Stakeholders and Scientists

Joanna Burger
Editor

Stakeholders and Scientists

Achieving Implementable Solutions to Energy and Environmental Issues

Editor
Joanna Burger
Division of Life Sciences
Environmental and Occupational Health Sciences Institute (EOHSI)
Consortium for Risk Evaluation with Stakeholder Participation (CRESP)
and Rutgers University
Piscataway, New Jersey 08854, USA
burger@biology.rutgers.edu

ISBN 978-1-4419-8812-6 e-ISBN 978-1-4419-8813-3
DOI 10.1007/978-1-4419-8813-3
Springer New York Dordrecht Heidelberg London

Library of Congress Control Number: 2011933225

Printed on acid-free paper

Springer is part of Springer Science+Business Media (www.springer.com)

For all the people in resource agencies,
regulatory agencies, tribes, industries,
and stakeholders who strive to solve our
energy and environmental problems.
May we leave future generations with a
balanced and sustainable environment,
and may we learn to incorporate
the concerns, views, and science of all
people in making better environmental
decisions

Joanna Burger, 18 January 2011

Foreword

Too often, discussion of energy focuses on the short-term costs, which is of primary concern to developers, investors, and politicians. The long-term costs associated with health effects of pollution and safety of facilities is often of more concern to neighbors living on the fenceline of these facilities, who are more likely to bear those costs. At the same time, the decision making that goes into approving facilities that produce the most damaging health effects often does not involve members of these affected communities. This is most true in environmental justice communities, those communities that are largely of low-income people or people of color and who live with disproportionate levels of environmental pollution when compared to the general population.

Bringing community groups together early may help to circumvent problems fenceline communities face. It helps identify major issues at play in a community, in particular to identify the sensitivities and situation which cause concern. Secondly, early engagement is a smart way to gauge community support for blocking or advancing changes in the local environment. Communities brought into the decision-making process early are often more willing to negotiate conditions and to recommend ways that certain facilities can be made acceptable. This may be as simple as changing proposed traffic patterns or hiring locally.

The community discussion process is often not easy, especially when siting energy facilities. However, in the end it may save money and keep the process of development moving rather than halted while tied up in endless legal battles. Discussion keeps hope alive for projects that can benefit communities. Community engagement can result in increased community support which can translate into public funding, regulatory approval, and public good will.

This book looks at different types of energy sources and discusses community engagement to block or approve siting of new energy facilities. Community engagement includes not only fenceline neighbors, but also other residents of the community, as well as regulators, interest groups, scientists, and public policy makers. I believe this book is a valuable tool not only for scientists, agencies, and regulators, but also for developers and fenceline communities looking to shape their own futures.

Dr. Mark Mitchell
Founder and President
Connecticut Coalition for Environmental Justice

Preface

The Nation and the World must move forward with development of a range of energy sources, all with attendant environmental problems. Solving these problems, and those remaining from past energy-related activities, will require transparency, iteration, inclusion, and collaboration with a wide range of stakeholders, including Tribes, U.S., State, and local governmental agencies, scientists, environmentalists, industries, concerned citizens groups, public policy makers, and the general public. While this is not a book about energy and energy-development, many of the chapters describe particular energy sources, as well as their advantages and disadvantages in terms of how stakeholders view the issues.

This book will describe and examine the interactions and integration of science and stakeholders to find solutions to some of the Nation's controversial environmental and energy-related issues. The initial chapter describes stakeholders, and discusses ways that stakeholders can be involved in decision making for environmental and energy-related problems. The second chapter examines in more detail the plight of minorities with respect to involvement in these issues, exploring some of the impediments to participation. And the third chapter examines sources of energy, and possible stakeholder involvement in decisions for these sources of energy. These three chapters serve as a basis for the other chapters, which are largely case studies dealing with a single energy source, with many sources or a comparison among types, or with general environmental problems.

The book uses case studies to explore the methods of integration and collaboration among diverse communities, and to develop a synthesis of true stakeholder involvement in energy-related issues that results in acceptable solutions that protect both human and ecological health. The focus of each chapter will be on problem definition, the process leading to the solution, and the mechanisms and collaborations among stakeholders that made the solutions (or lack thereof) possible. Many of the chapters are about place-based environmental management, but all of the chapters deal with how stakeholders have improved (or changed) the process. Some chapters will discuss failures, and how lack of stakeholder involvement contributed to these failures, and others will describe the role of the media and communication in stakeholder participation. This book is about stakeholder inclusion and its role in solving problems, not about the energy problems themselves, although obviously, this will be touched on in each chapter. Since it is about stakeholder participation, different people often collaborated to write each chapter.

The problems and issues discussed in the book range from contamination that resulted from the development and production of nuclear bombs during the Cold War, through public involvement in site selection and environmental management of a variety of energy sources, to Native American involvement in data-gathering, monitoring, and developing innovative solutions. The case studies are meant to illustrate the full range of ways that scientists and stakeholders can interact to find solutions to often contentious situations, or simply to improve the science or solutions.

You, the reader, and a wide range of other stakeholders, are the intended audience for the book. It will have served its purpose if the public, managers, public policy makers, and others can see a wide range of possible approaches to the environmental problems we face as a Nation and World. While the problems addressed in this book differ from one another, the main theme is that stakeholder involvement improved both the science and the solution, and that consensus can lead to both better science and to a solution that is effective in terms of alternatives, cost, effort, and time. The case studies are meant to provide not one template, but a range of templates for solutions that will be useful to a full range of stakeholders. In cases where no solution was reached, a firmer understanding of the alternatives and different viewpoints has led to a basis for further discussion and collaboration.

The genesis for this book came from my research as part of the Consortium for Risk Evaluation with Stakeholder Participation, a consortium of University scientists whose goal is to provide scientific expertise to the Department of Energy, and work with stakeholders in a number of different environmental arenas. I am more convinced every day that the inclusion of the widest possible group of stakeholders leads to better, more cost-effective, and environmentally sound decisions.

In my work it became apparent that many different groups of stakeholders, from agencies and regulators, to the public, did not always appreciate the ways to interact effectively and openly to work toward solutions to the problem of legacy wastes, waste from commercial nuclear power facilities, and other contamination and environmental issues. And it is increasingly clear that finding solutions to deal with legacy wastes, whether those remaining from the Cold War, those remaining from commercial nuclear power, or from other sources of energy, require the cooperation and collaboration of a wide range of agencies, organizations, and people. Space itself is a significant environmental problem faced by the siting of energy facilities, such as wind and solar.

Although the examples in the book mainly concern energy-related issues, the principles and frameworks provided apply equally to all environmental problems. We need to find solutions that are compromises between the views, concerns, and values of a wide range of people, from Native Tribes and First Nations to regulators, governmental agencies, and commercial interests. With increasing populations, it becomes even more important to conserve and preserve both ecological and societal resources and values within a context of human health and well-being. This book will hopefully contribute to the dialogue of how to meld these different views, as well as providing useful examples for the Tribal Nations, the general public, and private and governmental agencies.

Piscataway, NJ Joanna Burger

Acknowledgments

Over the years I have had many interesting and challenging discussions with people about the involvement of Native Americans and stakeholders in environmental evaluations, ecological assessments, and environmental decisions, including Michael Gochfeld, Caron Chess, Jim Clarke, Keith Cooper, Bernie Goldstein, Michael Greenberg, David Kosson, Charles Powers, and Jim Shissias. I have deeply appreciated my students and their thinking and enthusiasm, and I thank them now: Bill Boarman, Mandy Dey, Susan Elbin, Jeremy Feinberg, Tom Fikslin, Steve Garber. Chris Jeitner, Larry Niles, Brian Palestis, Carl Safina, Jorge Saliva, David Shealer, and Nellie Tsipoura. I could not have finished this work without the help of Taryn Pittfield and Chris Jeitner who helped with many phases of this book, and of course, without the authors themselves. Several ornithological colleagues have shared discussions over the years, including Bert Murray, Charlie Leck, Ted Stiles, Dee Boersma, B.A. Schreiber, and Patti Gowaty.

As always, my parents (Janette and Melvin Burger), and my brothers and sisters (Melvin Burger Jr., Christina Wiser, John Burger, Barbara Kamm, and Roy Burger, as well as their spouses), have provided encouragement, love, and at times, contentious discussions about environmental problems and their solutions. Brothers and sisters provide the continuity throughout our lives that no other people can, and I thank them now. My nieces and nephews have entertained me endlessly, and provided me with hope for the future: Kathy Drapeau, Eddie Burger, Michael, David and Daniel Wiser, Jacob and Andrew Burger, Ben Kamm, Erik, Beth, Emily, Allison, Alexis, and Amanda Burger, Jennifer Wolfson, and Douglas Gochfeld, as have my great nephew Greg and great niece Caroline Drapeau.

My husband, Michael Gochfeld, has been an inspiration throughout my career, always willing to help in any way he can, travel to out of the way places, discuss ideas, and challenge my mind. This book would never have been completed without his love, support, and attention. Our children, Debbie Gochfeld (and her husband Marc Slattery) and David Gochfeld, provided the hope for all our futures, and for a world where people with very different viewpoints can come together to make decisions that benefit the environment, as well as mankind. Finally, my parrot Tiko has been a constant reminder, standing behind me as I type, that we must work together to reach sound environmental decisions, before his kind, and indeed all biodiversity, perishes in the wild.

Over the years I have received funding from a wide range of agencies and organizations, and all have contributed to my thinking and work that lead to this book. These include NIMH, EPA, NIEHS (P30ES005022), the Department of the Interior, the Department of Energy (through the Consortium for Risk Evaluation with Stakeholder Participation, AI # DE-FG 26-00NT 40938 and DE-FC01-06EW07053), the Nuclear Regulatory Commission (NRC-38-07-502M02), the New Jersey Department of Environmental Protection (Office of Science, and Endangered and Nongame Species Program), Trust for Public Lands, New Jersey Audubon Society, the Jersey Coast Angler's Association, the Jersey Coast Shark Anglers, and the Environmental and Occupational Health Sciences Institute. The conclusions and interpretations reported herein are the sole responsibility of the author, and should not be interpreted as representing the views of the funding agencies.

Contents

Contributors

*Members of the Consortium for Risk Evaluation with Stakeholder Participation

Colin Apse The Nature Conservancy, North America and Africa Regions, 14 Maine Street, Suite 401, Brunswick, ME 04011, USA
capse@tnc.org

Fred Ayer The Low Impact Hydropower Institute, 34 Providence Street, Portland, ME 04103, USA
fayer@lowimpacthydro.org

Lauren C. Babcock-Dunning Edward J. Bloustein School of Planning and Public Policy, Rutgers University, 33 Livingston Avenue, New Brunswick, NJ 08901, USA
lbabcock-dunning@rutgers.edu

John H. Balletto ARCADIS U.S., Inc., 8 South River Road, Cranbury, NJ 08512-3698, USA
John.Balletto@arcadis-us.com

John Banks 12 WABANAKI WAY, Indian Island, ME, USA
john.banks@penobscotnation.org

Gina Bartlett Center for Collaborative Policy, California State University, Sacramento, 160 Delmar Street, San Francisco, CA 94117, USA
gina@ccp.csus.edu

Gabriel Bohnee Environmental Restoration and Waste Management (ERWM), Nez Perce Tribe, P.O. Box 365, Lapwai, ID 83540, USA
gabeb@nezperce.org

***Joanna Burger** Division of Life Sciences, Environmental and Occupational Health Sciences Institute (EOHSI), Consortium for Risk Evaluation with Stakeholder Participation (CRESP), and Rutgers University, 604 Allison Road, Piscataway, NJ 08854, USA
burger@biology.rutgers.edu

Marnie K. Carroll Diné Environmental Institute, Diné College, Shiprock, NM 87420, USA
mkcarroll@dinecollege.edu

Perry H. Charley Navajo Nation, Uranium Education Program, Diné College, Shiprock, NM 87420, USA
phcharley@dinecollege.edu

* **James H. Clarke** Department of Civil and Environmental Engineering, Consortium for Risk Evaluation with Stakeholder Participation (CRESP), Vanderbilt University, 2301 Vanderbilt Place, Nashville, TN 37235, USA
james.h.clarke@vanderbilt.edu

Laura Rose Day Penobscot River Restoration Trust, P.O. Box 5695 Augusta, ME 04336, USA
laura@penobscotriver.org

Mary R. English Institute for a Secure and Sustainable Environment, University of Tennessee, Knoxville, TN 37996, USA
menglish@utk.edu

Adam R. Fremeth Richard Ivey School of Business, University of Western Ontario, 1151 Richmond St. N, London, ON, N6A 3K7, Canada
afremeth@ivey.ca

Edward P. Glenn Environmental Research Laboratory, University of Arizona, Tucson, AZ 85705, USA
eglenn@ag.arizona.edu

* **Michael Gochfeld** Environmental and Occupational Medicine, EOHSI, Consortium for Risk Evaluation with Stakeholder Participation (CRESP), and UMDNJ-Robert Wood Johnson Medical School, 170 Frelinghuysen Road, Piscataway, NJ 08854, USA
Gochfeld@eohsi.rutgers.edu

* **Michael R. Greenberg** Edward J. Bloustein School of Planning and Public Policy, Consortium for Risk Evaluation with Stakeholder Participation (CRESP), Rutgers University, 33 Livingston Avenue, New Brunswick, NJ 08901, USA
mrg@rutgers.edu

Jonathan Hubert 14 Old Smith Rd, Tenafly, NJ 07670, USA
jon.hubert@gmail.com

* **James H. Johnson, Jr.** Department of Civil Engineering, Consortium for Risk Evaluation with Stakeholder Participation (CRESP), Howard University, Washington, DC 20060, USA
jjohnson@howard.edu

Amanda Kennedy 2 Landmark Square Suite 108, Stamford, CT 06901, USA
amandarutgers@gmail.com

* **David S. Kosson** Department of Civil and Environmental Engineering, Vanderbilt University, 2301 Vanderbilt Place, Nashville, TN 37235, USA
david.s.kosson@vanderbilt.edu

Gregory G. Lampman New York State Energy Research and Development Authority, 17 Columbia Circle, Albany, NY 12203, USA
ggl@nyserda.org

* **Karen W. Lowrie** Edward J. Bloustein School of Planning and Public Policy, Rutgers University, 33 Livingston Avenue, New Brunswick, NJ 08901, USA
klowrie@rutgers.edu

Alfred A. Marcus Carlson School of Management, University of Minnesota, 321 19th St. S, Minneapolis, MN 55408, USA
amarcus@umn.edu

Jonathan Paul Matthews Environmental Restoration and Waste Management (ERWM), Nez Perce Tribe, P.O. Box 365, Lapwai, ID 83540, USA
jonathanm@nezperce.org

Beverly Maxwell Diné Environmental Institute, Diné College, Shiprock, NM 87420, (Navajo Nation) USA
bevmaxwell@dinecollege.edu

Kenneth Morgan US Department of Energy, Ohio Field Office, Miamisburg, OH 45342, USA
morgan.ken@mac.com

James R. Newman Environmental Bioindicators Foundation, Inc. and Normandeau Associates, Inc. (Formerly Pandion Systems, inc.), 102 NE 10th Avenue, Gainesville, FL 32601, USA
jnewman@Nornandeau.com

Christian M. Newman Environmental Bioindicators Foundation, Inc. and Normandeau Associates, Inc. (Formerly Pandion Systems, inc.), 102 NE 10th Avenue, Gainesville, FL 32601, USA
cmnewman@nornandeau.com

Michael K. O'Neill New Mexico State University, Agricultural Science Center at Farmington, P.O. Box 1018, Farmington, NM 87499, USA
moneill@nmsu.edu

Jeffrey J. Opperman Global Freshwater Program, The Nature Conservancy, Chagrin Falls, OH 44022, USA
jopperman@tnc.org

Josiah Pinkham Cultural Resource Program, Nez Perce Tribe, P.O. Box 365, Lapwai, ID 83540, USA
josiahp@nezperce.org

* **Charles W. Powers** Department of Civil and Environmental Engineering, Consortium for Risk Evaluation with Stakeholder Participation (CRESP), Vanderbilt University, 2301 Vanderbilt Place, Nashville, TN 37235, USA
cwpowers@cresp.org

Joshua Royte The Nature Conservancy, Maine Chapter, Brunswick, ME 04011, USA
jroyte@tnc.org

John Seebach American Rivers, 1101 14th Street NW, Suite 1400, Washington, DC 20005, USA
jseebach@americanrivers.org

John Stanfill Environmental Restoration and Waste Management (ERWM), Nez Perce Tribe, P.O. Box 365, Lapwai, ID 83540, USA

Gary Stegner US Department of Energy, Ohio Field Office, Miamisburg, OH 45342, USA
Gary.Stegner@lm.doe.gov

Anthony Smith Environmental Restoration and Waste Management (ERWM), Nez Perce Tribe, P.O. Box 365, Lapwai, ID 83540, USA
asmith@nezperce.org

John M. Teal Professor Emeritus Woods Hole Oceanographic Institution, Teal Partners, 567 New Bedford Road, Rochester, MA 02770-4116, USA
teal.john@comcast.net

Mark R. Watson New York State Energy Research and Development Authority, 17 Columbia Circle, Albany, NY 12203, USA
mw1@nyserda.org

William J. Waugh DOE Environmental Sciences Laboratory, S.M. Stoller Corporation, Grand Junction, CO 81503, USA
Jody.Waugh@lm.doe.gov

Edward J. Zillioux Environmental Bioindicators Foundation, Inc., Zillioux Environmental, LLC, School of Public Health and Health Sciences, University of Massachusetts, Amherst, 207 Orange Avenue, Suite G, Fort Pierce, FL 34950, USA
zillioux@bioindicators.org; zillioux@schoolph.umass.edu

Chapter 1
Introduction: Stakeholders and Science

Joanna Burger

Contents

Abstract It has become fashionable to include stakeholders in environmental decisions, yet this inclusion often takes the form of one-way communication that involves imparting information or assessing concerns and perceptions. While risk communication and perception analysis is important to the process, a consensus can be reached in many contentious situations by the wide inclusion of stakeholders in a process whereby they actually participate in problem formulation, data acquisition and analysis, and in the final decision making. This chapter provides an introduction to stakeholder involvement, defines stakeholders, provides a template for the different types of stakeholder involvement, and suggests approaches to improve stakeholder participation in environmental and energy-related issues. Stakeholder participation includes Community-based participatory research, another method or description of collaboration between researchers and communites.

J. Burger (✉)
Division of Life Sciences, Environmental and Occupational Health Sciences Institute (EOHSI), Consortium for Risk Evaluation with Stakeholder Participation (CRESP), and Rutgers University, 604 Allison Road, Piscataway, NJ 08854, USA
e-mail: burger@biology.rutgers.edu

J. Burger (ed.), *Stakeholders and Scientists: Achieving Implementable Solutions to Energy and Environmental Issues*, DOI 10.1007/978-1-4419-8813-3_1,

1.1 Introduction

For many years environmental problems were "solved" with a top-down approach, whereby managers or governmental agencies defined the problem, conducted the science necessary to answer the question, and solved the problem. Solutions and plans were told to stakeholders, and sometimes at best they were asked their opinions about the problems or the solutions. However, stakeholders were not part of either problem formulation or the solutions. This led to solutions that did not reflect the wishes of the general public and a wide range of others, often including the scientific community.

Now many managers and public policy officials recognize the importance of including a full range of stakeholders in dealing with environmental problems in a manner that is informative, iterative, and interactive, and that solutions are sometimes dynamic, include feedback loops, and often involve ongoing management, now termed "adaptive management" (Walters and Hilborn 1978; Lee 1999), as well as community-based participatory research (O'Fallon and Dearry 2002; Wallerstein and Duran 2006). Adaptive management recognizes that there are few final solutions, and that adjustments must be made as new information and data appear. In some cases that information is from traditional ecological knowledge from Native Americans, Alaska Natives, or others close to the land (Berkes et al. 2000). Although initially decision makers and managers were reluctant to include stakeholders in their deliberations (Boiko et al. 1996), they gradually embraced them (PCCRAM 1997), particularly as citizen's advisory boards and committees (NRC 2008a; NRC 2008b). As discussed thoroughly in this book, "stakeholder" refers to everyone involved or interested in a particular site or problem, including Native Americans, governmental agencies (local, state, federal), regulators, scientists, social scientists, citizen's advisory boards, and the public (among others). The inclusion of stakeholders is critical for the development and acceptance of a sound and comprehensive energy policy, including further development of energy sources and environmental cleanup and restoration of contamination from the Cold War Legacy (DOE 1997; NRC 2008a; NRC 2008b).

The United States and the world are moving toward complex and diversified means of producing enough energy for the increasing needs of developed and developing nations related to growing populations and the growing per capita demand. The US population, for example, was just over 5 million in 1800, 76 million in 1900, and reached over 280 million by the year 2000 (US Census Bureau 2009, Fig. 1.1). Along with population increases worldwide, there has been a trend toward urbanization, creating densely populated cities (often coastal) with high energy demands.

Oil reserves are limited and vulnerable to production quotas, and the United States is dependent on foreign oil at a time when oil production is unstable, costly, and will inevitably decline. While other nations have moved toward greater use of nuclear energy (Anastasi 1998), the US nuclear industry has remained static for several decades, but is poised to move forward with upgrades and new power plant sites approved (NRC 2008a; NRC 2008b). The recent nuclear disaster at Fukushima (Japan) has led many nations and states to reconsider their commitment to increased nuclear power, or the building of new nuclear plants. Only with time

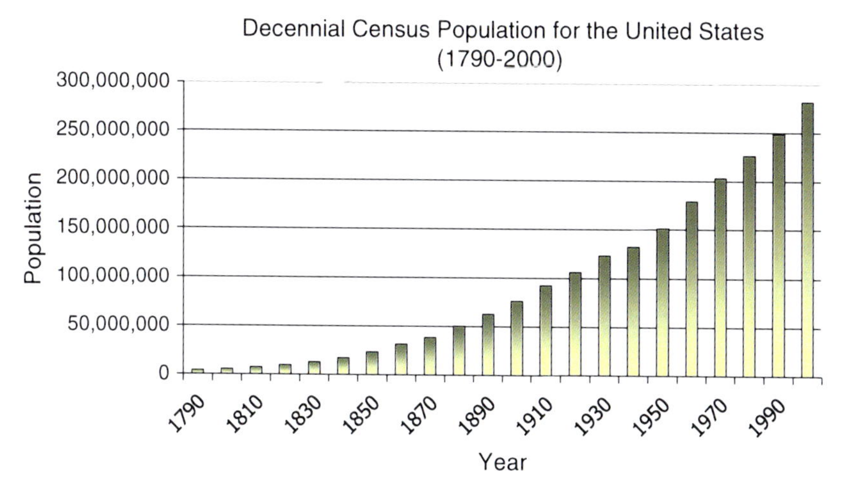

Fig. 1.1 Growth of population in the United States (US Census Bureau data base 2010)

Fig. 1.2 Vischer Ferry Plant of the New York power authority, shown is the power generation plant on the Mohawk River (Schenectady, New York) at Lock 7

will the full impact of the Fukushima nuclear disaster be clear on national energy policies. At the same time, many states and countries are moving toward diversification in alternative energy sources, including wind power, solar power, and hydropower (Anderson and Edens 2009; Singh 2009; Gochfeld 2011), and this has broad implications for stakeholder participation and collaboration. Hydropower has been an important source of electricity for some time (Fig. 1.2).

Many of the complex problems engendered by retooling, rebuilding, and siting new nuclear power-generating facilities, by cleaning up, remediating, and restoring Department of Energy (DOE) nuclear sites, and by siting other new energy-producing facilities (such as wind facilities), engender heated controversy. The controversies occur over the definition of the problem, data on characterization, remediation decisions, restoration options, short-term safety, and long-term protection of both human and ecological health around such facilities. While many of the solutions that served the nation in the past involved regulations, environmental protection laws, and top-down decisions, this approach may no longer work. The science, and the eventual solutions, may require inclusion of a wide range of stakeholders, including Native Americans and Alaska Natives. Both the problems and the solutions will depend on collaborations and interactions between the relevant parties. It is not a matter of placating a public, but of making better, more cost-effective management decisions, which can be shown both qualitatively and quantitatively (Burger et al. 2007b, 2009; Burger and Gochfeld 2009; Brody 2009).

Inclusion and participation of a wide range of stakeholders does not guarantee consensus and acceptance, for their may be circumstances where negotiation fails and irreconcilable views remain. One of the lessons learned from this volume is that such outcomes can be reduced or mitigated by early and comprehensive stakeholder involvement. Furthermore, the book includes cases regarding (a) existing situations that need attention such as legacy wastes, site remediation, and dealing with environmental problems at energy-generation facilities and (b) prospective projects such as wind turbines and solar panel arrays. Conditions for stakeholder involvement and the likelihood of a successful outcome (with "success" defined as carrying out the project largely as conceived by its proponents) are very different. Although there are similarities – the need to work within the regulatory framework, for example – cases of existing situations usually concern *how* to meet a widely agreed upon need for a goal (such as an existing location or ongoing mitigation). Prospective projects usually concern *whether* the proposed location is appropriate and *whether* the project is needed at all. These issues need to be clearly identified at the beginning of the project.

A note on the role of Indian Tribes is essential. There are over 500 federally recognized Tribes which are accorded Nation status and thus have the same status and roles as governments. Tribal representatives repeatedly point out that they are not "stakeholders" in the usual sense of the word. Some agencies treat Tribes and other governments as stakeholders (broad definition), so that one may see documents referring to "Tribes, governments, and stakeholders" as well as "Tribes, local governments, and other stakeholders." I recognize the Tribal sensitivity to their status as a Sovereign Nation and government. However, in this and most subsequent chapters, "stakeholder" is used in its broadest sense to include governments and Tribes, among others. I trust that readers will understand that this in no way undermines the special Sovereign Nation status accorded Indian (Native American and Alaska Native) Tribes by treaties.

1.1.1 Objectives of the Chapter

This chapter defines, discusses, and suggests that stakeholders and scientists need to collaborate to address the environmental problems resulting from energy, energy-related issues, and the environmental legacy from the Cold War (primarily radionuclide and chemical contamination). It provides a brief history of the changing views of stakeholder involvement and participation in solving environmental problems, as well as the evolving view of scientists and science in collaboration with a full range of stakeholders. Finally, it briefly describes the objectives for the book and provides a short summary of the types of problems addressed. The full range of energy sources, their advantages and disadvantages, and points of interaction with stakeholders, is discussed in Chap. 3 (Fig. 1.3).

Fig. 1.3 Wind power facilities are being developed in many areas of the US, Canada, and the world, and in many places, such facilities are compatible with agriculture

1.1.2 The Department of Energy

Some of the case studies in the book deal with cleanup and the DOE, the agency that faces the greatest environmental cleanup task in the world (Sink and Frank 1996; Crowley and Ahearne 2002). The environmental management (EM) division of DOE was created in 1989 to deal with the cleanup. While remediation of the DOE sites is often not energy related, for nuclear power to expand in the United States, safe and cost-effective methods for clean up and storage of nuclear wastes must be found. Thus dealing with remediation and related issues on DOE sites is a critical aspect of energy development in the Nation and the World. A history of the DOE's nuclear regulations can be found in Walker (2000).

The overall goal of cleanup is to protect human and ecological health during and after cleanup. Ecological principles and environmental evaluation methods and tools need to be integrated with environmental management of contaminated sites, as well as with stakeholders' concerns (Cairns 1994; NRC 1986; Suter 2001; Burger 2002, 2007; Burger et al. 2007a, b). Functioning ecosystems provide the goods and services that human populations require, including clean air and water, food and fiber, medicines, cultural products, protection from storms and inclement weather, recreational opportunities, aesthetic pleasures, cultural and religious experiences, and existence values (Harris and Harper 2000; Stumpff 2006; Burger et al. 2008a, b; Harper and Harris 2008). However, it is important to consider that intact ecosystems and environmental quality have intrinsic value, apart from the services they provide (McCauley 2006). It is particularly important to evaluate and integrate the remediation, protection of human health and the environment, and management with Tribal Nations and the full range of stakeholders.

The legacy of secrecy from the Cold War ended when DOE announced a policy of increased openness in 1993 (O'Leary 1997). This resulted in releases of documents, media accounts of actions and activity, and admissions of secret radiation information on radiation experiments with U.S. citizens. This led rather quickly to discussions of public participation (Creighton 1994), and to assessments by the National Research Council (NRC, arm of the National Academy of Sciences) on several aspects of DOE's operations and public participation policies (NRC 1994, 2000). Similarly, other agencies have recognized the importance of including stakeholders, and have issued directives or public involvement policies, such as the US Environmental Protections Agency (EPA 2003, Table 1.1).

Table 1.1 US environmental protection agency's guidance for public participation (after EPA 2003)

EPA's guideline steps
1. Plan and budget for activities
2. Identify appropriate stakeholders
3. Consider technical/financial assistance to facilitate involvement
4. Develop information and outreach for the public
5. Undertake public consultation
6. Review and use input, and provide feedback to the public
7. Evaluate public involvement activities

1.2 Role of Scientists in Environmental Decisions

Science usually forms the basis for environmental decisions, although it is the public, managers and public policy makers that make the decisions. Even so, scientists must provide the basic data that are necessary to inform the choices. Science can provide the data for different alternatives, or to address the main questions that are necessary before decisions can be made about cleanup, wastes resulting from energy-related activities, and siting of energy-producing plants or other facilities.

Scientific data are required for a number of different classes of questions dealing with EM. That is, environmental and risk management to reduce exposure and risks to humans and the environment requires information on a broad range of subjects, including attitudes and perceptions, behavior, exposure pathways and routes, hazards and finally, risk (Fig. 1.4, Burger and Gochfeld 2006). For many years, scientists and managers concentrated on collecting information and understanding exposure and the hazards from environmental contaminants, yet in reality, risk management requires understanding the perceptions and concerns of people, as well as their individual and group behavior. These categories of information are not mutually exclusive, but are interrelated. Recognition of the importance of each of these factors (and disciplines) is critical to finding solutions and managing risk (Grumbine 1997).

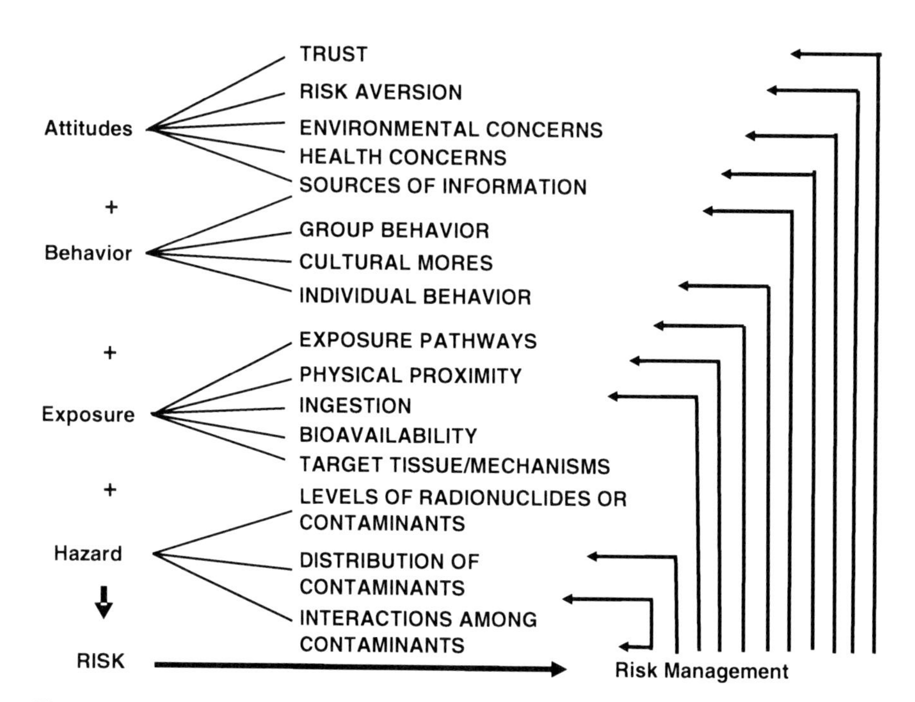

Fig. 1.4 Relationship between attitudes, behaviors, exposure and hazards that lead to risk and the need for risk management (after Burger and Gochfeld 2006)

In the last several decades, there has been a shift from research by individual scientists, to research by groups of scientists in the same discipline (e.g., a group of biologists), to collaborations among scientists in different disciplines (e.g., biologists, geologists, health physicists). While this has changed both the way science is funded and published, it has dramatically altered the way the public, public policy makers, managers, and others deal with science. These different kinds of collaborations among scientists need to be taken into account when stakeholders (defined broadly) consider what information is needed to move toward an environmental solution.

Finally, it is important to remember that science should be objective, but that scientists are stakeholders as well, and have his/her own views and concerns (Kraus et al. 1992). Scientists often view hazards and risks quite differently from the general public (Slovic 1987, 1993; Kunreuther et al. 1990). Scientists, as do other people, view risks based on their knowledge, experience, and perceived costs and benefits, although scientists may have a different knowledge base than others. It is thus the responsibility of scientists to provide the scientific data needed for a given environmental decision, without imposing his/her views. Where possible, personal biases, or those that derive from the scientist's work itself, need to be stated up front. For example, a person brought up on a farm (and thus exposed to pesticide use) might view pesticide use and hazards differently from a city dweller, and a person who devoted their life to the study of a particular illness might see that illness as inherently more important than others.

1.3 Role of Stakeholders in Environmental Decisions

In the past, managers and public policy makers made the decisions about environmental cleanup and energy-related issues. They decided the "what," "where," and "how" questions. This top-down approach, however, often resulted in solutions that were opposed by local communities, scientists, and other community members. This led ultimately to the realization that a full range of stakeholders needed to be included in both problem formulation and in the solving of the problems. In this section, I define and discuss the identification of stakeholders, the unique role of Native Americans and the early history of stakeholder involvement, as well as the fact hat assessment of perceptions is not participation. Finally, I provide a framework for different kinds of stakeholder involvement.

1.3.1 Identification of Stakeholders

Although the term "stakeholder" originally referred to a disinterested person who held the stakes and paid the money to the winner of the bet or contest, the modern usage has the opposite meaning – one who has an interest or share in land, property, treaty rights, or other aspects of a decision. It was the President's Commission (PCCRAM 1997) that elevated stakeholder engagement to a central role in problem solving and risk management (Fig. 1.5).

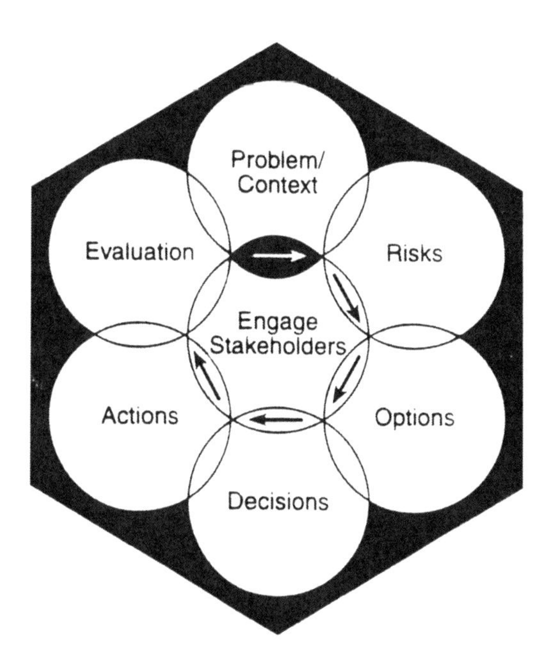

Fig. 1.5 PCCRAM diagram for the inclusion of stakeholders in environmental decision-making

In this book we define stakeholder broadly to include the full range of governments, people, agencies, and organizations that have an interest in solving environmental and energy-related problems. Thus, stakeholder includes managers, public policy managers, agency personnel, public officials (local, regional, state, federal), scientists, conservationists, conservation organizations, citizen groups, citizen advisory boards, national resource trustees, and the general public. While Native Americans are unique in that they have Sovereign Nation status and thus a government-to-government relationship with regard to environmental problems, we include them as stakeholders for brevity (but see below).

Broadly, stakeholders include anyone who is "concerned about or affected by the risk management problem" (PCCRAM 1997). I suggest that this means anyone who has one or several of the following: involvement, material benefit, direct use, indirect use, risk, aesthetics, or existence values (appreciates the mere existence of the good, service, or quality even though there is no intention to ever use or experience it, Brookshire et al. 1983; Larson 1993; Johansson-Stenman 1998). In short, anyone who has any interest can be considered a stakeholder, even if they live far away and will never come near the site. The latter point was well illustrated by the public uproar over the *Exxon Valdez* oil spill; most people in the United States never expected to go to the area in Alaska, but they were incensed by the oil spill's destruction of fisheries, animals, and the ecosystem (Burger 1997).

While it is relatively easy to identify the people who are directly at risk from an environmental exposure, such as those working, living, or recreating on-site or nearby (NRC 1983, 1993), it is equally important to identify stakeholders who have cultural, religious, medicinal, or aesthetic interests in the site or its resources. These other values can be as important as risks to health and safety (NRC 1996). Stakeholders' attitudes and concerns have been attributed to three value orientations: (1) self-interest, which often leads to exploitation of resources (Hardin 1968), (2) altruistic orientation toward human welfare, supporting a benefit to the general population (White et al. 1997), and (3) biospheric orientation, where individuals place a high personal value on the integrity of natural systems and the earth (Leopold 1949).

The identification of the relevant stakeholders is the first and the most critical step in addressing environmental problems (Burger et al. 2005a). The right people need to be brought to the table, and all views, opinions, and concerns need to be heard and addressed, although in any consensus process there will be disagreements and unresolved issues. However, these issues can usually be ameliorated if everyone is involved in the discussions from the start. Many of the examples in this book illustrate this point.

1.3.2 Unique Role of Native Americans

Native Americans are in a unique role because, by Treaty, their Tribes are recognized as Sovereign Nations. This means that they must be consulted when environmental decisions involve either their lands or their treaty rights (Nez Perce 2003). This consultation must involve government-to-government discussions and negotiations, although meetings and discussions can occur between relevant environmental officials and Tribal environmental agents. There are over 500 recognized Native American Tribes and Alaska Native groups in the United States (Federal Register 2008) (Fig. 1.6), and many of these hold lands or treaty rights to lands that are involved in legacy waste cleanup (DOE lands), siting of new energy-related facilities,

Fig. 1.6 Native Americans and Alaskan natives, such as these residents of St. Paul Island and Nikolski (Aleutian), have a stake in environmental decisions. The Aleuts and Pribilof islanders live on isolated Islands, hundreds of miles from mainland Alaska

or maintenance and operation of existing facilities. Their inclusion is not only advantageous, but required by law.

1.3.3 *Early History of Stakeholder Involvement*

Stakeholder involvement is not a new phenomenon, although the extent of active participation and collaboration is. Local and regional governments have conducted town meetings, and planning boards and environmental agencies have conducted public hearings, for many years. However, these meetings were often aimed at allowing the public to express their views or ask questions, and the decisions were left to governmental agencies and elected committees, all too often ignoring the public's suggestions Further, many agencies established citizen's advisory boards or committees circa 1990 (NRC 2008a; NRC 2008b), but these boards usually interacted only with the appropriate governmental agency (DOE, DOD, EPA), and were only advisory (but see Boiko et al. 1996).

Very early on it was recognized that the early involvement of a full range of stakeholders might lead to sounder and less costly solutions (Lynn 1987; Lynn and Busenberg 1995), but this recognition was not always taken to heart. Involving stakeholders is more than simply informing them of decisions. It is an iterative, inclusive, and interactive process that requires input at all stages, from defining the problem, identifying the suite of decision options, evaluating and selecting among them, determining the science needs, acquiring the information or data needed, and implementing the decisions, to the evaluation process itself. To be effective, the managers must understand the needs and perceptions of different stakeholders (PCCRAM 1997). While there are many examples where the views of stakeholders have been ignored, for example, the siting of locally unwanted land uses in multiple hazard neighborhoods (Greenberg et al. 1995), there are increasingly more examples where stakeholders have been active throughout the process (as shown by examples in this book). The social acceptance of various forms of energy, particularly renewable energy sources, will require social-political, community, and market acceptance (Wustenhagen et al. 2007), all of which will require stakeholder involvement. While I recognize that different stakeholders have different views that may be incompatible, the process of stakeholder involvement will allow these views to surface, and provide an opportunity for consensus (or at least a lack of disenfranchisement).

Moreover, stakeholder participation is beginning to be addressed within the business community, where three key levels are recognized: informative participation, consultative participation, and decisional participation (Green and Hunton-Clarke 2003). For companies, the questions often revolve around whether stakeholders should include only those people at some risk or whether they should also include anybody concerned with their activities (Crass and Greenbaum 2002). Unlike governmental agencies that are at least partly responsible to a public, companies are in a unique position because they also are responsible to shareholders, yet they still recognize that they need a network theory to appropriately include the full range of stakeholders in their organization's decisions (Rowley 1997).

1.3.4 *Assessment of Perceptions and Concerns is Not Participation*

In the 1980s, public policy makers, and to some extent scientists, felt that they needed to provide the public with information on environmental issues (Ruckelshaus 1983), and this included communication and understanding risk perceptions (Fischhoff 1995). Initially, stakeholder participation included two aspects: (1) presenting information to the public and (2) assessing the perceptions, concerns, and viewpoints of the public. The former was often done through public meetings, presentations, and making reports available (Pretyy 1995; O'Leary 1997), although this aspect has morphed into web-based information sources, interactive Web Sites, and even twittering (Rasmussen et al. 2007) And these programs led to methods to evaluate public participatory programs (Chess 2000) that included background documents, dedicated briefings and workshops, development of mental models, focused meetings, and mail (and e-mail) communication (Gregory et al. 2003). This form of public participation is extremely important in developing a dialogue, dispelling misconceptions, and starting with the same information base, although in many cases, public participation in meetings, workshops, or citizens advisory committees does not affect either process or outcome success (Chess and Purcell 1999).

Assessing perceptions and concerns is a second, and very common, form of "stakeholder participation" that has grown into an important discipline within risk assessment and risk management. There are literally hundreds of studies that describe what a particular group of people believe, perceive, and are concerned about environmental problems or issues (Fig. 1.7). However, stakeholder involvement has often been limited to the examination of public perceptions and attitudes about the siting of chemical plants, nuclear facilities, and hazardous waste sites (Kunreuther et al. 1990; Slovic et al. 1991; Slovic 1993; Flynn et al. 1994; Mitchell 1992; Williams et al. 1999; Burger 2004). In general, scientists view the risks from such facilities as less serious than does the general public (Barke and Jenkins-Smith 1993), and people are more willing to pay for management of risks that are direct, rather than diffuse (Fischer et al. 1991). Sophisticated weighting scales have been used to evaluate stakeholders' perceptions of environmental risks (Accorsi et al. 1999).

The fields of risk perception and communication have recognized developmental beliefs on the part of social scientists that led to failure, and these include the following (1) all we have to do is get the numbers right, (2) all we have to do is tell them the numbers, (3) all we have to do is explain what we mean by the numbers, (4) all we have to do is show them that they've accepted similar risks in the past, (5) all we have to do is show them that it's a good deal for them, (6) all we have to do is treat them nice, and (7) all we have to do is make them partners (Fischhoff 1995). Simply providing information is not enough to ensure that the "stakeholders" will form the "correct" view (what the communicators believe is correct). Some of the

Fig. 1.7 Survey research to address people's concerns and percptions is an important aspect of stakeholder involvement, and it often involves face-to-face interviews (here Burger conducts an interview). However, it is only one of several methods

controversies that develop come from mixed messages in risk communication. That is, "technical" people and others often use words differently, and understand the meanings differently (e.g., risk, safety, probability, association vs. causation, Jardine and Hrudey 1997). Thus, communication and information acquisition will always be an important part of the process of environmental decision making.

The two methods of involvement listed above (listening to information, assessing perception/concern) are no longer sufficient for Native Americans, Alaska Natives, and a full range of stakeholder, who want and deserve (by law in the case of Native Americans) to participate in environmental decisions. Many agencies consider these two levels to be the essential aspects of stakeholder involvement. For example, the EPA has given advice on stakeholder involvement that mostly involves information transfer and exchange (Table 1.1). It does, however, remain important to assess the perspectives of a full range of stakeholders in understanding their views of different energy technologies, especially for competing types (Reddy and Painuly 2004; Jardine et al. 2007; Wies et al. 2008, see chapters in this volume). Information and advisory panels or committees remain one of the most common forms of stakeholder participation (O'Connor et al. 2000) that fall under the category of providing information to the public, or seeking their advice.

Merely assessing the perceptions and concerns, and being sensitive to locally affected people (Supriyasilp et al. 2009) is no longer enough. It is no longer sufficient to simply: (1) listen to the public, or to evaluate whether the public feels that it is has been informed, (2) determine whether the public understands risk, (3) understand the public's satisfaction about their ability to make their views known, and (4) assess citizen's satisfaction with how concerns were dealt with (i.e., EPA's recent superfund evaluation, Charnley and Engelbert 2005). This level of stakeholder involvement is often perceived by the public as one sided and leaves stakeholders dissatisfied (Stave 2002).

True involvement, participation, and collaboration will be required to solve past environmental problems, such as from the Cold War legacy wastes, and to deal with future energy development issues. It is both sound policy and a requirement (laws, regulations) to include stakeholders in a participatory process for addressing many environmental and energy-related problems (Glicken 2000; PSNP 2008).

1.3.5 Conceptual Framework for Stakeholder Participation

It seems clear that an overall conceptual framework for stakeholder participation is needed to provide a template for moving forward, to describe options for entry for a range of stakeholders, and to form a basis to evaluate the efficacy of different forms of participation. I suggest that there are several levels of involvement, and each is important for solving environmental problems. The levels of stakeholder involvement include informational (providing information), acquisitional (gathering information), dialogue, intragovernmental, intragovernmental with outside scientists or advisors, stakeholder involvement, stakeholder-driven, and collaborative (Table 1.2).

Informational, acquisitional, and dialogue are all phases that have been extensively studied and generally come under the heading of risk communication and perception (see above, Reynolds and Seeger 2005). They are necessary and important parts of the stakeholder participation process, but are not the only essential parts. An information base is essential for all stakeholders, whether governmental, private, or Native American, to address issues with the same information. In this regard, the concerns and views of stakeholders ranging from children to the elderly need to be considered. Understanding concerns, however, preludes to the real task of working together collaboratively to solve environmental problems.

Intra governmental (with either other agencys' scientists or outside scientists) is a category that must be recognized because sometimes an agency works alone without either other agencies or outside scientists. Both types of involvement are essential for environmental problems that involve different ecosystems, types of organisms, human or ecological concerns, or geophysical concerns. Even so, the stove-piping of agencies has often resulted in their working alone, rather than together. The

Table 1.2 Levels of involvement possible for a full range of Native Americans, Native Alaskans, and stakeholders

Type	Definition	Example
Informational	Decision-makers provide information to the public, either in meetings, brochures or web-based. The information is uni-directional	Water quality monitoring data is often provided to the public Coliform counts are provided to beach communities and public officials
Acquisitional (information gathering)	Decision-makers, scientists or managers collect information on perceptions, concerns or resource use	Questionnaires aimed at understanding perceptions of future land use, health risks, or environmental concerns Can include phone, in-person or other survey methods
Dialogue	Open meetings or forums between managers, decisions makers, scientists, the public and others. Offers the opportunity for questions and answers and two way dialogue	Public meetings about environmental issues Public meetings about siting of facilities or cleanup decisions Citizen's advisory groups or panels
Intragovernmental	Collaboration or cooperation among governmental agencies to achieve some environmental goal	Environmental monitoring programs often involve several agencies, and may involve state and federal agency personnel. The environmental monitorin and assessment program (EMAP) program of the US environmental protection agency is such a program (EPA 2006a, b)
Intragovernmental with outside scientists (or other advisors)	Collaboration among governmental agencies, with the addition of outside scientists to address environmental issues	The Food and Drug Administration collaborated with EPA to devise fish consumption advisories to deal with environmental contamination Several state and federal agencies have developed human health or ecological health indicators with the help of outside scientists (Fox 1994)
Stakeholder involvement	Stakeholders other than governmental scientists or scientists are involved in forming policies and dealing with environmental issues	Some state agencies involve fishermen and fish consumers in devising fish consumption advisories
Stakeholder-driven	Stakeholders other than governmental agencies or scientists initiated the questions, or helped to formulate the problem	Watershed and riverine management has been directed by stakeholders, notably the public and environmental groups (PSNP 2008)

(continued)

Table 1.2 (continued)

Type	Definition	Example
Stakeholder collaboration	A full range of stakeholders participate in many phases of a project, notable in problem formulation, identifying the data gaps and science needs, participating in the research itself, and participating fully in decision-making. Community-based participatory research falls in this category	A full range of stakeholders, including Alaskan Natives (Aleuts), state and federal agencies, scientists, conservation organizations and others were involved in designing and implementing a science plan aimed at providing the information to close Amchitka Island (Burger 2011; Burger et al. 2005a, 2007b)
		Navajo students from Dine College were involved with scientists and others to examine and study the effects of manipulations of vegetation covers to contain radionuclides (see Chap. 6)
		Government officials, scientists, the general public, and fishers are collaborating to select indicator fish species for evaluation (and to collect the fish) to provide information on mercury contamination that might pose a human health risk (Burger et al. 2005b)

Stakeholders include managers, public-policy makers, scientists, governmental agencies (local, state, federal), and all others (after Burger 2009 and unpublished data)

most famous case of such stove-piping is what happened with security issues in the United States. Intelligence information was not shared between intelligence agencies and law enforcement agencies (e.g., CIA, FBI) (GAO 2004) There was a recognition following 9/11 that agencies must work together and pool information, databases, and expertise, rather than each working alone.

Stakeholder involvement refers to stakeholders taking part in some aspect of environmental decision making, but usually only in a limited sense. For example, fishermen may be involved in selecting or catching fish to be used for chemical analysis; the data are then used to develop fish consumption advisories. Similarly, the public may contribute observations on an endangered species residing near a nuclear power plant.

Stakeholder-driver refers to environmental decision making where one or more stakeholders identified the initial problem, or contributed ideas to the solving of the environmental problem. This type is quite well known in that often conservation or environmental groups alert the government to a specific problem. For example, people may notice that trees or other species are dying near a chemical or nuclear plant, and demand that local or regional officials examine the problem, and they continue to be involved in how the problem is addressed. Similarly, people may notice that fish are dying from contamination in streams and rivers near chemical plants, and demand action. It differs from collaborative research in that the stakeholders are driving problem formulation and the gathering of data to address the question, but they are not actually participating in the full design of the scientific plan or the gathering of data. They have changed the way scientists and agencies address the problem, but did not participate in the science process itself.

Collaborative environmental decision making involves actually collaborating with a full range of stakeholders, although the types of collaborations may vary. Some of the elements of collaborative decision making involve stakeholder involvement, knowledge-based, holistic, and proactive approaches, sharing of power, joint responsibility, and developing integrated solutions (Randolf and Bauer 1999), although not all stakeholders have to participate in each aspect. Collaboration is only possible if all parties agree to give up some of their power and sole decision power, and make substantive changes to the problem formulation and science acquisition phase. Community-based participatory research is a type of full collaboration. Several of the chapters in this book involve true collaborations among a full range of stakeholders, from government officials and regulators, to scientists, the general public, Native Americans or Alaska Natives, conservations groups, and others.

In addition to the levels of stakeholder participation, there are several elements necessary to carry on a successful program involving stakeholders, and these include (1) identifying a successful mission, vision, or objectives, (2) having sufficient resources for the program, (3) having an extensive planning process for both routine and rare events, (4) developing a list of process and information tools, (5) developing a lessons learned program that allows for iteration and improvements, and (6) identifying and overcoming obstacles as they occur (Greenberg and Lowrie 2001). Smutko et al. (2002) identified a suite of seven issues that lead to the need for collaborative stakeholder involvement, which include level of uncertainty, balance of

Fig. 1.8 Understanding concerns and perceptions of people ranging from children to elders is essential, as well as considering future generations. Shown are (from *left* to *right*) Emily, Melvin, Roy, Allison, and Beth Burger

information, risk, time horizon of effects, urgency of decision, distribution of effects, and clarity of the problem. Distribution of effects, clearly, must consider elements of environmental justice communities that often bear the majority of the risk, and have the least input into environmental decisions (Waller et al. 1999; EPA 2002; 2009; DOE 2008; 2009; Elliott et al. 2009).

Often complicated models are developed to deal with participatory and multiobjective development of energy projects that include a framework for planning, analysis and evaluation of ecological, social, and economic impacts, and value tree analysis (Marttunen and Suomalainen 2004). Beierle and Cayford's (2002) social goals for stakeholder participation included incorporating public values into decisions, improving the substantive quality of decisions, resolving conflicts, building trust in institutions, and educating and informing the public. These goals can only be met, however, if participants have the experience to participate fully (Alberts 2007). Regardless of the simplicity or complexity of the analyses, however, it is transparency, openness, consistency, quality, and degree of collaboration among stakeholders that will lead to solutions. And for many, it is a matter of considering not only current generations, and people of all ages, but also people of future generations (Fig. 1.8).

Finally, a note of caution is required; there are situations in which citizen participation in decision making may be costly, ineffective, and lead to unsatisfactory solutions (Irvin and Stansbury 2004). With increasing attention to reducing the costs of governmental programs, money spent on participatory processes will receive increased scrutiny. I view this as positive, since such scrutiny may identify stakeholder programs that have not been effective or successful, but caution is needed to assure that successful programs are fully funded. Delegating environmental decision making may result in a rollback of decades of environmental regulatory success in favor of economic considerations, and locally based citizen's groups may prefer to solutions that are not acceptable to the wider regional or national citizenry (Irvin and Stansbury 2004). While I might argue that the previous case occurs only when the identification of stakeholders does not include the wider citizenry, it nonetheless illustrates the importance of considering competing claims at the time of stakeholder identification and problem formulation.

1.4 Melding Scientists and Stakeholders

Often contentious situations are a result of contested or ambiguous goals or lack of scientific agreement on data needs or cause-and-effect relationships (Mccool and Guthrie 2001; Burger et al. 2005a, 2007b). It is in the latter case that the melding of scientists and stakeholders can facilitate solutions, or at least lead to a path forward toward solution. While scientists have been used to working on their own, often researching topics of personal and disciplinary interest, it is increasingly clear that environmental problems will be solved only when scientists can work with a full range of other stakeholders to address the data-based needs. The major questions include: What data are required to address issues of compliance, to answer the questions of stakeholders, to reduce risks to humans and the environment, and to assure long-term health of humans and the environment. Increasingly, qualitative and quantitative methods will be developed to rank, rate, or balance different stakeholder views when these views are particularly divergent (Apostolakis and Pickett 1998). Seymour et al. (2008) used a qualiltative assessment called Key Changes and Actor Mapping to illustrate and analyze stakeholder viewpoints on energy policy.

One of the salient features of true multi-stakeholder collaboration in environmental decision making is the opportunity to experience change in their subjective understanding of their relationship to each other and themselves (Poncelet 2001). Nowhere is this truer than of collaborations between scientists and other stakeholders. Scientists in the past, and some scientists today, view science as a solitary pursuit or one that involves collaboration only among other like-minded scientists. For some scientists, it was a leap to learn to deal with scientists from other disciplines, especially those in the social sciences. Yet, such interdisciplinary studies are often the ones that lead to major advancements. A greater leap for scientists, however, is to take part in stakeholder-driven research and to collaborate as much as possible with stakeholders. Such collaborations are in their infancy in most situations, and the examples in this book show the ways that a range of Native Americans, Alaska Natives, and stakeholders can collaborate with scientists in the research itself to provide the data needed to make environmental decisions, particularly about energy-related issues. Such community-based participatory research must be a true and open collaboration to be effective.

Some concerns have been raised about the quality of decisions that result from stakeholder involvement. However, in a study of 239 published case studies of stakeholder involvement in environmental decisions, the majority contained evidence that stakeholders improved decisions by adding new information, ideas, and analysis, while still using technical and scientific data (Beierle 2002). Indeed the data suggested that the more intense the stakeholder involvement was, the more likely was the resultant decision's high quality. And stakeholder involvement conveys a sense of "ownership" and enhances acceptance of outcomes. This certainly is true for several of the case studies reported in this volume. Collaborations of stakeholders with scientists, as happened at the Savannah River Site (Chap. 5), at Amchitka (Chap. 8), at monitoring and remediation sites (Chap. 4), and at uranium mill tailings site (Chap. 6), where stakeholders actually participated in the research and thus had partial ownership in the data, resulted in consensus solutions to otherwise contentious situations. Some of the chapters deal with contentious wind energy projects (Chaps. 11 and 12) or at hydropower sites (Chap. 13) where different stakeholders had very different goals. In some cases, the process was flawed (see Chap. 11), due mainly to differences in the goals and objectives of different stakeholders. Chapters 14 and 15 deal with evaluation among energy sources. Finally, one chapter deals with how tribes can affect environmental decisions (Chap.7), and potential failures during decommissioning and decontamination (Chap. 9), and communication and media issues (Chapters 16 and 17). The beginning chapters deal with the role of stakeholders (Chap. 1), the special role of minorities (Chap. 2), and energy diversification (Chap. 3). While the approaches differ, each chapter examines some aspect of involvement between scientists, stakeholders (used broadly), and contentious and complicated environmental and energy-related issues.

Acknowledgments I particularly thank Michael Gochfeld, Charles W. Powers, David S. Kosson, and James Clarke for valuable discussions about science, stakeholders, and environmental health problems. I also thank Caron Chess, Michael Greenberg, and Lisa Bliss for helpful discussions about science over the years, Mary English for insightful comments on the nature of stakeholder involvement, and Sheila Shukla, Chris. Jeitner and Taryn Pittfield for technical support. This research was funded mainly by the Consortium for Stakeholder Participation (CRESP) through a grant from the Department of Energy (DE-FC01-06EW07053) to Vanderbilt University and Rutgers University, as well as the Nuclear Regulatory Commission (NRC 38-07-502M02), NIEHS (P30ES005022), and the New Jersey Department of Environmental Protection. The conclusions and interpretations reported herein are the sole responsibility of the author, and should not in any way be interpreted as representing the views of the funding agencies.

References

Accorsi R, Apostolakis G, Zio E (1999) Prioritizing stakeholder concerns in environmental risk management. J Risk Res 2:11–29

Alberts DJ (2007) Stakeholders or subject matter experts, who should be consulted? Energy Pol 35:2336–2346

Anastasi C (1998) Nuclear power needs to be part of the sustainability debate. Electric. J 11:82–84

Anderson FR, Edens GE (2009) Alternative energy and the rebirth of NEPA. Nat Res Environ 23:22–24

Apostolakis GE, Pickett SE (1998) Deliberation: integrating analytical results into environmental decisions involving multiple stakeholders. Risk Anal 18:621–634

Barke RP, Jenkins-Smith HC (1993) Politics and scientific expertise: scientists, risk perception, and nuclear waste policy. Risk Anal 13:425–439

Beierle TC (2002) The quality of stakeholder-based decisions. Risk Anal 22:739–749
Beierle TC, Cayford J (2002) Democracy in practice: public participation in environmental decisions. Resources for the Future Press, Washington
Berkes F, Colding J, Folke C (2000) Rediscovery of traditional ecological knowledge as adaptive management. Ecol Applic 10:1251–1262
Boiko PE, Morrill RL, Flynn J, Faustman EM, van Belle G, Omen GS (1996) Who holds the stakes? A case study of stakeholder identification at two nuclear weapons sites. Risk Anal 16:237–249
Brody SD (2009) Measuring the effects of stakeholder participation on the quality of local plans based on principles of collaborative ecosystem management. J Plan Ed Res 22:407–419
Brookshire DS, Eubanks LS, Randall A (1983) Estimating option prices and existence values for wildlife resources. Land Econ 59:1–15
Burger J (1997) Oil spills. Rutgers University Press, New Jersey
Burger J (2002) Incorporating ecology and ecological risk into long-term stewardship on contaminated lands. Remediation 18:107–120
Burger J (2004) Recreational rates and future land-use preferences for four Department of Energy sites: consistency despite demographic and geographical differences. Environ Res 95:215–223
Burger J (2007) A Model for Selecting Bioindicators to Monitor Radionuclide Concentrations Using Amchitka Island in the Aleutians as a Case Study. Environ Res 105:316–323
Burger J (2009) Stakeholder involvement in indicator selection: case studies and levels of participation. Environ Bioindicat 4:170–190
Burger J (2011) Stakeholders, Risk from Mercury, and the Savannah River Site: Iterative and Inclusive Solutions to Deal with Risk from Fish Consumption. In: J Burger (ed) Stakeholders and scientists. Springer: New York
Burger J, Gochfeld M, Kosson DS, Powers CW, Friedlander B, Eichelberger J, Barnes D, Duffy LK, Jewett SC, Volz CD (2005a) Science, policy, and stakeholders: developing a consensus science plan for Amchitka Island, Aleutians, Alaska. Environ Manage 35:557–568
Burger J, Stern S, Gochfeld M (2005b) Mercury in Commercial Fish: Optimizing Individual Choices to Reduce Risk. Environ Health Persp 113:266–271
Burger J, Gochfeld M (2006) A framework and information needs for the management of the risks from consumption of self-caught fish. Environ Res 101:275–285
Burger J, Gochfeld M (2009) Changes in Aleut concerns following the stakeholder-driven Amchitka Independent Science Assessment. Risk Anal 29:1156–1169
Burger J, Gochfeld M, Powers CW (2007a) Integrating long-term stewardship goals into the remediation process: Natural resource damages and the Department of Energy. J Environ Manage 82:189–199
Burger J, Gochfeld M, Powers CW, Kosson DS, Halverson J, Siekaniec G, Morkill A, Patrick R, Duffy LK, Barnes L (2007b) Scientific research, stakeholders, and policy: continuing dialogue during research on radionuclides on Amchitka Island, Alaska. J Environ Manage 85:232–244
Burger J, Gochfeld M, Pletnikoff K, Snigaroff R, Snigaroff D, Stamm T (2008a) Ecocultural attributes: evaluating ecological degradation: ecological goods and services vs. subsistence and tribal values. Risk Anal 28:1261–1271
Burger J, Gochfeld M, Greenberg M (2008b) Natural resource protection of buffer lands: integrating resource evaluation and economics. Environ Monit Assess 142:1–9
Burger J, Gochfeld M, Pletnikoff K (2009) Collaboration versus communication: the Department of Energy's Amchitka Island and the Aleut community. Environ Res 109:503–510
Cairns J Jr (1994) Rehabilitating damaged ecosystems. CRC Press, Florida
Charnley S, Engelbert B (2005) Evaluating public participation in environmental decision-making: EPA's superfund community involvement program. J Environ Manage 77:165–182
Chess C (2000) Evaluating environmental public participatory methodological questions. J Environ Plan Manage 43:769–2000
Chess C, Purcell K (1999) Public participation and the environment: do we know what works? Environ Sci Technol 16:2685–2692

Crass W, Greenbaum A (2002) Reasoning about responsibilities: mining company managers on what stekholders are owed. J Business Ethics 39:319–335
Crowley KD, Ahearne JF (2002) Managing the environmental legacy of U.S. nuclear-weapons production. Am Sci 90:514–523
Creighton JL (1994) How to design a public participation program. Battelle Pacific Northwest Labs, Department of Energy (DOE- EM-22)
Department of Energy (DOE) (1997) Linking legacies: Connecting the Cold War Nuclear Weapons Production Processes To Their Environmental Consequences. Washington: Office of Environmental Management, Department of Energy http://www.em.doe.gov/Publications/linklegacy.aspx. Accessed February 3, 2010
Department of Energy (DOE) (2008) Environmental justice: five-year implementation plan. Washington (DOE/LM-1462)
Department of Energy (DOE) (2009) Environmental justice at the U.S. Department of Energy. http://www.1m.doe.gov/spotlight/ej3.htm. Accessed 18 Jan 2011
Elliott MR, Wang Y, Lowe RA, Kleindorfer PR (2009) Environmental justice: frequency and severity of US chemical industry accidents and socioeconomic status surrounding communities. J Epidemiol. Comm Health 58:24–30
Environmental Protection Agency (EPA) (2002) National Environmental Justice Advisory Council: fish consumption and environmental justice http://www.epa.gov/compliance/resources/publications/ej/fish_consump_report_1102.pdf Accessed Dec 2009
Environmental Protection Agency (EPA) (2003) Public involvement policy. http://www.epa.gov/policy2003/policy2003.pdf. Accessed 6 March 2005
Environmental Protection Agency (EPA) (2006a) Integrated Risk Information System data base for methylmercury. http://www.epa.gov/iris/. Accessed Jan 2010
Environmental Protection Agency (EPA) (2006b) A manager's guide to indicator selection. Washington, EPA Office of Research and Development (EPA/600/R-90/001a)
Environmental Protection Agency (EPA) (2009) Environmental justice: compliance and environment. http://www.epa.gov/environmentaljustice. Accessed Nov 2009
Federal Register (2008) Indian entities recognized and eligible to receive services from the Unites States Bureau of Indian Affairs. Federal Register 73:18553–18557
Fischer GW, Morgan MG, Fischoff B, Nair I, Lave LB (1991) What risks are people concerned about? Risk Anal 11:303–314
Fischhoff B (1995) Risk perception and communication unplugged: twenty years of process. Risk Anal 15:137–145
Flynn J, Slovic P, Mertz, C (1994) Decidedly Different: Expert and Public Views of Risks from a Radioactive Waste Repository. Risk Anal 6:643–648
Fox G (1994) Bioindicators as a measure of success for virtual elimination of persistence toxic substances. International Joint Commission, Canada
General Accounting Office (GAO) (2004) 9/11 commission Report: reorganization, transformation, and information sharing. GAO-04–1033T, Washington
Glicken J (2000) Getting stakeholder participation 'right': a discussion of participatory processes and possible pitfalls. Environ Sci Pol 3:305–310
Gochfeld M (2011) Energy Diversity: Options and Stakeholders. In: J Burger (ed) Stakeholders and scientists. Springer: New York
Green AO, Hunton-Clarke L (2003) A typology of stakeholder participation for company environmental decision-making. Bus Strategy Environ 12:292–299
Greenberg M, Lowrie K (2001) A proposed model for community participation and risk communication for a DOE-led stewardship program. Fed Fac Environ 12:125–142
Greenberg M, Schneider D, Parry J (1995) Brown fields, a regional incinerator and resident perceptions of neighborhood quality. Risk: Health Safety Environ 6:241–259
Gregory R, Fischhoff B, Thorne S, Butte G (2003) A multi-channel stakeholder consultation process for transmission deregulation. Energy Pol 31:1291–1299
Grumbine RE (1997) Reflections on "What is ecosystem management?" Conserv Biol 11:41–47
Hardin G (1968) The tragedy of the commons. Science 162:1243–1248

Harper BL, Harris SG (2008) A possible approach for setting a mercury risk-based action level based on tribal fish ingestion rates. Environ Res 107:60–68
Harris SG, Harper BL (2000) Using eco-cultural dependency webs in risk assessment and characterization of risks to tribal health and cultures. Environ Sci Pollut Res 2:91–100
Irvin RA, Stansbury J (2004) Citizen participation in decision making: is it worth the effort? Publ Admin Rev 64:55–65
Jardine CG, Hrudey SE (1997) Mixed messages in risk communication. Risk Anal 17:489–498
Jardine CG, Predy G, Mackenzie A (2007) Stakeholder participation in investigating the health impacts from coal-fired power generating stations in Alberta, Canada. J Risk Res 10:693–714
Johansson-Stenman O (1998) The importance of ethics in environmental economics with a focus on existence values. Environ Res Econ 11:429–442
Kraus N, Malmfore T, Slovic P (1992) Instuitive toxicology: expert and lay judgements of chemical risks. Risk Anal 12:215–232
Kunreuther H, Easterling D, Desvousges W, Slovic P (1990) Public attitudes toward siting a high-level nuclear waste repository in Nevada. Risk Anal 10:469–484
Larson DM (1993) On measuring existence value. Land Econ 69:177–188
Lee KN (1999) Appraising adaptive management. Conserv Ecol 3:3–18
Leopold AS (1949) Sand County Almanac. Oxford University Press, Oxford
Lynn PM (1987) Citizen involvement in hazardous waste sites: two North Carolina success stories. Environ Imp Assess Rev 7:347–361
Lynn PM, Busenberg JJ (1995) Citizen advisory committees and environmental policy: what we know, what's left to discover. Risk Anal 15:147–162
McCauley, DJ (2006) Selling out on nature. Nature 442:27–28
Marttunen M, Suomalainen M (2004) Participatory and multiobjective development of water course regulation – creation of regulation alternatives from stakeholders' preferences. J Multi-Criteria Dec Anal 13:29–49
Mccool SF, Guthrie K (2001) Mapping the dimensions of successful public participation in messy natural resources management situations. Soc Nat Res 14:309–323
Mitchell, J (1992) Perception of Risk and Credibility at Toxic Sites. Risk Anal 1:19–26
National Research Council (NRC) (1983) Risk assessment in the federal government: managing the process. National Academic Press. Washington
National Research Council (NRC) (1986) Ecological knowledge and environmental problem-solving. National Academic Press. Washington
National Research Council (NRC) (1993) Issues in risk assessment. National Academic Press. Washington
National Research Council (NRC) (1994) Building consensus through risk assessment and management of the Department of Energy's Environmental Remediation program. National Academic Press. Washington
National Research Council (NRC) (1996) Understanding risk: informing decisions in a democratic society. National Academic Press. Washington
National Research Council (NRC) (2000) Long-term institutional management of U.S. Department of Energy Legacy waste management. National Academic Press. Washington
National Research Council (NRC) (2008a) Public participation in environmental assessment and decision making. National Academic Press, Washington
Nuclear Regulatory Commission (NRC) (2008b) Expected new nuclear power plant applications (updated August 2008). http://www.nrc.gov/reactors/new.licensing/new-licensing-files/expected-new-rx-applications.pdf. Accessed December 2009
Nez Perce Tribe (2003) Treaties: Nez Perce perspectives. US DOE and Confluence Press, Richland
O'Connor RE, Anderson PJ, Fisher A, Bord RJ (2000) Stakeholder involvement in climate assessment: bridging the gap between scientific research and the public. Climate Res 14:255–260
O'Fallon LR, Dearry A (2002) Community-based participatory research as a tool to advance environmental health science. Envir Health Perspect 110:155–159

O'Leary HR (1997) DOE nurtures openness to mend past miscues. Forum Appl Res Publ Policy 12:102–102

Poncelet EC (2001) Personal transformation in multistakeholder environmental partnerships. Policy Sci 34:273–301

President's Commission (PCCRAM) (1997) Presidential/Congressional Commission on risk assessment and management. U.S. Government Printing Office, Washington

Pretyy JN (1995) Participatory learning for sustainable agriculture. World Develop 23:1247–1263

Puget Sound Nearshore Partnership (PSNP) (2008) Puget Sound Nearshore Partnership: stakeholder involvement strategy. http://www.pugetsoundnearshore.org/program_documents/stakeholder_strategies_dec08.pdf. Accessed 3 June 2009

Randolf J, Bauer M (1999) Improving environmental decision-making through collaborative methods. Pol Studies Rev 16:168–190

Rasmussen S, Mangalagiu D, Ziock H, Bollen J, Keating G. (2007) Collective intelligence for decision support in very large stakeholder networks: the future of US energy system. Paper presented at Law Soc Assoc, Berlin Germany http://www.allacademi.com/meta/p182321_index.html. Accessed 1 Jan 2011

Reddy S, Painuly JP (2004) Diffusion of renewable energy technologies – barriers and stakeholders' perspectives. Renewable Energy 29:1431–1447

Reynolds B, Seeger MW (2005) Crisis and emergency risk communication as an iterative model. J Health Comm 10:43–55

Ruckelshaus WD (1983) Science, risk, and public policy. Science 221:1026–1028

Sink CH, Frank CW (1996) DOE forges partnerships for environmental cleanup. Forum 11:65–69

Rowley TJ (1997) Moving beyond dyadic ties: a network theory of stakeholder influences. Acad Manage Rev 22:887–910

Seymour EH, Murray L, Fernandes R (2008) Key challenges to the introduction of hydrogen – European stakeholder views. Int J Hydrogen Energy 33:3015–3020

Singh S (2009) World Bank-directed development? Negotiating participation in the Nam Theun 2 hydropower project in Laos. Develop Change 40:487–507

Slovic P (1987) Perception of risk. Science 236:280–285

Slovic P (1993) Perceived Risk, Trust, and Democracy, Risk Anal 13:675–682

Slovic P, Layman M, Flynn J (1991) Lessons from Yucca Mountain. Environment 3:7–11, 28–30

Smutko LS, Klimek SH, Perrin CA, Danielson LE (2002) Involving watershed stakeholders: an issue attribute approach to determine willingness and need. J Am Water Res Assoc 38:995–1006

Stave KA (2002) Using system dynamics to improve participation in environmental decisions. System Dynamic Rev 18:139–167

Stumpff LM (2006) Reweaving the Earth: an indigenous perspective on restoration planning and the National Environmental Policy Act. Environ Pract 8:93–103

Supriyasilp T, Pongput K, Boonyasirikul T (2009) Hydropower development priority using MCDM method. Energy Pol 37:1866–1875

Suter GW II (2001) Applicability of indicator monitoring to ecological risk assessment. Ecol Indicators 1:101–112

US Census Bureau (2009) Resident population of the United States. http://eadiv.state.wy.us/demog_data/usdec_1790_00.html. Accessed 1 Aug 2009

Waller LA, Louis TA, Carlin BP (1999) Environmental justice and statistical summaries of differences in exposure distributions. J Expos Anal Environ Epedimiol 9:56–65

Wallerstein NB, Duran B (2006) Using community-based participatory research to address disparities. Health Promot Pract 7:312–323

Walker JS (2000) A short history of nuclear regulation, 1946–1999. NRC. Washington

Walters CT, Hilborn R (1978) Ecological optimization and adaptive management. Ann Rev Ecol Syst 9:157–188

White PCL, Gregory KN, Lindley PJ, Richards G (1997) Economic values of threatened mammals in Britain: a case of the Otter *Lutra lutra* and the Water Vole *Arvicola terrestris*. Biol Conserv 82:345–354

Wies TM, Ilinca A, Pinard J-P (2008) Stakeholders' perspectives on barriers to remote wind-diesel power plants in Canada. Energy Pol 36:1611–1621
Williams BL, Brown S, Greenberg M, Kahn MA (1999) Risk perception in context: the Savannah River Site stakeholder study. Risk Anal 19:1019–1035
Wustenhagen R, Olsink M, Burer MJ (2007) Social acceptance of renewable energy innovation: an introduction to the concept. Energy Pol 35:2683–2691

Chapter 2
Minority Participation in Environmental and Energy Decision Making

James H. Johnson Jr.

Contents

Abstract The most eloquent decisions are most often made when the decision-making process includes multiple perspectives. In the past, the lack of participation of minority and low income populations has lead to disproportionate impacts as a result of decisions affecting the environment and human health. In 1994, President Clinton signed Executive Order 12898 to require all federal agencies to develop a strategy to ensure fair treatment and meaningful participation of minority and low income populations. This chapter outlines the range of minority and low income stakeholders that should be considered in environmental and energy decision-making processes, their unique roles, and some of the key issues that should be included when identifying the stakeholders.

J.H. Johnson Jr. (✉)
Department of Civil Engineering, Consortium for Risk Evaluation with Stakeholder Participation (CRESP), Howard University, Washington, DC 20060, USA
e-mail: jjohnson@howard.edu

J. Burger (ed.), *Stakeholders and Scientists: Achieving Implementable Solutions to Energy and Environmental Issues*, DOI 10.1007/978-1-4419-8813-3_2,

2.1 Introduction: Problem Definition

The 1971 Council on Environmental Quality annual report acknowledged that racial discrimination adversely affects urban poor and the quality of their environment (CEQ 1971). The report observed the abnormally high levels of lead found in the blood of children from urban poverty neighborhoods in New York City. For the most part the residents of these neighborhoods were nonwhite families. This acknowledgement is said to have given birth to what is now known as the environmental justice (EJ) movement. Some of the highlight events on the timeline of the EJ movement includes the following:

- GAO Report states that three out of four hazardous waste facilities in EPA's Region 4 are in African-American communities and at least 26% of the residents are below the poverty level (1983). This was one of the first reports to present data to quantify the disproportionate exposure of minority and low income populations to the risk associated with an environmental facility.
- Robert Bullard's Book Dumping in Dixie (1990) is the first published textbook on EJ.
- President Clinton issues Execution Order (EO) 12898 requiring federal agencies to be accountable for EJ and calling for implementation plans. This EO directed federal agencies to analyze the environmental, human health, economic, and social effects of federal actions on communities, including minority and low income communities, as outlined in the National Environmental Policy Act (NEPA). It also directed federal agencies to provide opportunities for community participation to identify impacts and mitigating actions in the NEPA process (CEQ 1997).
- Nuclear Regulatory Commission (NRC) Licensing Board analyzes the applicability of EO 12898 in Louisiana Energy Services licensing application to privately own and operate a uranium enrichment facility (Discus 2002; NRC 1998). This was the first consideration by the NRC of EO 12898.
- Department of Energy issued its initial EJ Strategy in response to EO12898 in 1995. In 2008, it reviewed and updated its EJ Strategy (2008) and designated the Office of Legacy Management as the lead EJ program office.
- US Environmental Protection Agency issues its draft Plan EJ 2014 (2010a). One of Administrator Lisa Jackson's seven goals is to "Expand the Conversation on Environmentalism and Working for Environmental Justice." This draft represents EPA's response to EO 12898 and amplifies the administrator's commitment to outreach and fair treatment of historically underrepresented communities in EPA's decision-making processes. The USEPA is the designated lead Federal agency for EJ.

2.1.1 *The Need for Clinton's Executive Order*

Why the need for an Executive Order by President Clinton (1994) – an order that has been continued by every President since 1994? Why the need for a movement known as EJ?

Table 2.1 Range of minority stakeholders to be considered in decision processes and their unique roles

Type	Role
Potentially impacted population	Provides cultural, religious, medicinal, and aesthetic information
Interested population	Provides the viewpoint of the general population and helps to protect the integrity of the environment
Susceptible population	Provides a benchmark for the protection of human health against deleterious impacts
Trusted (minority) community organizations[a]	Provides information to help populations to understand what is being proposed and helps in the negotiation during the decision-making process

[a]Personal communication: Dr. Mark Mitchell, President, Connecticut Coalition of Environmental Justice, Harford, Connecticut

Simply, it is about fair treatment and *meaningful involvement in the decision-making process*. Meaningful involvement is characterized by not only when involvement takes place, but also the depth of the involvement (i.e., engagement) and the embracing of all the stakeholders. Table 2.1 lists the range of Minority Stakeholders that should be considered in the environmental decision-making process. The participation of each of the stakeholders listed in Table 2.1 is critical to reaching decisions that all parties feel are reflective of their concerns and equitably distributes the benefits and harmful impacts of the decision actions. For example, the lack of involvement of trusted community organizations can result in scientific information not being understood by minority and low income populations who often are below the average educational attainment levels. This lack of understanding could lead minority and low income populations to distrust scientific information and the overall decision-making process. This distrust is the result of minority and low income populations feeling as though they are at an information disadvantage or even feel talked down to. As a stakeholder participant in the decision-making process, trusted community organizations help the communities understand scientific information through educational and outreach programs about environmental and human health impacts, and assist in the preparation of testimony to state and federal agencies. In states like Connecticut, community organizations also assist in the understanding and implementation of state EJ laws.

2.1.2 Objectives of the Chapter

The objectives of this chapter are to first outline evidence of the unfair treatment of minorities in the siting of environmental and energy production facilities, and second to highlight efforts of the US EPA and Department of Energy to provide effective mechanisms for the participation of minority and low income populations in the decision-making processes to move toward equity in the impacts – both good and bad – of such decisions.

2.2 The Need for Minority Stakeholder Participation

There have been a host of studies demonstrating the disproportionate location of environmental facilities in minority and low income communities (see e.g., Dumping in Dixie, Bullard 1990; Bunyan and Mohai 1992). A few are highlighted below to provide a context for the basis of the action of EO 12898 and the subsequent EJ strategies and guidance developed by federal agencies.

2.2.1 Socio-Demographic Factors

Perlin et al. (1999) examined the socio-demographics of people living near industrial sources of air pollution in three areas of the United States. Using data from the 1990 Toxic Release Inventory (TRI) and the 1990 Census, the relationship between poverty status and race to the location of single or multiple air pollution emission sources was analyzed. The results for all three sites – Kanawha Valley, West Virginia, Baton Rouge – New Orleans corridor, Louisiana, and Baltimore Metropolitan area, Maryland were consistent. On average, African-Americans and households living below the poverty level lived closest to the nearest TRI facility and within two miles of multiple TRI facilities.

Elliott et al. (2004) took an expanded view of the relationship of the socio-demographics of people and the associated risks with respect to hazardous facilities. Their study considered the risk associated with the location of a facility and the risk associated with the methods of operation and standards of care that are used at a facility. Data for their study were obtained from EPA's RMP Info*database. RMP*Info database is set forth under Section 112 (r) by the 1990 Clean Air Act Amendments. With certain exceptions, facilities storing on-site at least one of 77 toxic or one of 63 flammable substances above a threshold quantity are mandated to file an assessment of hazards and accidents at the site. The study results indicated that facilities with more employees and facilities using a large number of regulated chemicals were more likely to be located in more heavily populated and more heavily African-American counties – location risk. In addition, facilities that were at a greater risk of accident and injury were in more heavily populated African-American counties – operation risk.

In 1979, Robert Bullard (Bullard 1990) was asked to conduct a study of the spatial location of municipal solid-waste disposal facilities in Houston. The information was to be used in a class action lawsuit against the city of Houston, the state of Texas and a garbage hauling and disposal company. The lawsuit was prompted by a plan to place a municipal landfill in a community where 82% of the residents were African-American. His study revealed that the siting of waste facilities in Houston were in predominantly African-American Communities. He also noted that the communities were in existence prior to the siting of the waste facilities in the communities.

Bullard (1990) later explored whether African-American communities in the South had been disproportionately exposed because of the location of waste or other types of environmental hazards. His exploration was done in the context of the following questions:

- Who are the most affected?
- Why are they affected?
- Who created this problem?
- What can be done to remedy the problem?
- How can the problem be prevented?

As shown in the discussion above, the *most affected* are the minority and low income populations. Beyond the impacts of location and operation risk cited by Elliott et al. (2004), the vulnerability of low income and minority populations must also be included. The consideration of vulnerabilities because of diet habits, multiple and cumulative exposure to risks, poor housing environments, employment in high-risk jobs, etc. are key to developing a holistic picture of the health impacts of decisions on minority and low income populations.

2.2.2 *Why Are Minority and Low Income Populations Affected?*

The bulk of the evidence suggests minority and low income populations were affected because of lack of empowerment to participate in the decision process. Bullard (1990) also cites limited housing and residential options combined with discriminatory siting of facilities as reasons minority and low income populations are disproportionately affected. The key point is fairness in the distribution of the benefits and harmful impacts of siting facilities of any type – disposal facilities (e.g., landfills and incinerators), industrial facilities, energy production facilities whether they be nuclear, wind, solar, or fossil. Empowerment involves the ability to intelligently and actively engage in the decision-making process from its beginning. Intelligent engagement many times will also include the need for resources for capacity building.

2.2.3 *What Can Be Done to Remedy the Problem?*

Table 2.2 lists some key issues that are often ignored when minorities and low income populations are asked to be involved in decision-making processes related to environmental and energy facilities. Each of the issues in Table 2.2 is important and must be considered in the discussion (Fig. 2.1).

Minority and low income populations are often thought to be monolithic (i.e., they have one point of view as a group). This is a myth. In the study by Elliott et al. (2004) it was found that counties with greater proportions of African-American populations tended to have slightly higher poverty rates suggesting a high degree of income inequality. Differences within these populations can also be found in educational level, thought processes, and perceived impacts. Therefore, representation of

Table 2.2 Key issues often ignored when including minority and low income stakeholders in environmental and energy-related decision making

- Complete community representation
- Prior history with the decision-making authority – trust issue[a]
- Capacity to engage in science decision
- Disruption on the community structure associated with an action
- Consideration of relevant public health data and industry data including multiple or cumulative exposure
- Faith and science intersection of the stakeholders[a]

[a] Personal communications: Ms. Melinda Downing, Office of legacy Management, DOE and Mr. John Rosenthal, National Small Town Alliance, Washington, DC

Fig. 2.1 Some key issues that are often ignored when minorities and low income populations are asked to be involved in decision-making processes related to environmental and energy facilities. These issues are important and must be considered in the discussion

all sectors of the minority and low income populations is critical to their engagement in the decision-making process.

A major barrier to community involvement is the prior history with the decision-making authority. This is the trust issue. A prior history of broken trust (see e.g., Benedek (1978): The Tuskegee Study of Syphilis) presents a barrier to current cooperation and support. One bridge to reinstatement of trust in a decision-making authority is finding trusted environmental, health, or other community organizations that are active in minority and low income communities. These organizations can help to explain the significance of what is being proposed to the communities and

Fig. 2.2 Minority and low income populations should build the capacity to directly engage in science decision making. This requires resources for training community leaders and providing materials in a language, format, and media

facilitate a more harmonious working relationship between minority and low income stakeholders and stakeholders from the science community, regulatory authorities, etc.

In addition to the engagement of trusted community organizations, minority and low income populations should build the capacity to directly engage in science decision making. This requires resources for training community leaders and providing materials in a language, format, and media that are both appealing and understandable by the populations (Fig. 2.2). An often untapped resource for capacity building is students. Students at all levels – K-12 and higher educational levels – provide fertile minds that are not biased by tradition and have unlimited capacity to learn. The engagement of students through educational activities in and outside of the classroom can lead to dialogues first with their parents, then parent to parent, and lastly parents with other stakeholders. The resources required to build capacity and provide educational materials for students are often not considered or provided.

The disruption of a community's structure or culture associated with the decisions can be significant and should be reviewed. For example, in one NRC case involving the licensing of a proposed centrifuge enrichment facility, there was a concern about the blocking of the route between a local community and a local church (Discus 2002). The community affected was composed of low income residents and many of the individuals did not own cars. The location for the proposed facility would block the existing walking route to the church and alternative routes would be significantly longer. The project was abandoned before this issue was resolved. This was the first consideration of an EJ situation for NRC. In Private Fuel Storage (NRC 2004), NRC stated that EJ as applied to NRC "means that the agency will make an effort under NEPA to become aware of the demographic and economic circumstances of local communities where nuclear facilities are to be sited and take care to mitigate or avoid special impacts attributable to the special character of the community." NRC is not obliged under Executive Order 12898 but considers EJ matters in the normal NEPA context. As such the focus is mainly on the identification and weighing of disproportionately significant and adverse environmental impacts on minority and low income populations.

Consideration of relevant public health and industry data are important in identifying disproportionate impact. The EJ Guidance under the NEPA (CEQ 1997) specifically mentions the examination of public health and industry data for the potential for multiple of cumulative human health or environmental hazards. The guidance further states these effects should be considered even when they are not in the control of the agency conducting the environmental assessment or environmental impact statement.

The role of faith in minority and low income populations can shape actions for stakeholders with respect to science. There are various areas of scientific exploration that many stakeholders will neither consider nor discuss based upon faith. In addition, there are other areas where faith drives fear of discussion and consideration. Therefore, faith can drive both inaction and action in terms of participation in decision making processes. The engagement of faith leaders early in the decision-making process can help extenuate fears of discussion and consideration of ideas and build trust in the interactions among stakeholders.

2.2.4 How Can the Problem Be Prevented?

Clearly disproportionate location of facilities can be prevented when potentially impacted populations, interested populations, trusted community organizations, and susceptible populations are actively engaged in the decision-making process. Executive Order 12898 requires EJ to be a part of each Federal Agency's Mission and emphasizes four issues pertinent to NEPA (CEQ 1997):

1. Each agency specifies an Executive Justice strategy.
2. Importance of research, data collection, and analysis particularly with respect to multiple and cumulative exposure.
3. Agencies should collect, maintain, and analyze information on the pattern of subsistence consumption of fish, vegetables, and wildlife.
4. Ensure effective public participation (see Chap. 1).

The later issues are particularly important and the focus of this chapter.

2.3 Stakeholder Participation in the Decision-Making Process

The Department of Energy and the US EPA have formulated Executive Justice Strategic Plans in compliance with Executive Order 12898. The Department of Energy has designated the Office of Legacy Management as its office responsible for implementing EJ strategies in its programs. The Environmental Protection Agency is the lead Federal Agency for EJ. The Interagency Group on EJ met for the first time in 10 years in September 2010.

The DOE published its revised EJ Strategic Plan in 2008 (DOE 2008). It defines Executive Justice as "... fair treatment and meaningful involvement of all people, regardless of race, ethnicity, culture, and income or education level with respect to development, implementation, and enforcement of environmental laws, regulations and policies." The plan recognizes that traditionally minority, low income, and Tribal communities have lacked access to the required information and technical advice to be informed participants. The vision for the plan includes public participation and trust. The following goals are articulated in the Plan:

1. Identify and address programs, polices, and activities that may have disproportionate human health or environmental impacts on minority, low income, and Tribal populations.
2. Enhance the credibility and public trust of the department by making public participation a fundamental component of all program operations, planning activities, and decision-making process.
3. Improve the research and data collection methods characterizing the human health and environment of minority, low income, and Tribal populations including the assessment of multiple and cumulative exposures.
4. All activities and processes should have an Executive Justice component related to human health and the environment.

In reference to Goal 2, DOE has identified several actions to enhance public participation (DOE 2008):

- Integration of Executive Justice with the Public Participation Policy
- Reflection of its advisory boards of the communities they represent
- Identification of community-based organizations, networks, and media that it should communicate with and through

The United States EPA recently published its draft Plan EJ 2014 strategic plan (EPA 2010a). EJ 2014 is a road map to help EPA to integrate EJ into its programs. EJ2014 is divided into the following three sections:

1. Cross-agency focus areas
2. Tool developments
3. Program initiatives

EPA has also developed an Interim Guidance on Considering EJ during the Development of an Action (EPA 2010b). The guidance is nonmandatory and seeks the answer to three basic questions including "How did the public participation process provide transparency and meaningful participation for minority, low income, and indigenous populations and Tribes?" Environmental concerns are identified as actions that may lead to disproportionate impacts such as lack of trust, lack of information, language barriers, sociocultural issues, and lack of traditional communication channels. Meaningful participation is clearly defined as going beyond the minimum requirement of standard notice and comment procedures and requires early engagement.

The EPA recently hosted a symposium on the Science of Disproportionate Environmental Health Impacts (EPA 2010c). The participants provided recommendations in the area of science, policy, capacity building, and promoting health of sustainable communities. Of particular note are the following recommendations and comments:

- Regulatory development process should engage communities early in the policy-making process.
- EPA should provide staff to effectively interact with communities.
- The Agency has taken the first step of engaging communities in the development of an Agency Action (EPA 2010b). The Guidance provides clear direction on how and when Executive Justice related questions and community participation are required for Tier 1, 2, and 3 decisions.

2.4 A Concluding Word

The participation of minority and low income populations should be developed on a case-by-case basis. Table 2.1 identifies the span of participants and Table 2.2 the issues often ignored when engaging minority and low income populations. The goal is to ensure fair treatment by all citizens through an active dialogue to have a conversation with all stakeholders that will lead to decisions that share both the beneficial and harmful impacts of environmental and energy activities.

Table 2.3 provides strategies to ensure inclusion of minority and low income populations in decision-making processes. The physical, intellectual, and communicative capacity of communities to effectively engage in the decision-making process

Table 2.3 Guidelines for ensuring inclusion of minority and low income stakeholders in the decision-making process

- Conveying issues in ways that are tailored (e.g., translation, timing, location) to each community
- Bridging cultural and economic differences that affect participation
- Using communication techniques that enable more effective interaction with other participants
- Develop partnerships on a one-to-one or small-group basis to ensure representation
- Develop trust between government and potentially affected populations
- Develop community capacity to effectively participate in future decision-making process

Sources: DOE EJ strategic plan (DOE 2008), EPA's EJ 2014 (EPA 2010a), and EPA's Draft EJ strategic plan

is key to sustaining a trusted relationship among stakeholders. The engagement of minority and low income populations should begin at the conceptualization stages of the decision-making process – not at the end when the decision is about to be made. The importance of trusted community organizations to provide the initial bridges between stakeholders to ensure communication modes and methods are effective, to assist with building capacity, and to help form partnerships is paramount to the overall success of the process.

References

Council on Environmental Quality (1971) Environmental Quality, The Second Annual Report of the Council on Environmental Quality, DC

GAO (1983) Siting of Hazardous Waste Landfills and Their Correlation With Racial and Economic Status of Surrounding Communities. RCED-83–168

Bullard DR (1990) Dumping in Dixie, Race, Class and Environmental Quality. Westview Press, Boulder, CO

Clinton W (1994) Executive Order on Federal Actions to Address Environmental Justice in Minority Populations and Low-Income Populations. White House, Washington, DC

Council on Environmental Quality (1997) Environmental Justice, Guidance under the National Environmental Policy Act. Executive Office of the President, Washington, DC

Dicus JG (2002) Growing factors in Radiological protection on the Environment. Office of Public Affairs, Washington, DC

NRC (1998) Louisiana Energy Services (Claiborne Enrichment Center). CLI-98-3. NRC77

U.S. Department of Energy (2008) Environmental Justice Strategy. U.S. Department of Energy, Washington, DC

EPA (2010a) Recommendations on the EPA's Environmental Justice Plan EJ 2014 and Permitting Charges. EPA Office of Enforcement and Compliance, Washington, DC

Bunyan B, Mohai P (eds.) (1992) Race and the Evidence of Environmental Hazards: A Time for Discourse. Westview. Boulder, CO

Perlin AS, Sexton K, Wong WS D (1999) An Examination of Race and Poverty for populations living near industrial Sources of Air Pollution. EPA National Center for Environmental Assessment, Washington, DC

Elliott MR Dr, Wang Y, Lowe RA, et al (2004) Frequency and Severity of US Chemical Industry accidents and the socioeconomic status of surrounding communities. Journal of Epidemiology and Community Health, 58(1):24–30

Benedek TG (1978) The Tuskegee Study of Syphilis: Analysis of Moral versus Methodological Aspects. Journal of Chronic Diseases, 31:31–50

EPA (2010b) Interim Guidance on Considering Environmental Justice during the Development of an Action. EPA Office of Policy, Economics and Innovation, Washington, DC

EPA (2010c) EPA Symposium on the Science of Disproportionate Environmental Health Impacts. Washington Convention Center, Washington, DC

NRC (2004) Policy Statement on the Treatment of Environmental Justice Matters in NRC Regulatory and Licensing Actions. U.S. Nuclear Regulatory Commission, Rockville, MD

Chapter 3
Energy Diversity: Options and Stakeholders

Michael Gochfeld

Contents

M. Gochfeld (✉)
Environmental and Occupational Medicine, EOHSI,
Consortium for Risk Evaluation with Stakeholder Participation (CRESP),
and UMDNJ-Robert Wood Johnson Medical School,
170 Frelinghuysen Road, Piscataway, NJ 08854, USA
e-mail: Gochfeld@eohsi.rutgers.edu

J. Burger (ed.), *Stakeholders and Scientists: Achieving Implementable Solutions to Energy and Environmental Issues*, DOI 10.1007/978-1-4419-8813-3_3,

Abstract Energy, climate, food, and economic development are intertwined at regional, national, and global levels. There are large international disparities in the availability of, and demand for, energy which will be exacerbated in the near future as world population increases and per capita demand grows. The traditional energy chains from fossil fuels to electricity, heat, and transportation are being diversified by increased reliance on renewable energy sources, as well as new technologies in varying stages of development. As fossil fuel reserves dwindle, thereby becoming more costly, and carbon and climate considerations grow, diversification to low carbon renewables becomes more attractive and cost-effective. Layers of stakeholders include owner–investors, workers, consumers, and regulators, each with different stakes in different energy chains. Issues of economic, ecologic and aesthetic consequences, footprint, emissions, and costs challenge stakeholders to agree on energy options. Often overlooked in stakeholder discussions, investors play a powerful role in the invention, design, demonstration, and implementation of new technologies. Often undervalued, workers facing health and safety hazards are stakeholders influencing the design, construction, and operation of energy chains. As energy dispersion (rather than large centralized power plants) becomes more popular or necessary, siting issues will confront larger numbers of neighbor stakeholders. Certain groups like the United Nations and the U.S. Department of Energy are positioned to facilitate stakeholder input on the international and national scale, achieving a diversity of energy chains.

3.1 Introduction

On a global scale, we are all stakeholders in the intertwined domains of energy, climate, food, and development. The "energy crisis" and energy "sustainability" engender a spectrum of opinions. For some the crisis lies in carbon and climate, for others in food security or national security, and still others focus primarily on equity issues of energy availability, distribution, and economic development. Many sectors and competing needs, demands, and desires seem to make the overall match between consumption and availability of energy intractable, even on a national, let alone a global, scale. As global population grows and developed countries experience growing per capita demand, the picture becomes gloomier for long-term energy sustainability. It is customary to describe energy as "renewable vs. nonrenewable," or in terms of carbon generation vs. neutrality. Another classification is energy capture (i.e., solar, wind) vs. release (i.e., fossil fuel or biomass).

From a media perspective, after decades of obliviousness and often heated scientific argument (Schneider 2009), the issue of climate change accelerated by carbon dioxide and carbon from energy production has caught public attention (Pearce 2005). Each type of energy (e.g., coal, oil, wind, solar) constitutes an energy chain from production to user, including, for fossil fuel and nuclear, the entire fuel cycle from mining to waste. Carbon emissions are now a major topic that stakeholders at all levels consider, and after the 1997 Kyoto conference, carbon has become a commodity as well (ICBE 2010).

The world is on the threshold of another great historical energy transition (Gore 2008; Smil 2010), following the wood to coal, and the coal to petroleum transitions of prior centuries. Stakeholders can view this as the fossil fuel to diversity transition, while wistfully awaiting a future transition to renewable, low carbon emission energy. Goldenberg (2007:808) argues that increasing global reliance on renewable energy will enhance sustainable development while prolonging fossil fuel reserves. Energy production (by capture or release) is both costly and profitable. Energy is closely tied to commerce, investment, and development (chapters in Hanjali'c et al. 2008), as well as equity (Johnson 2011), and quality of life. Conservation (use reduction) attracts great interest, but provides little individualized incentives. Exploitation, distribution and sale of relatively inexpensive fossil fuel has impeded the development of previously less profitable, and less energy-dense, renewable energy chains, while the vagaries of fuel prices challenge stakeholders to make sound investments. Intensive exploitation including nuclear energy, deep ocean drilling for oil, and fracture mining of natural gas, have become associated in the popular press and public mind with environmental degradation and threats to human health, with various stakeholder groups mobilized against these technologies.

Stakeholders are involved in invention and technology (R&D), investments, markets, policies, jobs, and behavior (both collective and individual). All energy chains have life cycle ramifications, land use and footprint considerations, and environmental consequences that transcend the facilities, whether centralized or dispersed. Stakeholders abound, and conversely, everyone everywhere is a stakeholder at some level. Each type of energy, whether centralized power plants or dispersed turbines and household solar panels, depends on the invention and development of conversion, generation and storage technologies and transmission systems, and supports secondary industries which produce, construct or erect, and maintain the generation capacity. All the ways in which we use energy, for example, heating, electricity, and transportation, must be designed to operate with, or take advantage, of diverse sources. For the enhancement of renewable sources, optimists emphasize individual and governmental choices. Pessimists emphasize technological limitations.

All energy chains from source to consumer involve infrastructure construction and operation to produce, transmit, and use the energy. Each step in the process has inefficiencies resulting in losses that challenge technology. Ultimately the energy used to light, heat, and power homes, vehicles, and institutions as well as the energy needed to provide food (and biomass fuel), are subject to these inefficiencies which can be reduced but not eliminated. Moreover, as with food and water, demands for electricity and transportation fuel increase with population and standard of living. Indeed the global population increase, projected to exceed 9 billion by 2050, coupled with the anticipated increased standard of living demanded by the developing world, is the ultimate driver of the growing demand for energy for transportation, electricity, heating, cooking and playing (Bockris 2008).

3.1.1 Temporal Scales

Decisions that were made (or not made) by stakeholders a generation ago plague us today, while decisions in this decade will impact stakeholders still unborn. The bumper sticker "we borrow the future from our children" captures the essence of transgenerational stakeholders. A current direction is to "hang on" as long as we can, squeezing every drop of energy and dollar, out of current fossil resources, trusting the future to find its own way. Willy Ley (1954) admonished that "nobody can predict what may be invented in the meantime." Alternative energy sources may extend the life of fossil fuels (for the uses to which they are best suited), by integrating other fuel sources where applicable (Goldenberg 2007).

3.1.2 Spatial Scales

Geographic or geopolitical scales are also important. Siting issues, a theme in this volume, extend relatively short distances beyond the proposed fence lines of, or transportation corridors to, proposed facilities. Investments in one or another energy proposal and their impacts on land use decisions, whether cutting forests for biomass, or selling corn for fuel rather than food, impact stakeholders on a broader scale. Whereas fossil fuels can be packaged and transported, some renewables, such as wind and solar energy, are limited by irregular availability and inherently inefficient storage and transmission systems which impose a limitation on their geographic distribution. International policy issues involved in petroleum commerce, including the United States' reliance on foreign oil, have been conspicuous for decades. However, today transnational stakeholders face decisions such as exchanges of carbon credits or negotiating the transfer of African solar generation to meet European electric demands.

3.1.3 The Energy Commons

The global pool of energy, including renewables, available at any point in time is finite, creating, in effect, a common pool resource, subject to all of the commons issues that have been written about by both pessimists (Hardin 1968) and optimists (Ostrum 2001). Ultimately many of our sources of energy were derived or are derived from the sun, and commercial efforts to control the various sources are probably the largest industry on earth (Berman and O'Connor 1996). Agreements may need to be at the United Nations level analogous to "Oceans and Law of the Sea" (UN 2010).

3.1.4 Objectives of This Chapter

The two previous chapters introduced the general issue of stakeholders and equity. This paper focuses on stakeholders for various energy options. It provides an overview of energy sources and consumption, and their relationship to stakeholders. The sources include both the major current sources (i.e., fossil fuel, hydropower, solar, wind) and less well-known and explored options. While the purpose is to provide a framework for energy sources, some attention is given to ways in which stakeholders can influence decisions regarding energy-related issues. Fuller descriptions of particular energy-related situations in which stakeholders have improved or impeded energy policies are discussed in the remainder of the book.

3.2 Stakeholder Overview

Certain categories of stakeholders are common to most or all forms of energy production (see Chap. 1). Primary stakeholders are those directly involved in an energy chain (including the fuel cycle, construction and operation of facilities, and delivery of energy) (Table 3.1). Also primary are the consumers and neighbors, the latter making up the "public" that needs to be involved in siting decisions regarding new facilities or installations. Secondary stakeholders are more removed from the chain including investors, inventors, R&D community, and academics on the one hand and regulators on the other.

On the production end of energy chains are functions such as construction, operations, and maintenance. On the consumer end are the facilities and individuals who use energy and the industries that support the development and production of devices that utilize energy. In between are transmission and distribution systems. Each of these in turn involves companies involved in design, production, and sales. Stakeholders are involved at various levels including planning and siting new facilities, or even in evaluating health impacts in surrounding communities (Jardinre et al. 2007).

Any "company" represents additional levels of stakeholders: owners, shareholders, employees, and various subcontractors that provide security, maintenance, and increasingly other operations. Indeed, large refineries and power plants are increasingly being operated by contract work forces, stakeholders in an evolving energy economy, who incur added health and safety risks (Gochfeld and Mohr 2007). These are stakeholders whose jobs rely on the health and growth of their energy sector. For most of these stakeholders, growth of consumption in their chain is more desirable than conservation.

Table 3.1 Stakeholders in the energy chain

At the source: mining, milling, processing: owners, workers, families, neighbors, local governments
At facilities (power plants): owners, workers, families, neighbors, local governments
Shipping or transmission: owners, drivers, route neighbors, local governments, regulators
Users: governments, researchers, institutions, corporations, commercial, individual
Worker groups and unions: from mines power plants to farms
Environmental stakeholders: regulators, natural resource trustees, researchers, public interest groups

Overlying all of these stakeholders is the ownership system: investors, entrepreneurs, industrial consortia, governments which develop, operate, or subsidize each energy chain. Governments, in this case are owners as well as regulators. The entrepreneurship system is an important variable as shown by the contrast between monopolitistic and free market development of wind energy in the Netherlands (Agterbosch et al. 2004). In addition, since energy chains are in competition for markets, customers, and even subsidies, the primary stakeholders in one chain are inevitably secondary stakeholders in other chains.

Environmental impacts of different energy chains involve additional stakeholders (regulators, nongovernmental environmental groups, natural resource trustees, and researchers concerned with waste disposal, environmental health, and ecosystem integrity)

3.2.1 Investors

Often overlooked are investors (or lack thereof) who have had potent influences on what and where energy sources are developed or available. On the international scale, the World Bank or Inter-American Development Bank (IBD) influence the selection and development of energy chains in developing countries. For example, IBD provided a $40 million loan to Nicaragua to expand a geothermal power project. The 72 MW facility has an estimated price tag of $177 million, with geothermal contributing 15% and renewables 86% of the country's electric generation (Renewable Energy News Aug 4, 2010). Venture capitalists play important roles in supporting the development of new technologies, and their choice of what to fund in the short run are *de facto* decisions in the long run.

3.2.2 Secondary Stakeholders

Secondary stakeholders have important roles in energy policy even if not directly employed in, affected by, or using an energy chain. Educators, communicators, and regulators contribute at many levels in an energy chain. A crucial stakeholder in the new transition, in addition to producers, investors, regulators, and consumers, are inventors. These include inventors and the R&D community whether located in industry, universities, or government. Inventors are busily developing new technologies and new applications of existing technologies to exploit specialized niches and improve both efficiency and profitability. Invention, marrying imagination with feasibility, is recognized as crucial for the storage/transmission enigmas. More importantly, inventors are required to develop the new motors that will operate on new energy sources. The vision that everything can be turned into electricity and used "as is," does not apply, for example, to transportation. Will airplanes be able to run on wind and solar, or will these new technologies, simply spare fossil fuel, while development of practical fuel cell transportation occurs?

Training programs in academia and industry must evolve rapidly to embrace new developments in energy and climate, and many new energy-related centers and departments are opening at various institutions. Most importantly, training in nuclear and radiation sciences and technology, which declined greatly over the last 30 years, corresponding to the dearth of new nuclear facilities in the United States, will have to be resurrected.

Governments, including the United Nations, become stakeholders at multiple levels. Energy is both cost and opportunity, and is closely linked to economic development, accelerating it in countries rich in energy-yielding resources. The United Nations Environmental Program (UNEP) has identified the following six priority areas: climate change, disasters and conflicts, ecosystem management, environmental governance, harmful substances, and resource efficiency, all of which intersect energy commerce. UNEP also addresses implicitly global equity issues: "Decoupling economic growth from environmental impact, and creating the 'space' for poor people to meet their basic needs, will require producers to change design, production and marketing activities. Consumers will also need to provide for environmental and social concerns – in addition to price, convenience and quality – in their consumption decisions," (UNEP 2010).

In the United States, the Department of Energy is extensively involved in energy research through its Office of Science (DOE-OS 2010) and its National Renewable Energy Laboratory (NREL). The NREL has a broad agenda of R&D priorities, including partnering with industry to bring new technologies online (NREL 2008).

3.2.3 Workers as Stakeholders for Health and Safety

At all levels of all energy chains there are workers, ranging from the Navajo uranium miners or West Virginia coal miners, to the utility employees repairing damaged power lines. "Safety" in the context of energy planning usually refers to facility security (from natural disasters, terrorism, or malfunction), rather than worker safety. The energy industry is not inherently safe for workers, although the principles of occupational health and industrial hygiene (Plog et al. 2004) provide the means to prevent injury and illness for all workers. In large corporate facilities, worker health and safety has a voice through its unions, while in distributed systems without such organizations, workers are seldom vocal about safety issues.

Mine safety with respect to explosions and collapses, trapped miners, and deaths is a highly visible issue, while exposure to radiation in underground mines receives less attention, but leads to high rates of disease (Roscoe et al. 1989). Even renewables, such as wind power, require an industrial production base (with attendant hazards), on-site construction workers to erect the towers and install the turbines, and maintenance crews (with occasional fatalities Gipe 2010). The Oregon Solar Energy Industries Association has published a comprehensive safety manual for worker safety during construction and erection of solar arrays (OSEIA 2006). Likewise, community stakeholders provide reports on the occupational and environmental hazards associated with solar construction (SVTC 2009).

"Agriculture ranks among the most hazardous industries" (NIOSH 2010), and worker health and safety on the industrial scale farms that will produce biomass for direct burning or conversion to liquid fuels are seldom addressed or considered.

3.3 Uses of Energy

Energy uses include agricultural and industrial production, heat, electricity, cooking, transportation, and recreation, to name a few (Table 3.2). Although electricity is a common pathway for many sources of energy, not all sources are equivalent for different uses.

Large-scale users such as government(s), military, industry, and large commercial firms (and farms) have great opportunities to achieve energy conservation goals through economies of scale, such as large-scale installation and purchases of energy-efficient transportation, lighting and heating, and wholesale energy purchases. More dispersed stakeholders such as residences, tourists, personal transport, and small farms have more limited opportunities to benefit from energy conservation incentives.

3.4 Sources of Energy

The sources of energy described are organized by the environmental medium: air, earth, water, and sun.

3.4.1 Air

Wind is generated by uneven solar heating of the planet surface, influenced by ocean surface and land topography. Wind is irregularly distributed in time and space, and in the absence of suitable storage, does not support 24/7 electricity demands. The North Atlantic region has favorable wind conditions, and plans for new construction are

Table 3.2 Users as stakeholders

Institutional (including government)
Industrial/commercial (including energy industry)
Household/individual
Recreation/tourism
Transportation
Agriculture
Military

Given are domains of energy use

Fig. 3.1 Wind turbines on the coast of Newfoundland (photo Joanna Burger)

in various stages. Figure 3.1 shows a wind facility on the Newfoundland coast. Wind is already a major electricity source in many areas of Europe with Spain getting over half of its energy from wind as of 2009 (Keeley 2009).

Just as energy alternatives influence climate, so climate change may influence energy options. Breslow and Sailor (2002) predicted that climate change would reduce wind speeds in the United States by a few percent during the twenty-first century. Siting wind facilities encounters opposition on aesthetic grounds (Saito 2004), and natural resource trustees voice concern about impacts on migrating bats and birds, particularly raptors (NRC 2007). But many environmentalists champion expansion of wind energy production.

3.4.2 Earth

In addition to the traditional earth-derived fossil fuels (coal, oil, natural gas), there are other materials such as peat as well as modified materials such as oil gas, extracted from the earth. Liquid natural gas has been developed to facilitate transport of gas where pipelines are not a cost-effective option. Oil shale and tar sands were already on the energy radar screen by the 1950s (Ley 1954:160) and are in various stages of development. A 2010 controversy concerns environmental impacts of tar sand oil extraction as well as a proposed 2,000-mile oil pipeline from the Alberta tar sands to Texas.

Fig. 3.2 Geothermal plant in southern California (photo Joanna Burger)

3.4.2.1 Geothermal and Hydrothermal

Geothermal energy offers a largely untapped though local energy source to provide electricity and heat. There are various types of geothermal energy pools dependent on the underlying geologic conditions. Commercialization depends on how deep one must drill to reach useful hot temperatures, often several kilometers below the surface. In some areas hydrothermal pools can be tapped directly. In other places surface water is pumped down under pressure, inducing fracturing, analogous to natural gas fracturing (see below). Until recently, geothermal plants were mainly found in areas of high tectonic activity and, for example, are conspicuous on the landscape of southern California (Fig. 3.2) R&D continues to enhance the efficiency of conversion and the capture of residual heat through expanding the plant cycle utilization efficiency, such that geothermal/hydrothermal heat can be tapped more widely. Enhanced geothermal technology may make this one of the leading energy sources by mid-century (Renner 2008).

Except for the construction phase and the fuel required to operate pumps, geothermal energy is considered highly sustainable. However, it does release some greenhouse gases, particularly CO_2 from the earth, although much less per kilowatt hour than fossil fuel burning (Renner 2008). Naturally occurring hot springs have been used for decades to provide heating and hot water to communities in a few favorable areas such as New Zealand and Iceland, and hot water from near surface geothermal is widely used for heating, greenhouses, and even snow melting. On the small scale, buildings including homes can tap the relatively fixed subterranean temperature about 5 m below the surface, using geothermal heat pumps

3.4.2.2 Coal

This is an old historic energy source. Developed countries have reduced reliance on coal, but it remains a major source of electricity production and air pollution. Although the finiteness of coal reserves was recognized a century ago (reported by Ley 1954), coal reserves are substantial, and it is expected to remain a major energy source for decades to come (Balat 2008). Coal burning is the major source of carbon dioxide pollution and other pollutants. Reducing coal use or emissions is the number one target for buffering climate change (IPCC 2010). However, "clean coal," eliminating sulfur, nitrogen and mercury impurities, will increase the cost of this component. Managing waste from coal-fired facilities is a serious problem, and coal ash can have significant radioactivity (Hvistendahl 2007) as well as offer a management challenge.

3.4.2.3 Petroleum

Petroleum ("Oil") is the major energy source for transportation, and is also used for heat and electricity. A current estimate is that known petroleum reserves will last until about 2025 (Goodwin 2008). In 1980 a Synthetic Fuels Corp. was established to produce petroleum from alternative sources (oil shale, tar sands, etc.). Petroleum is also a major feedstock for the chemical and plastics industries. International oil consortia have been exploiting oil reserves on many continents, and developed countries have often suffered from low royalties and high environmental contamination resulting, for example, in a protracted legal case in Ecuador claiming that Texaco contaminated rain forests with oil drilling waste. Thus Ecuadorian Indian Tribes, with very low petroleum use themselves, become stakeholders in the oil energy chain. Enhanced use of alternative energy chains will extend the life of petroleum reserves, allowing use for which oil is best suited. The 2010 "Deepwater Horizon" oil disaster underscored the patential hazards and limitations of deep ocean oil extraction.

3.4.2.4 Natural Gas

This is a major source for cooking and electricity, with increasing role in some transportation fleets. Natural gas reserves are sufficient to cover current usage levels for about a century (Goodwin 2008). Transportation of gas is somewhat problematic. Gas pipelines have been associated with PCB contamination (disposed of around pumping stations), and occasional explosions and fires.

Liquefied natural gas (LNG) is technologically problematic, with several spectacular explosions, resulting in public apprehension about siting LNG depots, but overall it probably elicits less fear than nuclear, despite periodic adverse events. Pipeline stakeholders compete with LNG stakeholders for sources and markets in terms of distance, feasibility, and costs, (Cornot-Gandolphe et al. 2003) if not safety (Fay 1980). The growth of the LNG subsector, for example, spawned a continually expanding industry for cryogenic plants and truck and ship transporters (Cornot-Gandolphe et al. 2003).

In the 1950s and 1960s, Operation Plowshare used nuclear explosions in a commercially unsuccessful effort to release and exploit natural gas deposits. Today hydraulic fracturing, injecting pressurized liquids into potential gas pockets, is a highly controversial method for exploiting additional natural gas sources.

3.4.2.5 Nuclear Energy

For over 50 years, nuclear energy has been a major source of electricity, with limited uses in transportation (Green and Kennedy 2008). In the aftermath of Three Mile Island and Chernobyl, new nuclear construction stagnated in the United States. However, the emphasis on carbon, climate, and security has fostered a cautious optimism about building new facilities and expanding nuclear energy production, thereby reducing reliance on imported oil. After a generation without new applications, the Nuclear Regulatory Commission has approved applications for new power plant construction, mainly on sites that already have nuclear power reactors (Greenberg 2009).

The 2011 Fukushima nuclear disaster cast a chilling effect on this optimism, slowing, if not halting nuclear planning. In the United States, siting nuclear facilities on virgin sites has required land use and demographic analysis (Greenberg and Krueckeberg 1974) and will require extensive environmental and ecologic assessments (Burger et al. 2011). Both optimism and skepticism abound in the popular press. The initial promise of nuclear energy in the 1940s–1950s was that it would be "too cheap to meter," a phrase never achieved in real life. Optimism about its future depended on stabilizing the "growth of the human population" (Hubbert 1956). Capital costs and life-cycle fuel costs conspired to keep the vision of cheap nuclear energy an elusive dream, until initial costs were amortized. New reactor design technologies promise safer, and perhaps cheaper, nuclear facilities.

Much current policy, dating back to President Carter, was focused on preventing nuclear proliferation and misuse of weapons-grade plutonium. Among stakeholders, technical people consider the lack of a permanent repository for spent nuclear fuel (SNF), coupled with the U.S. policy of not reprocessing SNF, as the major barriers to expanding nuclear power in the U.S. However, the "public" or lay stakeholders are more concerned about explosions, leaks, proliferation, and terrorism targets. This disconnect interferes with progress on how and how much nuclear energy will contribute to the U.S. energy portfolio. The uncertain future of the Yucca Mountain repository serves to perpetuate the impression that SNF is too dangerous to be disposed of safely. One partial solution to disposing of weapons-grade plutonium waste would be the development of Mixed Oxide (MOX) Fuel, blending oxides of plutonium and uranium for use in reactors.

There is another group of stakeholders who oppose nuclear energy on principle. These include people who associate nuclear energy with nuclear weapons as well as those who see any type of power plant as retarding investment in renewable energy. Environmental groups are divided, some opposing any expansion of nuclear energy, while others (emphasizing climate benefits) support nuclear growth.

Although nuclear plants are touted as "nonpolluting," this ignores the very extensive fossil fuel use required for exploration, mining, grinding, milling, enrichment,

Fig. 3.3 Indian Point nuclear plant on the Hudson river (photo Michael Gochfeld)

and transport, at the front end of the nuclear fuel cycle. Each of these stages brings additional stakeholders into the nuclear fuel cycle. Nuclear plants were built close to water (Fig. 3.3) to provide abundant water for cooling. Thermal pollution by releases of heated water and also tritium leaks have undermined confidence in nuclear power.

A future development may be dispersed, small, modular nuclear reactors, powering small isolated communities or industrial facilities, thereby reducing the challenge of distribution, while increasing concerns over nuclear security. The installation of Toshiba/CPEIR 4S ("super-safe, small, simple") 10 MW reactor proposed for Galena, Alaska, is awaiting review (scheduled for 2012) by the Nuclear Regulatory Committee (Bradner 2008). The isolated Galena Community, currently paying high energy costs, is an important stakeholder for this demonstration project. Finally, the spectre overshadowing new nuclear development is the proliferation of weapons-grade materials that will fuel fiction writers if not terrorists.

3.4.3 Sun

3.4.3.1 Solar

Long before photovoltaic devices were invented, solar collectors were developed to capture the sun's radiant energy (Ley 1954:167). Ley (1954:163) describes a solar motor dating to the 1600s, but solar power plants were not actually operated until the late 1800s, and these used the sun to heat water, just as passive solar heating systems operate today in many places. Today solar energy capture is in widespread use for local hot water, heating and cooling, and electricity production, although it provides a minute percent of the overall electricity generation. Because solar generation is directly related to the area covered by panels, new conversion technologies with higher efficiency per unit area are being developed and tested. Solar concentrators, such as parabolic trough technology, and rotating arrays offer a major step increase in efficiency. Basic photovoltaic research involves new materials, such as membranes that can be applied to buildings. Development, testing, and commercialization of new technologies and marketing to enhance acceptance are priorities and new nanomaterials, combined with development of high performance storage, will enhance the efficiency (and lower the cost and footprint) of solar generation (NREL 2008 web site). A major limitation to the development of solar energy has been the design of energy-efficient, cost-effective storage systems, although Metz (1978), long ago pointed out that this should not impede technologic development nor expansion of solar capture.

3.4.3.2 Biomass

Also dependent on the sun, "biomass" covers a broad range of fuels, including wood, crop wastes, waste-to-energy plants, and incineration of sewage sludge. The transition from wood to coal represented a major energy transition (Smil 2010), although wood continues to be a primary fuel in many developing areas and an optional fuel in developed countries as well. Biomass provides an opportunity to use domestic and sustainable resources to provide fuel, power, and chemical needs from plants and plant-derived materials. Trees, grasses, agricultural crops or other material which are now discarded, chipped, and composted, can be used as a solid fuel or converted into liquid or gaseous forms – for the production of electric power, heat, chemicals, or fuels (NREL 2008). Ethanol is one component of biomass fuel production, and is already commercially viable "and fully competitive with gasoline" in Brazil (Goldenberg 2007), although the negative ecological and agricultural consequences are substantial.

Biomass offers the promise of cost-effective energy "to reduce our nation's dependence on foreign oil, improve our air quality, and support rural economies" (NREL website 2008). Unspoken, however, is the competition for land, water, fertilizer, and labor, between biomass production and food production, and the tremendous price volatility and potential for famine, when wealthy energy-hungry countries raise the prices on purchasing biomass fuel that would otherwise have been consumed as food or competes for cropland.

Biomass energy introduces double jeopardy. Expensive oil drives up the price for agricultural production, while at the same time creating demand for biomass, rendering staple foods more scarce. The price increase in 2007 affected corn, wheat, and rice more in Africa than Asia (von Braun 2008), and was aggravated by drought (IMF 2007).

Wood and charcoal are still used extensively for cooking and heating in highly dispersed household units, mainly in developing countries. Firewood shortages are already serious problems in many places, and solar cooking devices are encouraged as a substitute. Auxiliary wood-burning stoves became popular in the United States during the energy crisis of the 1970s, but by the late 1980s were falling into disfavor because of air pollution, particularly from polyaromatic hydrocarbons. However, Basel, Switzerland is building a wood-fired cogeneration plant (Madlener and Vögtli 2008).

Biomass use for vehicles and electricity is still sparse. Given finite agricultural space and productivity, biomass burning must compete with biomass as a source of biodiesel or ethanol. And all of these must also compete with other land uses, particularly food production. Further, growing crops for biomass, particularly corn, soy beans, sugar cane, and Switch Grass, requires a variety of energy inputs (still mainly fossil fuel–based), including irrigation. The water requirement (liters per megawatt hour) is much higher for producing biofuel than for operating fossil fuel or geothermal facilities (Service 2009).

In 2007–2008, wholesale prices for corn, wheat, and rice rose rapidly, resulting in food shortages as these grains were diverted either to make biofuel or to feed livestock (Normile 2008). Although agricultural efficiency was enhanced during the Green Revolution, and presumably could see further efficiency, there is a limit to the carrying capacity of land imposed by solar radiation, fertilizer, and water, such that competition among these sectors will continue to pose problems.

3.4.4 *Water*

The movement of rivers and the ocean provide opportunities to harness renewable energy sources in the form of tides, waves, and currents. International ocean governance continues to be a challenge as nations compete for this vast but not infinite resource (Miles 1999).

3.4.4.1 Hydroelectric Power

Worldwide, hydropower is the most important renewable energy source. Many rivers provide adequate flow to generate electricity without dams (Fig. 3.4). However, large dam projects are still being constructed, for example China's Three Gorges and Brazils ITAIPU. Although "renewable," these hydro projects leave a large indelible footprint, flooding huge areas of otherwise productive land supporting agriculture, biodiversity, or both as well as aesthetic resources (Porter and Brower 1963). Indeed, dam removal projects are under way in some places to restore the natural flow of rivers. Bratrich et al. (2004) describe a so-called "Green Hydro" project in Switzerland, where ecological resources are considered in the design of the facility.

Fig. 3.4 Hydroelectric plant in Brazil (photo Joanna Burger)

3.4.4.2 Tidal, Wave and Current

These sources exploit the natural movement of ocean water, which provides many opportunities to tap its energy. Tidal generation and wave generation are already at the implementation stage in some localities. Ocean thermal energy conversion (OTEC) technology is extremely attractive because of the abundance of sea water and the substantial temperature gradient particularly in the tropics between sun-heated surface and deep water. First developed in the 1930s, OTEC has been abandoned and resurrected several times, but except for a few locations such as Hawaii, it has not proven commercially viable. Moreover, given a low theoretical maximum efficiency, it is questionable whether large-scale development is practical, except perhaps for small island nations with low total demand. Another use of water is pumped storage for load balancing using off-peak electricity to pump water to a high level, achieving partial payback at peak demand time as the water is allowed to fall back to the lower level.

3.4.4.3 Hydrogen for Fuel Cells

Hydrogen is ubiquitous and can be obtained from petroleum or water, with current plans leaning toward petroleum, which makes this a nonrenewable source. However, electrolysis of water is a more available, cheaper, source of hydrogen. Harnessing renewable electricity for electroysis of water would provide a carbon-free source, unlike the prevailing approach of steam methane reforming of natural gas (Rifkin 2003).

DOE's NREL states "Hydrogen and fuel cells are an important part of the comprehensive and balanced technology portfolio needed to address the nation's two most important energy challenges—significantly reducing carbon dioxide emissions and ending our dependence on imported oil." Unlike other renewable sources, fuel cells have great potential for powering vehicles. R&D has to focus on improved production, delivery, and storage of hydrogen and on safe, energy-efficient designs for the fuel cells themselves.

3.4.5 *Novel Sources of Energy*

Optimism regarding our energy future is often based on the assumption that novel, low-cost, low-emission technologies will be developed or are being developed. Ley (1954), anticipating solar power, suggested that new inventions would provide access to new energy sources. Here, inventors are a key stakeholder, as are the industries, governments, venture capitalists, and private investors who make decisions about investing in the R&D. Likewise, the academics who provide both the general and specific training leading to the new developments, and the academic institutions anxious to patent new technologies are important stakeholders. Several new technologies are at varying stages between imagination and pilot scale. The energy generated by walking, solar converting membranes, and the potential of microorganism cultures are among many options being explored. Other very practical technologies have stalled for various reasons, mostly related to economics and efficiency, including for example methane capture (Allison 2008) and ocean thermal energy.

Fusion energy using deuterium offers theoretical promise of high energy yield, and attempts to capture a positive net energy production from high temperature fusion technologies continue even after decades of frustrations. Despite the promise of a successful pilot on the horizon and the abundance of deuterium in sea water, the prospects of achieving a working fusion power plant are considered distant (Moyer 2010).

3.5 Energy Conservation

Energy conservation or the reduction in the per capita use of energy, particularly in rich countries, can play a role in averting the energy crisis. Numerous writers have emphasized the need for the developed nations, particularly the United States, to reduce per capita consumption widely viewed as profligate. Orders of magnitude separate the per capita consumption in the United States and the poorest nations, and there is a growing "middle class" of nations like Brazil, where consumption is rapidly increasing. "Energy-intensive services and luxuries are largely taken as entitlements in wealthier countries. People in poor countries need and aspire to the improved socioeconomic conditions that energy availability can facilitate" (Tester and Incropera. 2007), while at the same time adding to the global demand.

All consumers are stakeholders in conservation efforts. Conservation operates on several levels. The United States provides incentives (tax credits) for purchasing or installing energy-efficient appliances or renovations. The United States sets vehicle fuel-efficiency standards. Industry develops fuel-efficient devices in response to anticipated favorable market conditions. Ultimately though, individuals must choose to change behavior in favor of energy conservation, purchasing, for example, smaller fuel-efficient vehicles, installing newer efficient heating/cooling systems, driving less, or even making financial contributions to "green" organizations.

Even assuming that Americans and Europeans were willing to seriously conserve energy, it will take a huge reduction in their per capita use (at least 40%) to offset the approximately 40% increase anticipated in world population from 6.9 to 9.5 billion in 40 years, even assuming unrealistically that developing countries do not increase their current per capita demand. Population policy, though an essential component and ultimate driver of energy demand, is beyond the specific objectives of this chapter.

The great increase in oil prices in 1973 and 1979 brought about changes in both regulations (i.e., National Maximum Speed Limit Law of 1974) and consumer behavior. Many individuals and institutions abandoned oil heating for alternative fuels. However, individuals also conserved energy voluntarily, by driving less and installing and changing their thermostats, resulting in an estimated 1.2 million barrel per day reduction for home use (Schipper and Ketoff 1985). About half of the reduction was considered permanent and the rest subject to price changes. Indeed, the 55 MPH national speed limit was relaxed by 1987 and repealed in 1995.

Today, motor vehicle use of petroleum in the United States already exceeds domestic production and continues to grow (NREL web site 2008). This reliance on foreign oil raises national security concerns, which engage the public and reinforce the financial benefits of buying a fuel-efficient vehicle. On the design end, fuel economy involves tradeoffs between vehicle weight (and safety), engine performance, and vehicle type (Bezdek and Wendling 2005). The popularity of hybrid vehicles, even when the excess purchase price exceeds the likely lifetime fuel savings benefit, attests to the willingness or even the enthusiasm of some stakeholders to sponsor this aspect of conservation.

New approaches to building construction, new materials, and new thermal efficiency designs allow buildings to conserve energy and even to generate it. At the local and national level changes in building codes and appliance standards encourage energy efficiency. A combination of economic incentives/disincentives is necessary. Efficiency in heating, lighting, and operation point toward the goal of "Zero Energy Buildings." Heat pumps, using the constant temperature subterranean earth for cooling (summer) or heating (winter), have high installation, low operational costs. New solid-state lighting and electro chromic windows offer promise. But although this is a major area for conservation (Wing 2008), zero energy/zero emission buildings remain a distant vision.

Several spectacular blackouts have highlighted the vulnerability of electricity distribution. In some places where energy generation is inadequate or too expensive to be provided 24/7, involuntary blackout periods are imposed.

3.6 Energy Challenges for the Future

3.6.1 Challenges to Energy Production

In addition to the actual form of energy, there are issues that relate to inefficiencies in all energy chains, which include increasing the efficiency of capture or release, of storage, transmission, and use (Table 3.3).

3.6.2 Political Challenges and Stakeholders

Political considerations commonly influence energy decisions. In 1954 the U.S. Atomic Energy Act was amended to allow private ownership and operation of nuclear power plants. Faced with competition from relatively inexpensive fossil fuel power plants, the nuclear industry embarked on design pathways that were cheaper rather than most efficient or safer. A number of political issues face the development of different energy sources, and it is these challenges where stakeholders can have the greatest impact on policy and decisions. The ways in which stakeholders influence these issues are covered in subsequent chapters.

Site neighbors frequently invoke aesthetic issues for opposing the construction of particular facilities. This becomes part of political and policy negotiations. Sometimes safety and congestion are the primary issue; sometimes they are surrogates for primarily aesthetic objections. Some stakeholders object to any nonrenewable energy expansion as "nonnatural" and as delaying the anticipated transition to renewable energy. The nuclear industry has recognized that building new facilities on or adjacent to current nuclear plants will engender more public support than opposition (Greenberg 2009). It is uncertain how quickly such support will return after Fukushima (March 2011).

Two politically charged issues are (1) review of Yucca Mountain as a SNF repository (halted by President Obama in 2009, but still subject to legal review, Marshall 2010) and (2) the future of SNF reprocessing, halted by President Carter in 1978. In 1981 President Reagan lifted the "ban," while in 1993 President Clinton, in effect, reinstated it (Andrews 2008).

Table 3.3 Technical and policy challenges for diversifying energy

Capture or release: increased efficiency
Store (i.e., batteries)
Transmit (smart grids) efficiency and security
Disposal (emissions, production waste, ash, spent nuclear fuel)
Local vs. distant
Increasing reliance on dispersed sources
Reuse or recycle of materials

3.6.3 Energy Transition and Sustainability

Although it is widely acknowledged that some transition from current energy production/use patterns is desirable, diverse stakeholders have divergent preferences regarding "transition to what." Transition targets include "sustainability" (undefined), more nuclear (Cravens 2007), more hydrogen (Rifkin 2003), more renewables, and less carbon. Except on small spatial scales, energy transitions are long, complex processes, requiring "a sequence of scientific advances, technical innovations, organizational actions, and economic and political and strategic circumstances" (Smil 2010:20). "To achieve sustainable energy, we must make informed decisions among competing policies and technologies. Ideally options will be selected because their behavior fulfills enough expectations of enough stakeholders to create a broad consensus."(Tester et al. 2005:88).

Although there is widespread agreement on the desirability of expanding renewable energy sources, as well as regulatory targets for utilities to increase their own stake in renewables, there is substantial disagreement over the short-term and long-term feasibility of transitioning from fossil to renewable energy (Smil 2010). The major U.S. governmental initiatives are based in the U.S. Department of Energy's. National Renewable Research Laboratory (NREL 2008), and focus on biomass, geothermal, hydropower, wind, solar energy, and the ocean thermal gradient. This includes supporting R&D and building industry partnerships for demonstration projects.

3.7 Distributed Energy Sources

The iconic image of an American farm with its windmill pumping water or the Dutch windmill-studded landscape are examples of traditional distributed energy. Distributed sources are a growing part of the energy portfolio, serving as an alternative to huge centralized gigawatt power plants. But distributed facilities raise their own siting problems, repeated many times over. Land uses footprint, and aesthetics are common issues. Some communities have banned solar panels on roofs as unsightly. Distributed sources may provide peaking power or backup power, or depending on location may serve the cooling and heating needs of a single facility. The micronuclear generator mentioned above has yet to be installed anywhere, but may provide local power to isolated communities. These sources of energy also allow consumers to be part of the electric power market as well as assure backup power in the event of a blackout. Aside from wind and sun, "the primary fuel for many distributed generation systems is natural gas, but hydrogen may well play an important role in the future" (Salminen et al. 2008).

3.8 Stakeholders and Disasters

Stakeholders, particularly site neighbors, are sensitive to the potential for disasters, high-impact but rare events. Giant explosions or fires are rare, but media coverage assures that news spreads beyond the confines of local neighborhoods. Thus neighbors of proposed nuclear facilities are aware of the Three Mile Island and Chernobyl and Fukushima disasters, even though those sites are quite remote in space and time. It is likely that nuclear siting will be constrained by those memories for another generation. Likewise, the Tennessee Valley Authority's Kingston Fossil Plant coal fly ash slurry disaster (December 2008), may influence siting or environmental control decisions around coal-fired power plants. Explosions involving LNG are invoked to block siting LNG facilities. The 2010 Gulf Oil spill has slowed deep ocean drilling. Refinery explosions may influence response to new petroleum facilities.

Many of the public concerns and policy decisions governing several energy chains are strongly influenced by fears of terrorism, both fuel security and facility security. This includes the theft of weapons grade material, deadly conflagrations, or the vulnerability of the energy grid (Rifkin 2003). Terrorism looms as a major factor in the siting, design, (or even continued operation) of nuclear facilities. Integrity assessments of nuclear facilities have been performed in the past (e.g., Chelapati et al 1972), but since 2001, the simulations and design have assumed deliberate rather than accidental crashes. Since 2009, NRC requires all new nuclear plant designs to ensure the reactor core would remain cooled and containment intact in the event of a commercial jetliner crash (Holt and Andrews 2010). Terror-proof standards greatly increase the costs of construction and operation. These apply on a lesser scale to other fixed facilities such as petroleum power plants and LNG facilities (gasification, shipping, degasification).

Terrorism is one aspect of the Department of Energy's *Grid 2030 vision* which "calls for the construction of a twenty first century electric system that connects everyone to 'abundant', affordable, clean, efficient, and reliable electric power anytime, anywhere." The smart grid would increase efficiency and reduce carbon emission per kwh (DOE 2010). Fukushima is a reminder that natural disasters must be considered as well.

3.9 Conclusions

Energy policy and design, including the selection among an increasing number of source options, hinge on dialogues among many classes of stakeholders. Some classes, particularly those seeing direct and immediate financial benefit or harm, or those objecting to nearby facilities (current or future), speak loudly. However, many other stakeholders stand to benefit or lose in the short or long term by decisions made today. Moreover, many energy decisions are actually tacit "nondecisions," acceptance of status quo, and failure to invent or invest in new options that will achieve the nearly universal goals of reliable, affordable, clean, and efficient energy for all persons and purposes.

Acknowledgments I have benefited greatly from the discussions with colleagues in the Environmental and Occupational Health Sciences Institute and the Consortium for Risk Evaluation with Stakeholder Participation (CRESP), particularly Joanna Burger, David Kosson, Paul Lioy, Chuck Powers, and Michael Greenberg. Part of this synthesis was funded by the Department of Energy through a grant to CRESP (DE-FC01-06EW07053) and National Institute of Environmental Health Sciences grant (P30ES005022). I very much appreciate Joanna Burger's patience, advice, and encouragement.

References

Agterbosch S, Vermeulen W, Glasbergen P (2004) Implementation of wind energy in the Netherlands: the importance of the social-institutional setting. Energy Pol 32:2049–2066

Allison E (2008) Methane hydrates. In: Letcher TM (ed) Future Energy: Improved, Sustainable and Clean Options for our Planet. Elsevier, New York

Andrews A (2008) Nuclear Fuel Reprocessing: U.S. Policy Development. Congressional Res Service Report RS22542, Washington

Balat M (2008) Clean coal. In: Letcher TM (ed) Future Energy: Improved, Sustainable and Clean Options for our Planet. Elsevier, New York

Berman DM, O'Connor JT (1996) Who Owns the Sun? People, Politics and the Struggle for a Solar Economy. Chelsea Green Publ Co, White River Jct

Bezdek RH, Wendling RM (2005) Fuel efficiency and the economy. Amer Sci 93:132–139

Bockris J (2008) Preface. In: Letcher TM (ed) Future Energy: Improved, Sustainable and Clean Options for our Planet. Elsevier, New York

Bradner T (2008) Toshiba continues efforts for Galena nuclear power plant The Alaska Journal of Commerce. Apr 27, 2008. http://www.alaskajournal.com/stories/042708/hom_20080427006.html. Accessed 13 Oct 2010

Bratrich C, Truffer B, Jorde K et al (2004) Green Hydropower: a new assessment procedure for river management. River Res & Applic 20:865–882

Breslow PB, Sailor DJ (2002) Vulnerability of wind power resources to climate change in the continental United States. Renew Energy 27:585–598

Burger J, Clarke J, Gochfeld M (2011) Information needs for siting new, and evaluating current, nuclear facilities: ecology, fate and transport, and human health. Environ Monit Assess 172(1–4):121–134

Chelapati CV, Kennedy RP, Wall IB (1972) Probabilistic assessment of aircraft hazard for nuclear power plants. Nuclear Engin Design 19:33–364

Cornot-Gandolphe S, Appert O, Dickel R et al (2003) The Challenges of Further Cost Reductions for New Supply Options (Pipeline, LNG, GTL). 22ND World Gas Conference (Tokyo 2003), CEDIGAZ, Paris. http://www.cedigaz.org/Fichiers/pdf_papers/challenge of further.pdf. Accessed 10 Oct 2010

Cravens G (2007) Power to Save the World: The Truth About Nuclear Energy. A.A. Knopf, New York

DOE (2010) Grid 2030 Vision. http://www.oe.energy.gov/smartgrid.html. Accessed 14 Oct 2010

DOE-OS (2010) Office of Science-Energy Research. http://www.er.doe.gov/. Accessed 10 Oct 2010

Fay JA. 1980. Risks of LNG and PNG. Ann. Review Energy 5:89–105.

Gipe P (2010) Deaths DataBase: Wind-works. http://www.wind-works.org/articles/Deaths Database.xls. Accessed 17 Oct 2010

Gochfeld M, Mohr S (2007) Protecting Contract Workers: Case Study of the US Department of Energy's Nuclear and Chemical Waste Management. Am J Pub Health 97:1607–1617

Goldenberg J (2007) Ethanol for a Sustainable Energy Future. Science 315:808–810

Goodwin ARH (2008) The future of oil and gas fossil fuels. In: Letcher TM (ed) Future Energy: Improved, Sustainable and Clean Options for our Planet. Elsevier, New York

Gore A (2008) Speech on renewable energy. July 17, 2008. Washington, DC. http://www.npr.org/templates/story/story.php?storyID=92638501

Green S, Kennedy D (2008) Nuclear energy (fission). In: Letcher TM (ed) Future Energy: Improved, Sustainable and Clean Options for our Planet. Elsevier, New York

Greenberg M (2009) NIMBY, CLAMP, and the Location of New Nuclear-Related Facilities: U.S. National and 11 Site-Specific Surveys. Risk Anal 29:1242–1254

Greenberg MR, Krueckeberg DA (1974) Demographic analysis for nuclear power plant siting: A set of computerized models and a suggestion for improving siting practices. Comp Op Res 1:497–506

Hanjali'c K, De Krol RV, Leki'CA (2008) Sustainable Energy Technologies: Options and Prospects. Springer, Netherland

Hardin G (1968) Tragedy of the Commons. Science 162:1243–1248

Holt M, Andrews A (2010) Nuclear plant security and vulnerabilities. Congressional Research Service Report for Congress.7–5700, Washington

Hubbert MK (1956) Nuclear Energy and the Fossil Fuel Drilling and Production Practice, Report. American Petroleum Institute: Shell Development Company. http://www.onepetro.org/mslib/servlet/onepetropreview?id=API-56-007&soc=API&speAppNameCookie=ONEPETRO. Accessed 11 Oct 2010

Hvistendahl, M (2007) Coal ash is more radioactive than nuclear waste. Scientific American 12/13/2007. http://www.scientificamerican.com/article.cfm?id=coal-ash-is-more-radioactive-than-nuclear-waste. Accessed 17 Oct 2010

IMF (2007) Biofuel demand pushes up food prices. International Monetary Fund. http://www.imf.org/external/pubs/ft/survey/so/2007/RES1017A.html. Accessed 17 Oct 2010

IPCC (2010) Intergovernmental Panel on Climate Change. Chapter 9.2.2 Coal. http://www.ipcc.ch/ipccreports/tar/wg3/index.php?idp=359. Accessed 17 Oct 2010

Keeley G (2009) Spain's wind turbines supply half of the national power grid. The Sunday Times (London). Nov 9, 2009. http://business.timesonline.co.uk/tol/business/industry_sectors/natural_resources/article6910298.ece. Accessed 15 Oct 2010

ICBE (2010) Carbon as a commodity. International Carbon Bank and Exchange. http://www.icbe.com/emissions/commodity.asp. Accessed 15 Oct 2010

Jardinre CG, Predy G, Mackenzie A (2007) Stakeholder participation I investigating the health impacts from coal-fired power generating stations in Alberta, Canada. J. Risk research 10:693–714

Johnson JH (2011) Minority Participants in Environmental and Energy Decision Making. In: J Burger (ed) Stakeholders and Scientists. Springer: New York

Ley W (1954) Engineers' Dreams: Great Projects that Could Come True. Viking Press, New York

Madlener R, Vögtli S (2008) Diffusion of bioenergy in urban areas: a socio-economic analysis of the Swiss wood-fired cogeneration plant. Biomass Bioenergy 32:815–828

Marshall E (2010) Republicans Charge 'Impropriety' in Halting Yucca Mountain Safety Review. Science Insider website. 14 Oct 2010. http://news.sciencemag.org/scienceinsider/2010/10/republicans-charge-impropriety.html. Accessed 21 Oct 2010

Metz WD (1978) Energy storage and solar power: an exaggerated problem. Science 200: 1471–1473

Miles EL (1999) The concept of ocean governance: evolution toward the 21st century and the principle of sustainable ocean use. Coast Manag 27:1–30

Moyer M (2010) Fusion's false dawn. Sci Am. 302(3):50–57

NIOSH (2010) Agriculture Health and Safety http://www.cdc.gov/niosh/topics/agriculture/#agcenters. Accessed 14 Oct 2010

Normile D (2008) As food prices rise, U.S. support for agricultural centers wilts. Science 320:303

National Research Council (2007) Environmental Impacts of Wind Energy Projects. National Academies Press, Washington
NREL (2008) National Renewable Energy Laboratory. U.S. Department of Energy. http://www.nrel.gov/. Accessed 15 Oct 2010
OSEIA (2006) Oregon Solar Energy Industries Association. Solar Construction Safety. http://www.coshnetwork.org/sites/default/files/OSEIA_Solar_Safety_12-06.pdf. Accessed 10 Oct 2010
Ostrum E (2001) Reformulating the commons. In: Burger J, Ostrum E, Norgaard RB, Policansky D, Goldstein BD, (eds) Protecting the Commons. Island Press, Washington
Pearce F (2005) Act now before it's too late. New Sci Feb 12, 185:8–11
Plog B, Niland J, Quinlan PJ (2004) Fundamentals of Industrial Hygiene. National Safety Council, 4th edition, Washington
Porter E, Brower D (1963) The Place No One Knew: Glen Canyon on the Colorado San Francisco: Sierra Club
Renner JL (2008) Geothermal energy. In: Letcher TM (ed) Future Energy: Improved, Sustainable and Clean Options for our Planet. Elsevier, New York
Rifkin J (2003) The Hydrogen Economy: After Oil, Clean Energy From a Fuel-Cell-Driven Global Hydrogen Web. EMagazine.com 14(1) Jan – Feb 2003. http://www.emagazine.com/view/?171. Accessed 13 Oct 2010
Roscoe RJ, Steenland K, Halperin WE et al (1989) Lung Cancer Mortality Among Nonsmoking Uranium Miners Exposed to Radon Daughters. J Am Med Ass 262:629–633
Saito Y (2004) Machines in the ocean: the aesthetics of wind farms. Contemporary Aesthetics vol 2. http://www.contempaesthetics.org/newvolume/pages/article.php?articleID=247. Accessed 14 Oct 2010
Salminen J, Steingart D, Kallio T (2008) Fuel cells and batteries. In Letcher TM (ed) Future Energy: Improved, Sustainable and Clean Options for our Planet. Elsevier, New York
Schneider SH (2009) Science as a Contact Sport: Inside the Battle to Save the Earth's Climate. National Geographic, Washington
Schipper L, Ketoff AN (1985) The international decline in household oil use. Science 230: 1118–1125
Service RF (2009) Another biofuel drawback: the demand for irrigation. Science 326:516–517
Smil V (2010) Energy Transitions: History, Requirements, Prospects. Praeger, Santa Barbara
SVTC (2009) Towards a Just and Sustainable Solar Industry. A Silicon Valley Toxics Coalition White Paper. http://www.svtc.org/site/DocServer/Silicon_Valley_Toxics_Coalition_Toward_a_Just_and_Sust.pdf?docID=821. Accessed 14 Oct 2010
Tester JW, Drake EM, Driscoll MJ et al (2005) Sustainable Energy: Choosing Among Options. MIT Press, Cambridge
Tester J, Incropera F (2007) Sustainable Energy- Spring 2007. Massachusetts Institute of Technology: MIT (10.391J OpenCourseWare) License: Creative Commons BY-NC-SA. http://ocw.mit.edu. Accessed 01 Mar 2010
United Nations (UN2010) Oceans and Law of the Sea. http://www.un.org/Depts/los/index.html. Accessed 13 Oct 2010
United Nations Environmental Programme (UNEP2010) Six Priorities. http://www.unep.org/. Accessed 13 Oct 2010
Von Braun J (2008) High and rising food prices: why are they rising, who is affected how are they affected, and what should be done. Internatl Food Policy Research Institute. www.ifpri.org/sites/default/files/publications/20080411jvbfoodprices.pdf. Accessed 20 Oct 2010
Wing RD (2008) Smart energy houses of the future – self-supporting in energy and zero emission. In: Letcher TM (ed) Future Energy: Improved, Sustainable and Clean Options for our Planet. Elsevier, New York

Chapter 4
How Clean Is Clean? Stakeholders and Consensus-Building at the Fernald Uranium Plant

Kenneth Morgan and Gary Stegner

Contents

K. Morgan (✉)
US Department of Energy, Ohio Field Office, Miamisburg, OH 45342, USA
e-mail: morgan.ken@mac.com

J. Burger (ed.), *Stakeholders and Scientists: Achieving Implementable Solutions to Energy and Environmental Issues*, DOI 10.1007/978-1-4419-8813-3_4,

Abstract This is an account of the Fernald Uranium plant, the pollution from the plant, and its impact on the community. The Fernald Feed Materials Plant provided uranium metal to the United States nuclear weapons program from 1951 to 1989. In 1984, public awareness and concern over environmental releases began to grow, culminating in a lawsuit against the operators of the plant. Public reaction to the site was so negative that it became difficult, if not impossible, to operate or remediate the site. Systematic application of a program of public participation restored institutional credibility. Stakeholder input dramatically improved the quality of decision making, resulting in reduced costs and an accelerated environmental restoration.

4.1 Introduction

On October 31, 1988, the Fernald uranium plant made the cover of *Time* magazine (Magnuson 1988). The cover showed four people standing in front of a chain link fence with the smoking stacks of the Fernald plant silhouetted behind them. The headline read, "The Nuclear Scandal: The Clawsons of Ohio blame the Fernald uranium plant for cancer in their family." Inside the magazine, the article led with: "They lied to us." The Fernald plant was a facility of the U.S. Department of Energy (DOE). By implication, the United States government was harming public health and lying about it.

Time was not alone in its criticism. *The Cincinnati Enquirer* pilloried the plant in reporting, editorials, and political cartoons. Fernald was located just beyond the Cincinnati metro outer-belt (Fig. 4.1), so it was naturally the turf of Cincinnati's major newspaper. But, as the *Time* magazine story shows, concern reached far beyond Ohio. Those who felt victimized by Fernald gave full vent on television in an episode

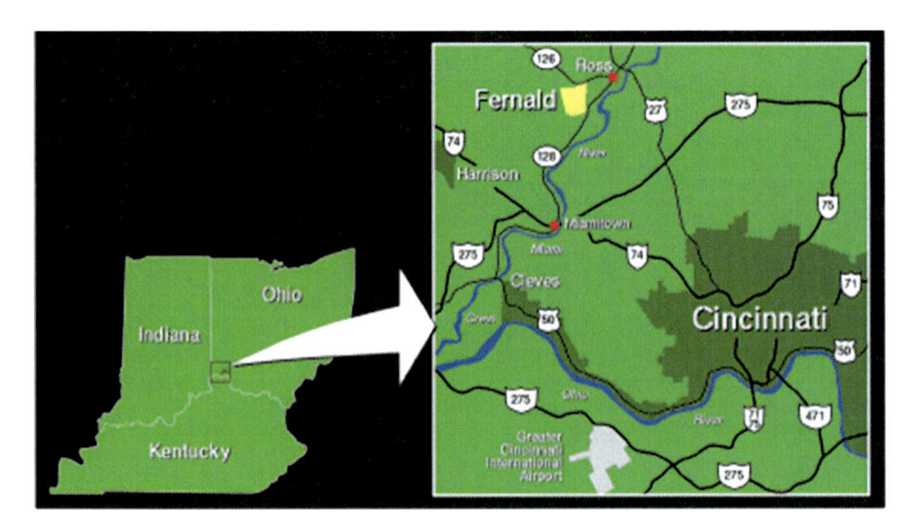

Fig. 4.1 Fernald location map

of *The Phil Donahue Show.* Eventually, the BBC spent a day at Fernald. It became a worldwide story. It is sad that any public institution should get into such a situation. It is sorrier still that it took such infamy to provoke change. The situation did change, and to the better, but the form and consequences of that change were not inevitable.

This chapter is an account of the 23-year drama of Fernald's metamorphosis by two participant observers (Ken Morgan and Gary Stegner). It chronicles the key role that stakeholders played in the ultimate cleanup and restoration of Fernald. Morgan was the DOE, Director of Public Affairs at Fernald from 1992 to 1995 and subsequently Director of Public Affairs for the DOE Ohio Field Office from 1995 until 2004. The Ohio Field Office had oversight of five environmental restoration sites: Fernald Ashtabula, Columbus; Mound and West Valley. Gary Stegner was a Public Affairs Specialist at Fernald from 1992 to 1996 and Director of Public Affairs for the Fernald project from 1996 until site closure in 2006.

4.2 Fear and Outrage

The drama of the Fernald uranium plant offers lessons about the merit and methods of public participation and consensus building. The drama began with fear and outrage. The outrage quickened citizen activism against what was perceived by some as an indifferent, if not malicious, government that could not be trusted. Some residents began to fight their own government through the courts and in public speech.

From inside the plant, things looked hopeless at Fernald. No one had intended to pollute the air or groundwater, but it happened. When this was announced to the public, it came as a complete surprise. A great many of the people who lived near the site had no idea what it did. Although the mission of the plant was not a federal secret, it had been the policy of the government to obscure as much of the operations of the plant as possible. The justification was national security. When the plant began operations at the height of the Cold War, it was a state-of-the-art facility, but in the early 1950s, people were more afraid of communists than carcinogens. The focus of the plant was uranium metal production and keeping the Soviet Union in the dark about the scale of that production. *Silent Spring* had not been written. The ideas of environment and ecology had not yet formed in the American consciousness. An adage at the Fernald plant was that "the only way uranium could hurt you is if you dropped it on your foot." Most of the workers at the plant were proud of their work and viewed it as an act of patriotism.

Inadvertently, because of a focus on a foreign threat, the government and its contractor, National Lead of Ohio, seemed to have lost touch with the mood of many in the nation it served. Fernald's mission to create uranium metal was in the public record. It was reported in the newspapers when the plant was constructed and hiring began for its workers. Certainly, thousands of people who worked there over the plant's lifetime understood very well what the plant did. However, they were instructed not to talk about their work. By 1986, the secretive nature of operations at the plant had obscured what it did and gave the impression that what it did had been deliberately disguised. Two other factors gave the public an entirely wrong impression of what the plant did: the sign outside the gate and the water tower everyone could see. The sign said, "Fernald Feed Materials Production Center" and gave the impression to some people that animal food was made there. This impression was reinforced by a large water tower that had been painted with red checkerboard squares, much like the brand logo of the pet and livestock food producer, Purina Mills. Although the paint job had been done purely as a safety precaution to alert low-flying aircraft, the paint job added to the impression that the government was trying it hide what was done at the plant. While some people in the community knew very well that the plant made uranium metal, others said, with some rankle, that they thought the place was making dog food.

The mostly secret operations of the plant suddenly became public in 1984 when a failure in a dust collector system released nearly 300 pounds of uranium oxide to the environment. That same year, plant officials disclosed that, in 1981, three off-property wells had been contaminated with uranium. One of the people who had used one of these wells, Lisa Crawford, became a public figure. She and other concerned citizens organized Fernald Residents for Environmental Safety and Health (FRESH) to monitor Fernald activities. From that point on, national and worldwide attention only increased. The press told a story of a government plant that had become a threat to the community around it. In 1986, two waste silos vented gas into the atmosphere, and a crack appeared in a plant built to treat waste. The State of Ohio filed a claim against DOE for injury to natural resources under Section 107 of CERCLA (The Comprehensive Environmental Response, Compensation, and Liability Act, known as Superfund). In reaction to the situation, Fernald discontinued using waste pits for storage and began placing waste in drums. The DOE also changed contractors. The Westinghouse Material Company of Ohio took over the Fernald site, replacing National Lead of Ohio after 33 years.

Negative publicity did not stop. Two summer camps, Fort Scott Camp, located two miles from Fernald, and Camp Ross Trails, a local Girl Scout camp that sat on a hill overlooking Fernald, closed both citing concerns about Fernald. In 1990, a class action lawsuit was brought against National Lead of Ohio. It was eventually settled in 1996 in a multi-million-dollar settlement in favor of the plaintiffs.

Within the DOE, and among many at Fernald, things seemed hopeless. Fernald's reputation within the DOE kept falling. Morale declined. No clear technical remedy was at hand for eliminating that threat, and the institution that managed the site had lost the moral and technical credibility needed to apply a remedy. The scope of the problem was immense. The site contained 259 buildings, many of which were contaminated. There were 40 acres of waste pits, some as deep as 30 ft, and thousands of barrels full of waste, some of which were leaking. There was a warehouse full of radioactive thorium and no other place to put it, and tons of uranium metal and uranium hexafluoride. No one really knew the full scope of the problem. A final inventory of the huge quantity of waste products is given in Fig. 4.2. Even in the best of circumstances, no one in the country was inclined to have nuclear waste treated, stored, or even moved through their backyards. For some, "my backyard" was essentially anywhere on the planet.

Waste	Volume	Final Disposition
High Activity Low-Level Radioactive Waste held in two concrete silos.	8,900 cubic yards.	Hydraulic mining and transfer to storage tanks. Waste blended with flyash and concrete to reduce leachability and decrease moisture content, Package and shiped off site for disposal by Waste Control Specialists, Inc;, in Texas
Low-level waste held in one concrete silo.	5,100 cubic yards	Pneumatic retrieval with remotely controlled and manned mechanical excavators. Material conditioned to reduce dispersability, package and ship off for disposal at Envirocare; Inc. in Utah
Six waste pits ranging in size from a baseball diamond to a football field and depths up to 30 feet. Low-level radioactive waste byproducts of uranium and thorium.	979,000 tons	154 unit trains transported the waste over a six years moving 9,100 rail cars 580,000 miles without incident for disposition at Envirocare, Inc, in Utah.
Soil and debris with contamination levels higher than the on-site waste acceptance criteria	212,896 tons	33 unit trainspulling 2,043 railcars shipped waste to Envirocare, Inc, in Utah.
Contaminated soil and debris from foundations and below-grade piping	2.95 million cubic yards	Construct on-site waste cell to contain waste.
Ground water remediation. Extract uranium from the aquifer to reach level at or below 30ppb.	225 acre area	As of July 2006, the project had extracted more than 19.2 billion gallons of aquifer water, treated more than 11.9 billion gallons and removed more than 7,500 pounds of uranium from aquifer
Low-level radioactive, hazardous and mixed waste site inventories stored in barrels and tanks	174,912 gallons low-level liquid mixed waste and over seven million cubic feet of low level mixed waste	Waste packaged and shipped to the Nevada Test Site, the Waste Isolation Pilot Project in New Mexico, and
Uranium product	Over 40 millions pounds,	Transfer to other DOE sites for programmatic use, storage, and sale to private sector.

Fig. 4.2 Quantities and kinds of waste at Fernald and their final disposition

In the face of the huge technical problems, and with a large part of the public frightened, disgusted, or angry, the Secretary of Energy began directing greater resources to Fernald. For the first time in its history, the plant gained a permanent DOE staff, including an office of Public Affairs. This DOE public affairs staff, along with the contractor's public relations staff, brought a new approach to the relationship between the plant and the public. There was a strong intent by DOE Fernald to change its mode of communication with the public. Instead of relying upon communication through television, radio, and newspapers, the new strategy was to favor symmetrical two-way communication from person to person.

The experience of the Fernald staff included time spent at DOE Hanford (Washington State), US Army Jefferson Proving Ground (Indiana), and the Department of Interior (Colorado). The Director of Public Affairs at Fernald was a charter member of the International Association of Public Participation. The new approach came from this past experience and the lessons of a series of theorists. The first exposure to this theoretical base came from a pamphlet written by Caron Chess for the New Jersey Department of Transportation (Chess is now at Rutgers University). Then came James Creighton's experience with public participation at the Bonneville Power administration, and Hans Bleiker's *Systematic Development of Informed Consent*. Creighton, a private consultant, was founding President of the International Association for Public Participation, and Bleiker is a private consultant specializing in helping public institutions build consensus for public policy. All of this was tied together by the concept of opinion leaders as articulated by Pat Jackson, a widely respected public relations professional. Fernald began to use communication techniques that fitted substantially within James Grunig's *Excellence Theory of Communication Organizational Effectiveness* Grunig 1992 that relies heavily on symmetrical two-way communication.

4.3 Creating a Path Forward

The approach to restoring trust and developing a solution unfolded gradually. Guidance for action came partly from theory but much of it from intuitive impressions about what to do next. Caron Chess once observed that sometimes public participation "is like knowing whether to kiss on a first date." Of course, it is not about kissing, but rather being deeply sensitive to the mood and concern of the people who are affected by your actions.

The actions sketched below are in a rough sequence, but it is a sequence of emphasis rather than a set of completed tasks begun and completed one after the other. All of these tasks had to be done simultaneously. However, activity higher in the list required more time and resources early on, and after substantial progress was made on those things, more time and resources could be applied to activity lower in the list.

4.3.1 Establish Two-Way Personal Communication with Primary Stakeholders (Hereafter Opinion Leaders)

Stakeholders are people who, in some way, are affected by the actions and decisions of an organization or an activity. If there are many stakeholders, it becomes difficult or impossible to have personal communication with them. It is easy to start thinking of them as a class or group and drift into the idea that mass communication will suffice. Mass communication techniques allow little scope for two-way, reciprocal communication, and tend to disintegrate into advertising, self-promotion, or propaganda. The idea of the opinion leader offered a theory for how symmetrical two-way communication with relatively few individuals can ultimately shape the opinion of an entire community. It presumes that social groups contain certain persons who are especially respected. When these people become committed to an idea or action, they influence other members of their community. Thus, one-to-one iterative dialogue with opinion leaders can lead to consensus among a population.

This one-on-one communication eventually effects what is newsworthy. Good reporters tend to find opinion leaders and interview them to create the substance of their stories. Most good journalism is a reflection of the state of mind of opinion leaders. So, for the Fernald staff, the first step to building a public consensus was not to persuade a reporter, but rather to identify opinion leaders and begin a dialogue with them.

It took some time to get a sense of the "groups" or communities of interest that existed in relation to Fernald. It took more time to discover who the opinion leaders were. However, some opinion leaders were immediately obvious. Direct interaction with obvious opinion leaders led to the discovery of other opinion groups and their leaders. For instance, from 1992 on, Fernald staff sought to improve its relationship with the site's greatest and most influential critic – Lisa Crawford, the president of Fernald Residents for Environment, Safety, and Health (FRESH). For starters, instructions were given that if anyone received a call from Lisa Crawford, that call should be treated as if it was a call from an Assistant Secretary of Energy – which is to say, respectfully and responsively. This was a beginning of a general attempt to elevate all members of the public to the status of important people.

The next effort was to see if Crawford and her colleagues would consent to letting Fernald staff attend the regular meetings of FRESH. At first, this request was met with skepticism and caution. We were asked to limit our visit to 5 min, in which we could introduce ourselves. We were then expected to leave. However, over time, the relationship became better and more reciprocal. Within a year, Fernald staff routinely attended entire sessions, provided briefings, and answered questions. DOE or contractor staff might be asked to bring an overhead projector, or they might help set up chairs in the church basement where FRESH met.

Interaction with FRESH also began to inform Fernald staff of other communities of opinion, some of whom did not agree with FRESH. A list of stakeholders drawn up in 1993 is shown in Fig. 4.3. This list is not comprehensive, nor could one be drawn up given the long duration of the project. Opinions and interest in Fernald

Fernald Stakeholders.This table lists organizations and individuals affected by Fernald operations. From within these groups Fernald staff identified over 89 opinion leaders with which Fernald staff made a point of developing some kind of personal direct-contact relationship. The list is extracted from a handbook on Fernald's Public Involvement Program published in 1993 by the Department of Energy's, Fernald Field Office

AGENCY OR GROUP	ROLE AND INTERESTS
Fernald Residents for Environment Safety and Health (FRESH)	Citizen watchdog organization formed in 1986 to monitor public health, safety, and environmental issues at Fernald.
US Environmental Protection Agency	Federal agency responsible for human health and the environment, and that has jurisdiction for environmental restoration under CERCLA
Ohio Environmental Protection Agency	Ohio state agency responsible for human health and environment and which has regulatory jurisdiction delegated to it under RCRA
Morgan Township Trustees Crosby Township Trustees Ross Township Trustees Reilly Township Trustees	Township trustees are elected officials responsible for local affairs including fire and hazard protection for the community. Portions of Crosby, Ross and Morgan townships were within a three mile radius of Fernald
Labor Unions Fernald Atomic Trades and Labor Council Greater Cincinnati Building and Construction Trades Council	These two labor organizations represented the hourly work force at Fernald. The Atomic Trades Council represented operations workers while the Building and Construction Trades Council represented hourly workers of contractors who would do dismantling of the site. The two councils sometimes had conflicting aims, and were in some ways in competition with one another.
Fernald Citizens Task Force, later called the Fernald Citizens Advisory Board,	DOE-appointed body to provide advice on environmental issues and cleanup at Fernald.
The Ohio and Indiana Congressional Delegation	Seven separate congressional offices were engaged in the controversy surrounding Fernald. Indiana delegates had constituents who worked at Fernald

Fig. 4.3 Fernald stakeholders (DOE 1993)

Hamilton and Butler County Commissioners	Fernald affected two Ohio counties,
County Public Health Officials • Administrator, Ohio Department of Health, Radiological Health Program • Chief of Environmental Services, Butler County Health Dpartment • Deputy Director, Hamilton County Health Department • Director, Butler County Emergency Management	County officials who have greatest technical responsibility for issues relating to Fernald..
Executive Directors American Red Cross, Hamilton County American Red Cross, Cincinnati	Non governmental official with an interest in emergency and public health issues related to Fernald.
Local Businesses	As November 1993, Fernald's stakeholder list included the managers of nine businesses who's operations could be affected by Fernald operations,
Civic and local development agencies	Fernald kept in contact with fourteen regional development and civic organiztions, including Chambers of Commerce, League of Women Voters, building and zoning departments, and regional development organizations.
Private Individuals	A dozen or so individuals who had no organizational affiliation were included as Stakeholders because of they specified because they expressed a desire to be considered a Stakeholder.
Centers for Disease Control (CDC)	U.S. federal body that monitors diseases, and public health.
Schools	Fernald kept in close communication with administrators and teachers at fourteen local public schools. The schools were within the three mile radius of Fernald or the students lived with that radius.
Newspaper, Radio and Television	Fernald staff developed relationships with many reporters local, national, and international. However, reporters were not given "first informant" status. It was Fernald policy to first consult with stakeholders most affected by an issue or event. Fernald staff encouraged reporters to contact informed stakeholders when writing their stories.

Fig. 4.3 (continued)

tended to wax and wane among individuals and groups, depending on the issues that were salient at a particular time and the life experiences of individuals.

4.3.2 *Create an Advisory Board with a Commission to Find a Solution*

The scope of problems at Fernald were so great and so complicated that solving them required a forum for sustained dialogue with the most important opinion leaders. DOE and Fluor-Daniel (contractors for Fernald) managers believed that neither the government nor Fluor-Daniel could answer the key questions for Fernald. Any remedy would have to be satisfactory to the community. The key issues were these:

1. What should be the future use of the Fernald site?
2. What residual risk and remediation levels should remain following remediation? (How clean is clean?)
3. Where should the waste be disposed?
4. What should be the priorities among remedial actions?

What we sought was a board that could credibly answer these questions. It had to be a board where every concerned person in the community could see on the board at least one person he or she trusted to stand up for their concerns. If such a board could be convened and it could come to consensus among its members, there was a chance that its recommendations would be accepted by the community and that the U.S. Congress would provide funding to implement those recommendations. That was the hope.

Our experience with advisory boards as well as the counsel of Jackson, Creighton, and Bleiker raised concern that an advisory board might become a burden and actually create more problems than it solved. The history and efficacy of citizens' boards is mixed. Sometimes they become captured by a faction that represents only a part of the community. Sometimes they become battlegrounds between factions. In other cases, boards go on fishing expeditions looking into trivial issues that consume time and resources of the staff that would be better spent elsewhere. In other instances, boards become just another critic and spend their time sniping at the institution that formed the board, undermining its credibility and ability to accomplish its mission. In every case, boards can consume a great deal of time and energy.

Despite misgivings, both DOE and Fluor-Daniel were persuaded that a board was the only way to obtain the kind of iterative dialogue needed to discover remedies for the problems at Fernald and build credibility for those remedies. To avoid the pitfall of a potentially an ineffective or counter-productive board, the Fernald staff went carefully, using a set of assumptions and steps based on past experience and the counsel of James Creighton and Hans Bleiker. This advice included what kind of board to form and how it might be formed.

The Fernald Advisory Board rapidly gained a reputation for its effectiveness and kept that reputation throughout its existence. Several comparative studies by DOE's Office of Environmental Management confirmed the effectiveness of the board (Fig. 4.4). This effectiveness may have arisen from serendipitous circumstances, but we believe it was a result of the principles under which it was organized.

Unfortunately, at the time the board was needed, DOE staff did not have enough familiarity with the community to identify the opinion leaders necessary for a quality board. Further, it did not have the credibility to form a board. It was felt that any effort by the DOE to appoint members to a board would tend to discredit those appointed, or at least give the impression that DOE was trying to control the outcome of the deliberations. To avoid this, the DOE engaged a convener. An effective convener is a person with so much objectivity, credibility, and gravitas that his or her decisions will be accepted. Fernald's convener was Dr. Eula Bingham, a person who admirably fit the criteria. Dr. Bingham had been Assistant Secretary of Labor for Occupational Safety and Health during the Carter Administration. She is also a distinguished professor of environmental health at the University of Cincinnati, and past Vice President and University Dean for Graduate Studies and Research (1982–1990) at the University of Cincinnati.

DOE's instructions to Dr. Bingham were to interview people in the community to discover who was respected within the community or communities most affected by Fernald. She was also asked to draft a provisional charter and find a person with the skills to serve as chair of the board. During a series of private interviews with members of the community, Dr. Bingham explained that she was looking for people who could communicate to DOE the concerns of the community. When she asked people who they would trust, certain names were repeated. The way these names surfaced in conversation tends to support the Opinion Leader theory. Dr. Bingham interviewed the people named, pointed out to them their respect in the community, and asked them to serve on the board. Her appointment of these very people to the advisory board simultaneously tended to confirm Dr. Bingam's reputation as a wise person as well as give credence and authority of the board. Dr. Bingham organized the Fernald advisory board under the name, The Fernald Citizen's Task Force.

The Fernald Citizens Task Force required a modest budget to offset the cost of member travel to DOE-sanctioned national site-specific advisory board conferences, and to pay for the services of a professional facilitator. Funding was provided through the project's public affairs budget.

The need for outside facilitation became apparent very early. The Task Force required a large volume of technical information that had to be assembled and presented in a manner easily understood by lay people. Immediately following its organizational meetings, the Fernald Citizens Task Force chairman, Dr. John Applegate, approached DOE and requested the assistance of a non-DOE/Fluor-Daniel staff person. The Task Force felt that to avoid any appearance of conflict of interest and to insure objectivity, it was important that the information be prepared and presented by an individual with no direct association with DOE or its contractor. The person selected to be the facilitator would not only develop and present information, but would also establish agendas, facilitate meetings, and draft the final report and recommendations for the Task Force. The facilitator, Douglas Sarno was selected through a formal solicitation process and ultimately served throughout the life of the Task Force, which concluded its operations in 2006. What follows are the steps we used to ensure the formation of an effective board.

4.3.2.1 Determine What Kind of Board Is Needed

Two key issues are the need to identify the problem you expect an advisory board to solve and determine what kind of board is needed to solve that problem. Hans Bleiker (Bleiker and Bleiker 1995) provides a very good decision-making matrix showing the types, character, merits, and drawbacks of different kinds of boards.

4.3.2.2 Get Commitment

The top decision maker for the problem being solved must resolve to be an active participant. Advisory boards will fail if they do not have the continuing attention of decision makers.

4.3.2.3 Make Sure the Affected Community Understands the Problem

It is critical to make sure the community accepts that the problem must be solved and understands that the board will help solve it. This is a case where the news media can be a big help. The press loves bad news. Defining a problem is a wonderful way to give it to them. When a critic announces that a government institution has a problem, it tends to undercut the credibility of the institution. When an institution announces it has a problem and wishes to solve it, ironically, this can raise its credibility. Democratic government exists to solve problems. When someone outside government finds problems, the government is seen as not on the job or in denial. When the government identifies problems, it is seen as doing its job. Thus, the government should always be the first with bad news.

4.3.2.4 Define and Recruit Membership

Assess the community affected by the problem to determine who should serve on the Board. Ask people within opinion groups who they respect and invite respected voices to participate on the board. Make sure that everyone in the affected community can see at least one person on the board they trust – a person who will stand up for their issues and concerns. Do not rely on volunteers or make a general call for volunteers. A purely volunteer board will automatically be skewed in its composition and not fully represent the full spectrum of the community. Ask specific people to serve. Some of the best people will be very busy and reluctant to volunteer. Appeal to their patriotism or community spirit.

4.3.2.5 Create a Provisional Charter

Develop a provisional charter that defines the problem to be solved, the scope of inquiry, the number and characteristics of the membership, the nature of the deliverable, a time frame for resolution, and provisions for dissolution of the board.

4.3.2.6 Recruit a Chair

Select a chairperson who has the necessary qualifications: credibility, dignity, objectivity, parliamentary skill, and commitment to the process.

4.3.2.7 Find a Facilitator to Assist the Chairperson

Every board meeting should be run efficiently – agendas, presenters, presentations, and handouts take a great deal of time to assemble. An efficiently run monthly board meeting can easily occupy at least one talented person full-time 40 or more hours a week.

4.3.2.8 Ratify the Charter

The first meeting should be organizational. The board will need to discuss the charter, adjust it as necessary, ratify it, develop any additional bylaws, and set an initial agenda.

4.3.2.9 Support the Board

An advisory board is not a silver bullet that will solve all problems. Not only must the board be provided all the resources it needs, its work must be featured within the context of fully developed public information program. The decisions the board makes will not act as a catalyst for public consensus unless they are widely known.

4.3.3 Create a Transparent Decision-Making Process

Jim Creighton frequently pointed out that part of the reason that people could not trust or understand the decisions of large organizations like the government is that they could not understand the decision-making process. Once an organization attempts to explain its own decision-making process to itself, it can begin to have logical access points for the public to obtain information and insert information and opinion. The decision-making process becomes rational.

CERCLA and RCRA (The Resource Conservation and Recovery Act) processes provided a framework at Fernald for a rational decision-making process, but as the regulations are often applied, they provide too little information too late. Stakeholders need to be involved much earlier than the regulations seem to suggest, before alternative solutions are developed. Information has to become accessible. That means documents must be written in plain English. Videos and illustrations must be made. Opinion leaders must be consulted repeatedly to validate the quality and clarity of information products. The creation of the Citizen's Task Force, with ex officio members from DOE and its regulators, brought citizens to the table with decision makers early in the process.

For managers and administrators, all of this public interaction seems like it will just slow down and complicate progress. The contrary is true. It speeds things up because less effort is needed to defend and justify decisions after they are made, and in the worst case, projects are not brought to a stop midway, and done over.

4.3.4 *Identify as Many Opinion Leaders as Possible and Establish Relationships*

Throughout the process, the number of people to be communicated with continually grows. Actual human contact, even relatively brief contact, brings credibility to the whole institution and its processes if those contacts are sincere, open, friendly, and helpful. To this end, in 1994, Fernald developed the Envoy program to help extend the reach of its person-to-person message. DOE and Fluor-Daniel selected staff from all fields and disciplines, including engineering, management, construction, labor, and support organizations, that were respected both at work and in their community. They were themselves opinion leaders at work and at home. Their task was to create and extend relationships in the community. Fernald Envoys were instructed not to promote or defend Fernald or its mission, but merely let people in the community know they worked at Fernald and would be happy to provide information if it was wanted. They served as Fernald's eyes and ears for local business leaders, social groups, labor unions, school officials, environmental groups, regulatory agencies, and elected officials. Their job was to provide timely information about Fernald to the community and relay public concerns and ideas back to decision-makers. Envoys met in a monthly meeting with Public Affairs staff, which were frequently attended by senior DOE and Fluor-Daniel managers. The Envoy program proved a far more useful system of garnering public opinion than reading newspapers or doing surveys.

4.3.5 Move Toward Person-to-Person Communication and Avoid Using the Mass Media as the Principle Form of Communication

Fernald staff did not rely upon mass media to win public acceptance. The experience of Fernald staff was that the media tend to get their sense of a story from stakeholders rather than spokespersons. Thus, our primary effort was to inform opinion leaders first. Fernald staff strove to give reporters the status of stakeholders rather than a special class of utmost importance for communicating with the public. Although timely and accurate information was provided to reporters, they were encouraged to get their story from opinion leaders. To the degree possible, reporters joined rather than preceded stakeholders in receiving information. DOE and Fernald used chains of personal relationships to pass along urgent or time-sensitive information. In general, the opinion leaders most affected or concerned by an issue had information about it before or at the same time as any reporter had it. The goal was that opinion leaders were never surprised by a call from a reporter. To the degree that institutional spokespersons were needed to provide information to the media, they were sought from as low a level as possible. The people closest to the work have far more credibility than a plant manager or a public relations person.

4.3.6 Create a Responsive Communication Organization to Get Questions Answered and Issues Resolved Quickly

Fernald attempted to get answers back to any member of the public within one day. An accurate answer was not always possible to obtain in one day, but an explanation was. For instance, if, in a public meeting, or phone call, or e-mail there was a question that could not be answered immediately, staff would get back promptly to the questioner with an explanation for why no answer was immediately available and give a time frame for how long it would take to get an answer. The important thing was to be responsive and live up to any promises made.

4.3.7 Open Public Meetings so That They Are More Interactive, Informative, and Responsive to the Public

Public meetings were originally quite formal with a panel of people in suits seated in front of a public who grilled them with questions. Both the public and the suits tended to go away frustrated. A big breakthrough came when a game was developed that simulated the problems of plant management. It was called Cleanupoly (Fig. 4.4). Plant personnel, union leaders, environmental activists, local residents, and state and federal officials were seated in small mixed groups around circular tables, parlor style, to play a Monopoly-like game where each person was challenged

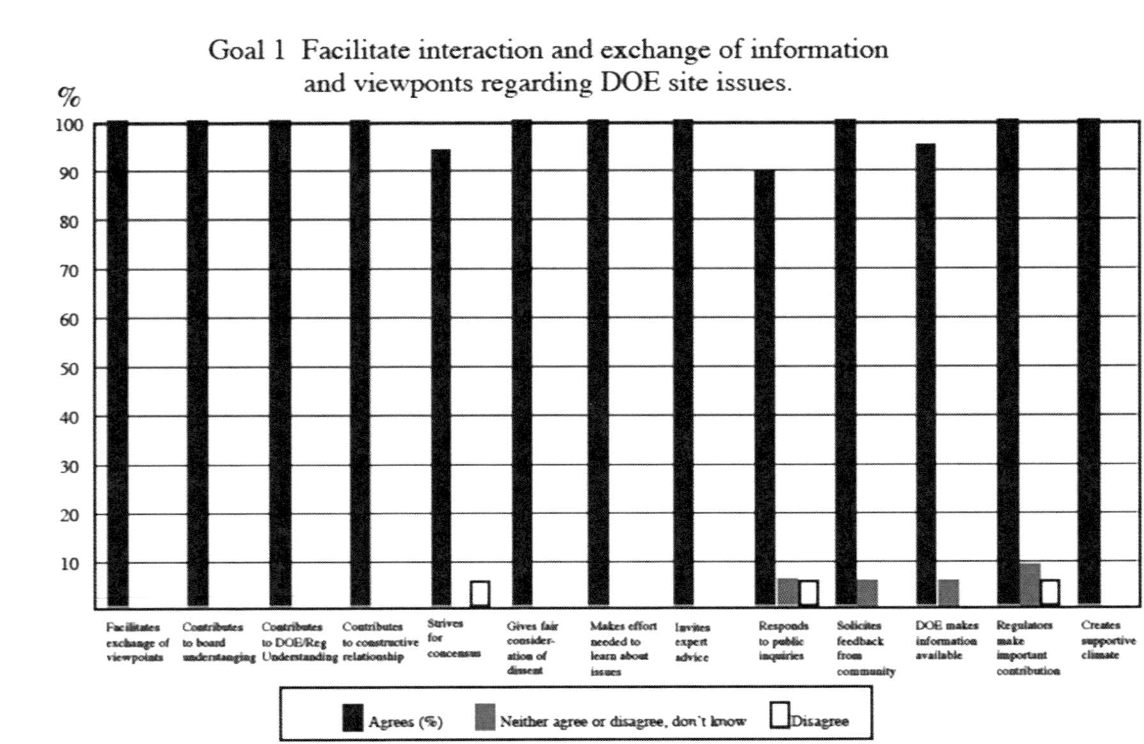

This figure is extracted from the August 1996, *SSAB Supplementary Appendix* to the *Site Specific Advisory Board Initiative Evaluation Survey Results*, The US Department of Energy Office of Public Accountability. June 1996. The survey evaluated seven DOE Advisory Boards, including Fernald. The survey was done by the Pacific Northwest National Laboratory support team: Dr. Mary Zalensy, Ms. Kristi Branch, Dr. Judith Bradbury with assistance and of an Advisory Panel: Dr. Jim Creighton, Dr. Carol Wise, and Dr. Charles Hulin.

Fig. 4.4 Stakeholders playing "future site," a simulation for solving waste disposition problems

to make decisions about how to manage the plant. This experience began to break down the boundaries among points of view. Public meetings evolved into more dynamic experiences often organized around poster sessions where people could move at their own pace around a room and have a chance to talk with someone about issues of most interest to them. The Citizen's Task Force used the simulation concept as well, and developed a game called Future Site. It helped not only the member of the Task Force but also members of the engaged public to grasp the issues and weigh alternatives.

4.3.8 Create a Full-Spectrum Communications Program

There is a danger that public participation can occur in a closed community of insiders. To avoid that, a variety of communication programs and techniques were applied at Fernald to provide personal experience of Fernald to a wider and wider public. This included Open Houses and Education Outreach programs that brought Fernald environmental science into local class rooms. The DOE and Fluor-Daniel sponsored a Science Bowl competition for High School students. Mass communication techniques were also employed, especially media events that brought reporters to the site and in contact with plant staff and stakeholders. It was important to keep the general public aware of the existence and work of the Citizen's Task Force as well as progress made at the plant.

As required by CRCLA and RCRA, a pubic reading room was opened to provide public access to documents. Fernald provided more documents on more topics than was required under the law. There is some risk involved in such transparency, as will be shown later in the case of Danger & Deceit. Easily accessible information can be used or interpreted in many ways. Still, Fernald was committed to transparency in its operations and decision making. Trust cannot be built or sustained for long without transparency.

4.4 The Fernald Citizen's Task Force Reaches Consensus and Makes Its Recommendations

It is impossible in this space to adequately summarize the long and methodical process of the Fernald Citizens' Task Force. Most meetings began early on Saturday morning, went for about four hours, and concluded with a meal provided by Fluor-Daniel. The meal was important. From ancient times, if disputing parties can break bread together, it seems to help bring them together. It is hard to demonize someone you actually know and eat dinner with. The members of the Task Force served without any other remuneration than some traveling expenses and this simple meal. They listened to long briefings by DOE contractors and when they thought it necessary, sought other expertise and did their own research. The Task Force began its meetings in September of 1993 and issued it recommendations a year and a half later, in July of 1995. It published these in a formal report of 51 pages, with another seventy pages of appendices. A summary of their recommendations taken directly from their report (FCTF 1995) is as follows:

4.4.1 Waste Disposition

The Fernald Citizens Task Force evaluated the political and logistical considerations involved in disposing over three million cubic yards of contaminated material and determined that a balanced approach, in which some waste was disposed on-site and some was disposed off-site, was most prudent. Of paramount importance was that the highest-level wastes be taken off site for safe disposal and that no new wastes come to Fernald for disposal. *Author's note*: The term "highest-level" here should be understood to mean highest activity or highest hazard and should not be confused with the terms high-level and low-level waste defined by the Nuclear Regulatory Commission (NRC). *The NRC terms do not have a direct correspondence to risk or hazard but are labels arising from the source of the waste. All radioactive waste at Fernald would be classified under the NRC terms as low-level, but some of it was very hazardous.*

Therefore, the Task Force concurred with existing DOE decisions that the most highly contaminated materials be disposed off-site. The Task Force recommended that an on-site facility be constructed to store materials with low levels of contamination from only the Fernald site. One Task force member, Daryl Huff, objected to this recommendation, preferring that all contaminated material be removed from Fernald and disposed of off-site.

4.4.2 Priorities on Remediation

The Fernald Citizens Task Force recommends that Fernald adopt an accelerated remediation schedule to provide rapid protection of human health and the environment, and to control overall costs. The recommendation calls for DOE to focus on remediation by reducing nonremediation costs as quickly as possible and to eliminate redundant requirements. Specific sequencing of activities within that accelerated schedule was viewed to be less important. However, the Task Force made specific recommendations for higher risk wastes awaiting shipment to be removed immediately.

4.4.3 Recommendations on Future Use

The Fernald Citizens' Task Force focused its future use recommendations on creating a broad understanding of how the Fernald site could best be used after remediation, rather than identifying specific detailed ideas for future use of the property.

The Task Force recommended that residential and agricultural uses of the property be avoided. However, it was also important to the Task Force that the land be used productively. For this reason, the remediation levels recommended for the site provide for all uses other that residential or agricultural. The Task Force also recommended that a sufficient buffer be provided between the on-site disposal cell and any other uses of the property. Ultimately, the Task Force recommended that, within the guidelines set forth, specific uses of the property would be best determined closer to the time of reuse by the people most impacted by that use.

The recommendations of the Fernald Citizens' Task Force are notable for their wisdom, their inclusiveness for the best interest of all affected stakeholders, and their persuasiveness. In several instances, they were ahead of the thinking of DOE and the regulators. This brought cohesiveness not only to the outlook of the community but among DOE, its contractor, and the regulators. The Task Force was not parochial in its outlook. The members recognized that a great deal of public money was involved and weighed carefully the merits of cleaning up the site against the other needs of society. By being reasonable in their expectations and looking sharply at how regulatory requirements and operating efficiencies could be streamlined, the cost of the project was reduced by many billions of dollars. Perhaps most importantly, Congress and DOE headquarters were persuaded to support the recommendations.

4.5 Stakeholders Upheld the Process

Two incidents reveal how solidly the community began to grow together to the point that they would defend the process and those that were part of the process.

4.5.1 Yes, They Can Be Trusted

In any long-term consensus building process, there is a risk that new stakeholders may suddenly emerge who are hostile to an alternative adopted before the new stakeholders became interested. People who had no interest in the issues and the process suddenly become very interested when they discover that some alternative, about to be adopted, will affect them. However, at Fernald it was found that because diligent effort had been made to engage every affected stakeholder, decisions tended to stick. This was revealed one evening during a public session about the new on-site waste disposal cell.

The cell was then in the planning stages, and because of its massive size it was going to be a dramatic alteration of the topography. A picture of the constructed cell is shown in Fig. 4.5. Some real estate developers had acquired the Girl Scout Camp overlooking Fernald and, confident of the progress of the clean up, were developing it as a site for some very nice homes. When they saw that the waste cell might

Fig. 4.5 Fernald waste cell

interfere with the view from the property, they were not happy. They came to the meeting to object. There was an exchange among stakeholders. The gist of it was that the real estate developers raised a variety of objections to the waste cell and concluded with the assertion that "You can't trust them," meaning DOE and its contractors. Immediately, folks who had once been harsh critics of Fernald jumped to the defense of the people and process that had brought the community this far. Trust seems to be won mostly by reliable behavior over time. The stakeholders were not inclined to let someone new enter the dialogue and make unfounded accusations when those new people had plenty of opportunity to participate earlier in the process. The real estate developers gave up and went away, apparently accepting a community decision, even though they didn't particularly like it.

4.5.2 The "Danger and Deceit" Crisis

Washington sometimes seems to work under the system of "management by newspaper." Every day the Secretary (of Energy, Interior, Defense, or whatever agency) gets up and reads the morning newspaper. He or she usually has an emotional response and starts immediately to develop public policy. Actually, he or she reads a stack of clippings from major newspapers brought every morning by an aide. At the same time, all the Assistant Secretaries diligently read the same stack of clippings so that they know what kind of day they are going to have.

A series of bad days began Saturday afternoon of February 11, 1996. Much to our surprise, a television news reporter sought an interview about an upcoming expose of Fernald that was to appear in the Sunday edition of the Cincinnati Enquirer. The Enquirer was a Gannett-owned newspaper, and the reporter was from the local Gannett-owned television station. We had no clue what this could be about, but it appeared that Gannett was using the evening television news as a "promo" to push newspaper sales.

To our eyes, Fluor-Daniel and DOE had been working diligently and competently. They were being prudent with the public treasury, and they were keeping the public informed and building consensus for the recommendations of the Citizens Task Force. On Sunday morning, the headline and top half of the fold story was "Taxpayers bilked in Fernald cleanup." There followed a barrage of stories over the next 10 months, many on the front page. Headlines ran, "Fernald workers' safety threatened," "Agency failed to detect problems," "Secret plan inflates cost," and many others with a similar ilk. We were deeply worried that all of our carefully built confidence building would now be undermined and consensus would fall apart. Yet, it turned out that community solidarity was, in the end, strengthened. It became clear to nearly everyone closely associated with the site that something peculiar was going on at the Enquirer.

This series of stories were usually branded under a logo composed of a radiation-warning symbol and the words Danger & Deceit. It was written under the byline of a new reporter at the Enquirer named Michael Gallagher. The stories were also entirely different in approach to their sources than any previous journalism we had seen. Our previous experience of conventional journalism was that reporters went to authorities and informed opinion leaders. We well understood the reasons for previous negative reporting about Fernald. It may have sometimes been sensationalized and not always entirely accurate, but it broadly reflected the general scope of problems at the site and how many in the community felt. The Danger & Deceit series was different. Gallagher made little or no use of the observations of informed stakeholders like Lisa Crawford of FRESH, or John Applegate, the president of the Citizen's Task force, or other informed opinion leaders. Instead, Gallagher's informants were nearly always undisclosed. Gallagher also made narrow interpretations of documents, quoting things without a proper context. The documents he quoted were readily available to any member of the public who took the trouble of visit the public reading room, yet, the reporter did not mention this, but rather gave the impression the documents had been "obtained" by some special diligence. While the documents might, at first glance, seem to justify the reporter's conclusions, a more informed understanding of the documents revealed they did not.

Washington reacted to the Enquirer's series with alarm. The local congressmen asked the Government Accountability Office (GAO) to investigate. At this writing, 14 years later, a web search can bring up some of the Cincinnati Enquirer's stories or a web site with information based upon them. As of February 2010, a web page from globalsecurity.org had this to say based upon the Enquirer's story, "Poor site management, unsafe practices, and improper financial conduct uncovered by the Cincinnati Enquirer prompted a GAO investigation..." Newspaper allegations tend to be their own "proof" of the validity of those allegations. "Where there is smoke there is fire", goes the adage.

Despite all of the tempestuousness of the Enquirer's reportage, the reaction of Fernald stakeholders was mellow. The Fernald envoys asked their contacts in the community whether they were concerned about the issues raised by the Enquirer. They were not. Some were angry with the Enquirer. Some rolled their eyes with disdain for the paper.

Another interesting thing occurred. All other reporters from newspapers, television, and the AP stuck to general journalistic practice and sought out informed opinion leaders for comment. The result of these conversations seemed to undercut the Enquirer's position. The story remained an Enquirer exclusive. Shortly after the Danger & Deceit series broke, Business Week contacted Fluor-Daniel. Fluor made their case to the Business Week reporter and then suggested he call a few stakeholders in the community. After a few days, the reporter called back to let the folks at Fluor-Daniel know that the story was not what Business Week was inclined to cover.

After the Enquirer's assault, all other local media seemed to take a friendlier stand toward Fernald and its cleanup mission, including the Gannett-owned television station. Only one outlet picked up the Fernald story, but it did so in an entirely different vein. Randy Katz of *Everybody's News* wrote a series of articles deconstructing the Enquirer's story. The March 8 1996 story ended with the remark, "The reporting style of The Enquirer's ... series contains its own element of hazardous waste, and that is unfortunate for readers concerned about the issues raised." DOE headquarters began to relax when the story did not seem to be spreading. For six long months, the site weathered the storm of the Enquirer's wrath, yet developed closer ties and greater confidence within the community.

The adage that "Time will tell" may apply here. The GAO found very little to corroborate the Enquirer's claims. Neither did a DOE HQ investigation. Despite the Enquirer series alleging grave safety violations and gross mismanagement of funds, Fluor-Daniel achieved all of the objectives specified by DOE. They did this on time or ahead of schedule, within budget, and with an exemplary safety record.

It may be of interest to follow the reporter who printed the "hard-hitting" series on Fernald. On June 28, 1998, a story appeared on the front page of the Enquirer, top half of the fold:

> "The Cincinnati Enquirer today issued an apology to Chiquita Brands International Inc. for articles published May 3, 1998, that were based on illegally obtained voice mail messages that questioned Chiquita's business practices. In a statement, Enquirer Publisher Harry M. Whipple and Editor Lawrence K. Beaupre said that facts obtained since publication have convinced them that the lead reporter had deceived them and others involved in the preparation of the articles. As a result, they said that the newspaper has renounced them. The reporter was identified as Michael Gallagher. The newspaper said he was terminated Friday for misconduct.... It said it has also agreed to other terms – including a payment in excess of $10 million – in exchange for settlement of claims against it by Chiquita."

In October 1998, the reporter Michael Gallagher pleaded guilty to felony charges of unlawful interception of communications and unauthorized access to voice-mail systems in the Court of Common Pleas in Cincinnati.

4.6 A Continued Commitment

The Fernald Citizens' Task Force accomplished the scope of its charter when it issued its recommendations to DOE in 1995. However, one of its recommendations was that the Task Force continued to operate. This recommendation was accepted, and the Task Force continued to prove valuable as shown in the Danger & Deceit crisis. The Task Force changed its name to the Fernald Citizens Advisory Board (CAB) and met less often, but it played a central role in planning details for the final land use of the site and dealt with a variety of smaller issues.

A completely unanticipated and crucial development was made possible by the Fernald CAB. It opened up Nevada. Although Fernald was keeping the vast majority of its waste on site, as shown in Figure 4.2, a great deal of waste still needed to be removed from the site, and a place had to be found for everything that left the site. Nevada was not inclined to become the nation's dumping ground. Resistance to nuclear waste shipments to Nevada arose because Congress had unilaterally designated Yucca Mountain, in Nevada, as the sole depository for all high-level nuclear waste in the country. The State of Nevada and its citizens were using every legal means to prevent Yucca Mountain from ever opening or receiving nuclear waste. At this writing (2010), Yucca Mountain has yet to open. The members of the Fernald CAB recognized the problem. Several members went to Nevada to meet with stakeholders there. They explained the situation at Fernald and the effort to keep the vast majority of waste at Fernald. They made the case that the DOE's Nevada Test Site was the best available place to put Fernald's large stock of radioactive thorium. The Nevada stakeholders acknowledged the willingness of Ohioans to accept the greatest share of the nuclear waste burden. They also seemed to appreciate that Fernald was asking rather than demanding. The Nevada stakeholders allowed that Fernald was a special case, and the specified waste was shipped to Nevada without incident.

4.7 A Final Accounting

The Fernald Environmental Restoration Project was brought to substantive completion in 2006. In 1992, forecasts of the best case for project completion were 27 years at a cost of $12.2 billion. Because of the Fernald CAB's recommendations and efficiencies found by Fluor-Daniel, 12 years were peeled off that schedule, with a cost saving of $7.8 billion. The final cleanup cost for Fluor's work came to approximately $4.4 billion. It is hard to imagine a more successful outcome. A methodical pump-and-treat project to reduce uranium in the Great Miami aquifer to drinking water levels continued and are nearing completion as this is written (2010).

On October 31, 2006, a formal announcement was made of project completion. Lisa Crawford had this to say "I thought this day would never come. We were all very upset about what plant operations did to our community. But we saw that DOE

Fig. 4.6 Fernald wet lands

and Fluor were just as committed to fixing what had happened as we were. Over time we came to trust each other. We didn't always agree, but they opened the process to us, they listened and even followed our guidance when we proposed a better way. Fluor and DOE delivered on their promise of site closure. Together, we made a difference!" (Business Wire 2006). Ohio EPA Director Joe Koncelik agreed, "The progress made at Fernald would not have been possible without the effective partnership of informed citizens, a committed contractor, and strong regulatory oversight" (Business Wire 2006).

The Fernald site is now a wildlife preserve. The buildings have been removed, and have been replaced by meadows and forest, planted with native species. The waste pits have been excavated and are now ponds and wetlands for wildlife (Fig. 4.6). Because of the unique variety of protected habitat, many kinds of plants and animals rarely seen in the region have appeared on the site, a popular place for birders who want to see a bobolink, a dickcissel, or a mute swan. Less obvious might be an Indiana bat. Near the center of the site is a museum preserving the cold war history of patriotism and sacrifice. Hidden in the woods are burials sacred to Shawnee, and Miami nations where the remains of their ancient ancestors are protected.

Only time will tell if the decisions made by the Fernald community will be found truly wise. If you are ever in the Cincinnati area, you might visit the Fernald preserve. Find a place and sit for a while. Get a sense of it. It is a peaceful place.

4.8 Conclusion and Acknowledgments

This chapter has focused on the important role of public participation policy and practice. However, success at Fernald ultimately arose from the commitment and concern of many people – activists, site workers, news reporters, concerned citizens, managers, engineers, and government officials. An attempt to list everyone who made this happen would exceed the space available. It was the effort of a whole community.

We believe that building trust across that community was essential. Perhaps it could have been done by top-down decision-making, but would the outcome have been trusted? Top-down decision-making is what created the crisis. Would suspicion remain about the health of the community? Could the cleanup have been done so economically? The legitimacy of government and industry ultimately rests upon the quality of its service to the community. Is it not a presumption for business and government leaders to think they know what is best?

Dr. John Applegate, chairman of the Fernald Citizen's task force said it, "When we talk about trust at DOE sites like Fernald, we are usually talking about the trust the public has for the department, but there is another important part of that and that is the trust the department has for the public.... There is a view that all the public adds to environmental decisions is emotion and selfishness. I think that the Fernald Citizens' task force showed that is simply not true."

References

Bleiker H, Bleiker A (1995) Citizen Participation Handbook for Public Officials and Other Professionals Serving the Public. Ninth Edition. Institute for Participatory Management and Planning. Monterey California

Business Wire (2006) Fluor Declares Fernald, Ohio Cleanup Complete. http://www.4-traders.com/FLUOR-12614/news/FLUOR-CORP-NEW-Fluor-Declares-Fernald-Ohio-Cleanup-Complete-Awaiting-DOE-Concurrence-245991/. Accessed 15 August 2010

Fernald Citizens Task Force (1995) Recommendations on remediation levels, waste disposition, priorities, and future use. Ross, Ohio

Grunig JE (1992) Excellence in public relations and communication management. Lawrence Erlbaum Associates, New Jersey

Magnuson, E (1988) They lied to us. Time. http://www.time.com/time/magazine/article/0,9171,968800-1,00.html. Accessed 9 Jul 2010

US Department of Energy, Ohio Field Office (1993) Public Involvement Program for the Fernald Environmental Management Project. Fernald, OH

Chapter 5
Stakeholders, Risk from Mercury, and the Savannah River Site: Iterative and Inclusive Solutions to Deal with Risk from Fish Consumption

Joanna Burger

Contents

J. Burger (✉)
Division of Life Sciences, Environmental and Occupational Health Sciences Institute (EOHSI), Consortium for Risk Evaluation with Stakeholder Participation (CRESP), and Rutgers University, 604 Allison Road, Piscataway, NJ 08854, USA
e-mail: burger@biology.rutgers.edu

J. Burger (ed.), *Stakeholders and Scientists: Achieving Implementable Solutions to Energy and Environmental Issues*, DOI 10.1007/978-1-4419-8813-3_5,

Abstract Many states issue consumption advisories to provide information, mainly to anglers, on the risk from eating fish from contaminated water bodies. The Savannah River passes between South Carolina and Georgia, yet, in 1999, the state-issued consumption advisories for self-caught fish were not in agreement. This chapter examines a stakeholder-driven process that involved state and federal regulators, wildlife biologists, Center for Disease Control, Department of Energy (DOE), fishers themselves, and others to reduce risk for people eating self-caught fish from the river adjacent to the Savannah River Site (a DOE facility). The process included problem formulation, stakeholder identification, identification of the scientific data needed to answer the key questions, development of studies to address these questions, refinement based on stakeholder collaboration, and then development of a mechanism to advise potentially affected persons of the risk . In sum, data on fishing behavior, consumption patterns and mercury levels in fish indicated that people who ate fish frequently were at risk from excess mercury exposure from eating some fish, and an information brochure embraced by the several regulatory agencies and jurisdictions was developed that specifically addressed these issues for people fishing in the Savannah River. This solution sidestepped competing jurisdictional issues between the two states and allowed all parties to create a Fish Fact Sheet brochure that could be distributed annually to those fishing along the Savannah River.

5.1 Introduction

The threat to humans and ecosystems from anthropogenic contaminants is a complex and persistent by-product of our industrialized and urbanized society. Among the several environmental exposures that remain from the Cold War Legacy are stockpiles of radionuclide and chemical wastes, contaminated buildings, and contamination of soils, sediment, and rivers. Mercury can remain in these media for decades and centuries. For mercury, the key feature is that inorganic mercury can be converted by anaerobic bacteria to methylmercury, can bioaccumulate in organisms, and can then be amplified up the food chain to top-level predators, such as sharks, eagles, wolves, lions, and people (Downs et al. 1998). Since mercury is a neurotoxin that can adversely affect eggs, chicks, and young animals, as well as developing fetuses and children, it is important to reduce mercury exposure in biota (including humans) and ecosystems. Excessive mercury exposure from consuming fish high in mercury can even adversely affect adults (Hightower and Moore 2003). In this chapter, all analytic results refer to total mercury, of which on average about 90% would be methylmercury (Jewett et al. 2003).

Contamination can be reduced in the environment by reducing or eliminating the source of the contamination, and by cleaning up sites that are already contaminated by released contaminants. However, sometimes it is not possible to remove the source, or to clean up the contamination, and immediate responses to address the risks are necessary. In some cases, the short-term solution is to manage the exposure through education or intervention or by physically blocking exposure pathways (such as erecting fences). In all cases, sufficient scientific information is necessary to document the extent of the potential exposure, locate exposure pathways, and thus define the potential risk to humans and the environment from the exposure. Obtaining such data is often difficult, time-consuming, and expensive, but the information is needed nonetheless.

Many sites, especially streams, rivers, and lakes, are contaminated with mercury (EPA 2009a) from both local and distant sources. While the method of eliminating the risks from mercury should be source reduction and mercury removal in ecosystems, interim early response risk reduction strategies for humans and the environment are essential tools needed by conscientious public policy makers, governmental agencies, health professionals, and the public at large. In many cases, however, the extent of mercury contamination on- and off-site is unknown: both the potential exposure of consumers (including humans) and the potential risk to consumers are unknown, making it difficult to devise a reasonable strategy for reducing risk to humans and the environment. Moving forward with a risk management strategy requires defining the problem, obtaining site-specific information on contaminants and exposure pathways, determining exposure information, examining risk, and devising a strategy for reducing risk to the public (in this case, from fish consumption). Optimally, the process should be iterative, interactive, inclusive, and collaborative with as many of the interested and affected people, organizations and agencies as possible.

5.1.1 Objectives for This Chapter

This chapter addresses the issues posed by mercury contamination in a riverine system adjacent to a U. S. Department of Energy (DOE) site. At the start of the research, both the extent of the contamination in fish and the extent of human consumption of fish were unknown. Thus the potential risk was unknown. And if there was a risk, both the source and possible risk-reduction solutions were unclear.

In this chapter, I describe the process of examining and managing the potential risk from mercury to people who eat fish from the Savannah River that flows adjacent to the DOE's Savannah River Site (SRS), including (1) Defining the problem, (2) Identifying the stakeholders, (3) Deciding what information was necessary to examine the potential risk from fish consumption, (4) Obtaining data to determine the risk, (5) Evaluating the data, (6) Determining a path forward to reduce risk to people who eat fish, and (7) Evaluating whether the strategy was effective in informing the public and reducing potential risk (Fig. 5.1). In this chapter the popularly used, gender-neutral term "fishers" is used to refer to fishermen, fisherwomen, fisher children, and their families. Throughout this whole process, a range of stakeholders was involved, and the major focus of the chapter is examining the role of stakeholders in improving the science that led to a reasonable solution to reduce exposure from fish consumption.

The main environmental problem addressed in this chapter is thus how to identify and deal with mercury levels in fish that might pose a health threat to people, and

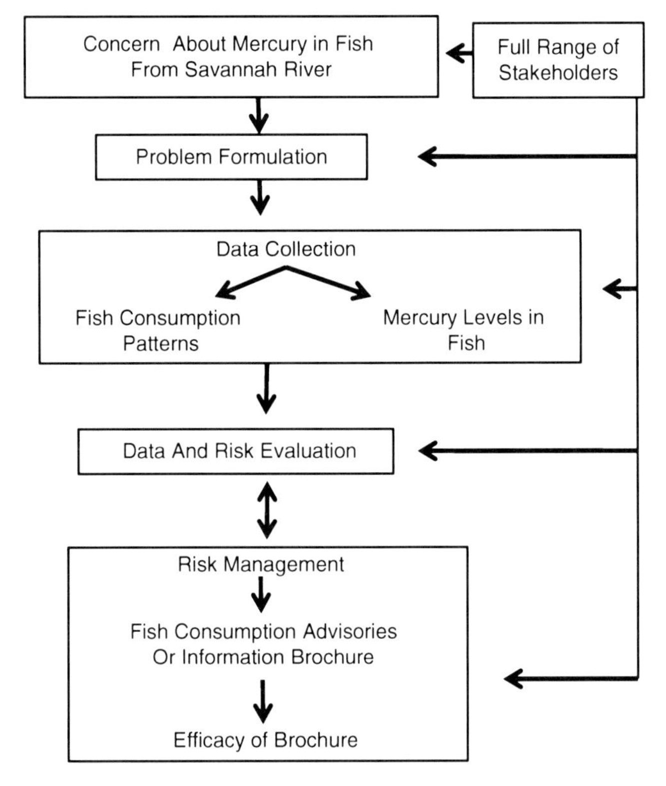

Fig. 5.1 Overall process for developing a path forward to resolve discrepancies and conflicts over the potential risk from mercury in fish consumed from the Savannah River (flowing between South Carolina and Georgia)

the focus is on how to integrate and collaborate with a full range of stakeholders to obtain sufficient data to evaluate the risk, and to devise methods to reduce risk. Stakeholders were integral to all phases, and greatly improved both the science and the management of the issue. I use the term "stakeholders" broadly to include any individuals or organizations that are interested in mercury in fish in the Savannah River (see Table 5.1).

Briefly, consumption information was needed from the stakeholders themselves to determine whether there was a risk to them, and it was necessary to build credibility with these same stakeholders (people possibly at risk) so that when the data and its interpretation were available and had been translated into a Fish Fact statement, the stakeholders (fishers) would be sufficiently trustful to follow the Fish Fact Sheet advise once they understood it. I will show that the desired risk reduction could likely not have been obtained without the stakeholder involvement in the data development process. Thus, the process of securing the information from them (and for them) was at the same time a preparatory step for building the kind of relationship with them and their consciousness that led to acceptance of the Fish Fact Sheet.

Table 5.1 Key stakeholders involved in examining risks to people from mercury in fish in the Savannah River

Agency or group	Role and interests
Fisherfolk and people who eat the fish they catch	People who fish along the Savannah River eat their own fish, distribute it to friends and neighbors, and give it to churches and others for fish fries. Their interests are in knowing what fish are safe to eat, which fish to avoid, and how to optimize fish consumption with risk reduction
General public	People who occasionally come to the Savannah River to fish or recreate, and people who are interested in biota and ecosystem health. Interested in understanding contaminant levels in fish
U.S. Department of Energy (DOE)	Landowner of the Savannah River Site (SRS) that runs adjacent to the Savannah River, and is responsible for some mercury contamination in Steel Creek (on site) and in the river. Interested in reducing risk to fish consumers
U.S. Environmental Protection Agency (EPA)	Federal agency responsible for human health and the environment, and that has some jurisdiction for waters jointly held by two states. Interested in risk reduction and in having uniform fish consumption advisories, as well as having oversight responsibility for CERCL cleanup on these lands
Citizen's Advisory Board (for DOE) (CAB)	DOE-appointed body to provide advice on environmental issues and cleanup at the SRS. Interested in monitoring DOE's activities and providing advice to them about issues such as mercury in fish
Savannah River Ecology Laboratory (University of Georgia, on SRS) (SREL)	Ecology laboratory of the University of Georgia with facilities adjacent to the site, and largely funded by DOE at the time. Interested in sound science and interactions with both DOE and the public about this issue
Centers for Disease Control (CDC)	U.S. federal body that monitors diseases and had conducted limited studies of mercury contamination on the river
South Carolina Department of Health and Environmental Control (SCDHEC)	State agency responsible for issuing consumption advisories (for South Carolina residents). Interested in mercury levels in fish from both an ecological and a human health risk perspective. Also interested in efficacy of fish consumption advisories and in materials for the public
Georgia Department of Natural Resources (GDNR)	State agency responsible for issuing consumption advisories (for Georgia residents). Interested in mercury levels in fish from both an ecological and a human health risk perspective. Also interested in efficacy of fish consumption advisories and in materials for the public
Consortium for Risk Evaluation with Stakeholder Participation (CRESP)	DOE-funded, independent research entity that conducts original research, reviews documents and reports, and involves stakeholders in research and management processes

The Savannah River flows between South Carolina and Georgia

When data were then turned into advice that depended on shaping their behavior, the groundwork for their willingness to do so had been laid. This case study shows that stakeholder inclusion in the data development process lays the foundation for successful risk management.

5.1.2 Mercury in the Environment

Mercury enters the environment from both natural and anthropogenic (human-generated) sources, and it is the latter that are of greatest concern because of both historic and recent changes in mercury levels. The increasing demand for electric power generation that releases mercury into the atmosphere from burning coal results in both local and regional pollution through atmospheric transport and deposition. In addition, mercury exposure can come from point-sources that release mercury into streams and rivers.

Mercury occurs naturally in seawater, and coastal waters receive mercury runoff from land, input from rivers, and airborne deposition. Biomethylation of mercury occurs in sediment, allowing for food chain biomagnifications (Montiero et al. 1996; Downs et al. 1998). Mercury in fish tissue may be six orders of magnitude higher than the mercury concentration in the water column (Scudder et al. 2009). Anthropogenic sources of mercury account for about 80% of the annual inputs of mercury to the environment. For many areas of the world, atmospheric deposition, both regional and global, is the primary source of mercury (Fitzgerald and Mason 1996; Driscoll et al. 2006). The global contribution of mercury to the atmosphere is unevenly distributed. The Asian countries contribute about 54% of the mercury to total atmospheric sources. China alone contributes 28% to the total emissions, followed by Africa (18%) and Europe (11%, Pacyna et al. 2006). These sources are primarily from electric power generation, although regionally mercury inputs can be due to other industrial processes, as well as from gold mining extraction (Mueezzinoglu 2003). Hospitals have traditionally contributed to local mercury sources (SFWMD 2007 and their earlier reports).

Mercury is a persistent toxicant that bioaccumulates in animals (including humans; Nichols 2001), making it critical to identify the sources of mercury, evaluate exposure to humans and the environment, and determine the risk it poses to humans and the environment. Armed with such knowledge, managers and regulators can lower mercury emissions to a level that reduces adverse effects on individuals and populations, and other risk managers (including health professionals) can develop strategies to reduce mercury exposure. For example, in the Everglades of south Florida, knowledge of the high levels of mercury in predatory fish led to enacting controls on emissions by local power plants and other industries, which ultimately led to a drastic reduction in the mercury levels in the fish eaten by birds and mammals, including people (Davis and Ogden 1994; Lange et al. 1994; SFWMD 2007). Even in the Everglades, however, it was necessary to issue fish consumption advisories to reduce human exposure to mercury.

5.2 Background on Fish Advisories

5.2.1 Fish Consumption Advisories

In the absence of an effective cleanup strategy, or way to rapidly reduce mercury in biota (such as fish) and in ecosystems to immediate risk issues, fish consumption advisories are often issued. To be sure, there is a general distrust of governmental agencies, even in other countries (Wentholt et al. 2009), and thus fishing advisories are often ignored (Burger and Gochfeld 2006). Still, although mitigation of the mercury contamination and its source is clearly preferable for many reasons, fish advisories and their dependence on getting affected parties to change their normal patterns of behavior are sometimes the only effective near-term risk management alternative. This approach to risk reduction has shifted the burden from pollution reduction to personal behavior (Jakus et al. 1997; Halkier 1999).

In general, states are responsible for issuing contaminant advisories to the public, should they become necessary, although the Environmental Protection Agency (EPA) and the Federal Food and Drug Administration (FDA) can issue them. The U.S. EPA posts a "National Listing of Fish Advisories" every 2 years (EPA 2009a), and its most recent edition notes that all 50 states, the District of Columbia , American Samoa and Guam (U.S. territories), and five Native American Tribes have fish consumption advisories for some local waters. Fish advisories are not regulations, but are recommendations to help protect human health. Fish advisories are developed for the general population, and for sensitive populations (such as pregnant women and children). Approximately 18 million lake acres (43% of U.S. lakes) and 1.4 million river miles (39% of U.S. total river miles) were under advisory in 2008 (EPA 2009a), so the problem is by no means local only to the Savannah River. States may be hindered in their issuance of fish advisories by lack of information on either consumption rates or contaminant levels in fish, organizational problems, or by jurisdictional issues when two or more states share the same waters (Chess et al. 2005).

All 50 states have advisories for mercury, and 80% of all fish consumption advisories were issued in part because of mercury. Mercury accounts for the greatest number of advisories for lakes (about 17,000 acres) and rivers (over 1.3 million miles, EPA 2009a). Mercury thus remains one of our most important environmental issues, both from point-source and from atmospheric sources.

As noted, compliance with fish consumption advisories is sometimes low, leading to questions about the efficacy of such advisories as a public health policy (Connelly and Knuth 1998; Burger 2000; Jardine 2003). However, Hispanic fishermen from Newark Bay (New Jersey) showed a willingness to change their consumption behavior when presented with clear risk information (Burger et al. 1999a, b; Pflugh et al. 1999), and others have shown a decline in fish consumption among pregnant women following a federal mercury advisory issued in January 2001 (Oken et al. 2003). People have to know about advisories to follow them, and often people are not aware of them (Burger 2005; Burger and Gochfeld 2006). And agencies need to know whether the advisories are effective and result in changes in behavior.

5.2.2 Why Fish Advisories are Necessary: The Role of Fish in Diets

Fishing provides a nutritious source of protein and is very popular in both urban and rural areas of the United States, and elsewhere in the World (Toth and Brown 1997; Burger et al. 1992, 1993, 2001b, c; Ramos and Crain 2001). Fishing not only provides fish and shellfish to eat, but it also provides a range of social benefits that include interactions with family and friends, allows people to get away from the stresses of life, and provides opportunities for people to commune with nature (Fleming et al. 1995; Toth and Brown 1997; Burger 2002; Burger and Gochfeld 2006). Increasing attention to health and nutrition in the media has increased the public's consumption of fish, even among those who never fish themselves (NOAA 2004).

In the United States, there has been a general upward trend in seafood consumption since the 1960s, despite an increase in price and warnings about contaminants. The trend has waxed with nutrition advice and waned with hazard advisories (FOA 1998), but continues to increase, now exceeding 7.4 kg/capita per year (NOAA 2004). Increases in contaminant levels in fish have led to the necessity to issue consumption advisories to deal with the risk.

Fish are a healthy source of protein, provide omega-3 (n-3) fatty acids that are generally accepted to reduce cholesterol levels, and reduce the incidence of heart disease, stroke, and preterm delivery (Anderson and Wiener 1995; Patterson 2002; Albert et al. 2002). Iribarren et al. (2004) showed a positive relationship between consumption of fish with high n-3 fatty acids and a lower likelihood of high hostility in young adults.

However, contaminant levels in some fish are sufficiently high to potentially cause adverse human health effects, making it necessary to consider both the risks and benefits of fish consumption. (Gochfeld and Burger 2005) Adverse effects from contaminants in fish include counteracting the cardioprotective effects (Guallar et al. 2002), damaging unborn babies and young children (IOM 1991; Neuringer et al. 1994; ATSDR 1996; Iso and Rexrode 2001; Olsen and Secher 2002; Moya 2004), and adversely affecting adult behavior and physiology (Hightower and Moore 2003; Hites et al. 2004). There is a positive relationship between mercury and polychlorinated biphenyl (PCB) levels in fish, fish consumption by pregnant women, and deficits in neurobehavioral development in children (IOM 1991; Sparks and Shepherd 1994; Schantz 1996; NRC 2000; Schantz et al. 2003). There is a decline in fecundity in women who consume large quantities of contaminated fish from Lake Ontario (Buck et al. 2000). There is also a suggestion that mercury affects blood pressure (Vupputuri et al. 2005). Generally, there is a positive relationship between mercury levels in people and fish consumption (Knobeloch et al. 2005; Johnsson et al. 2005). These largely epidemiology studies have demonstrated a relationship between mercury levels in fish they eat and mercury levels in people, and between mercury levels in human tissue and adverse effects. And although these relationships are correlational, they are persuasive in identifying the need to issue consumption advisories.

The extensive discussion about what the "safe" level of exposure is may be partly political; it is surely controversial, and will continue to be so for some time (Stern 1993; NRC 2000; Stern et al. 2004). Even the role of occasional peak exposures vs. chronic lower level exposures to methylmercury requires closer attention, and, depending on how the science of that question evolves in the near term, it may also be essential to develop single-meal fish consumption advisories for fish species high in methylmercury (Ginsberg and Toal 2000).

The bottom line, however, is that fish are an important and nutritious form of protein, and people should not stop eating fish. However, they should be aware of the relative benefits (in terms of omega-3 levels) and risks (contaminant levels) and optimize their eating patterns to reduce the risks while increasing the benefits I have cited. Obviously, they cannot do so without understandable, credible, and persuasive information.

5.3 Mercury, Fish Consumption and Stakeholders from the Savannah River

The general problem faced in this example is how to deal with mercury in fish in the Savannah River, and the solution involved a multistep process (Fig. 5.1). Each of the major steps are described below, with further refinements of the process and how stakeholders were involved. That mercury in fish from the Savannah River might pose a human health risk had previously been identified by all the state and federal agencies with responsibility for human health.

The controversy surrounding the issue involved the DOE's assertion that no one fished on the site itself (because of access restrictions and inability to get on site to fish in Steel Creek) and that no one fished adjacent to the SRS. Thus, the Department contended there was no risk to the fishing public. The state agencies and the federal Environmental Protection Agency were unsure whether people fished along the river at the edge of SRS, whether there was indeed any specific risk from the fish that resided in the waters along the SRS, and whether the resultant fishing/consumption patterns and mercury levels in fish from this stretch of the river posed a human health risk.

5.3.1 Background on Contamination and Conflicts

5.3.1.1 Site Description

The Savannah River originates in the southern Appalachians of North Carolina, passes through South Carolina and Georgia, and flows into the Atlantic Ocean. It winds through several large reservoirs, past various industrial sites (including chemical facilities), nuclear power plants, and the SRS. The original mercury contamination of the Savannah River came from chemical plants upstream from SRS.

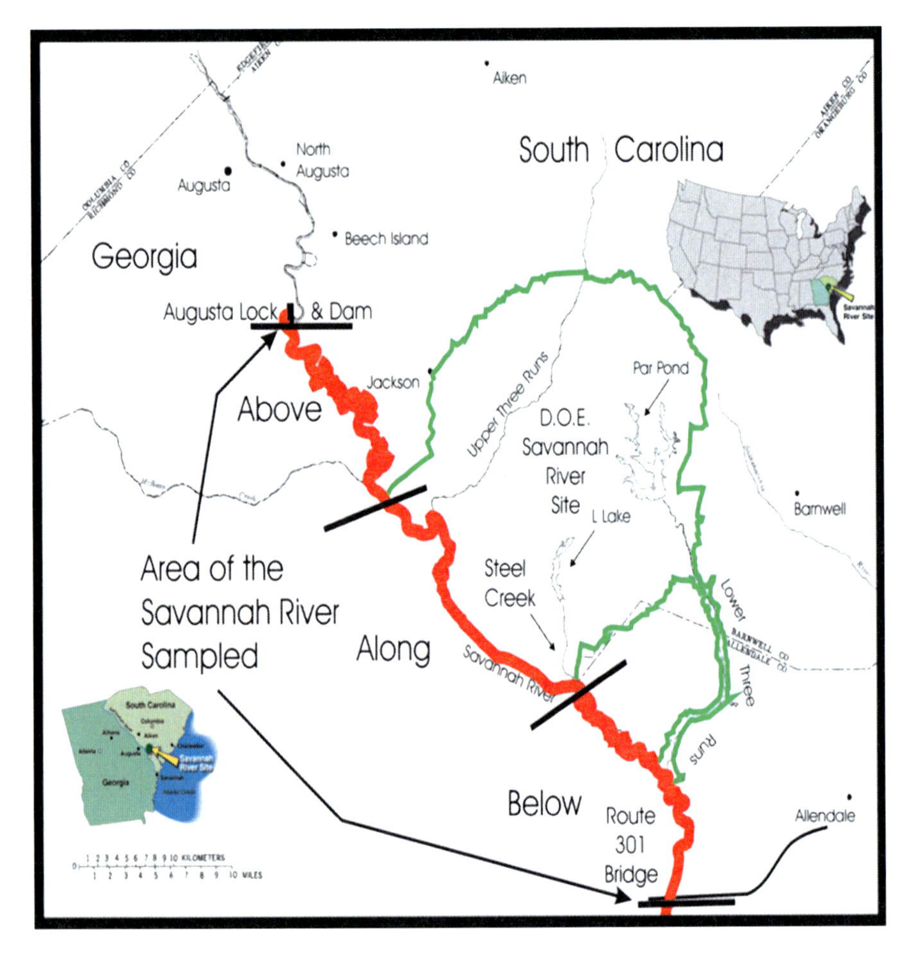

Fig. 5.2 Map of the Savannah River showing the location of the Department of Energy's Savannah River site

The SRS (33.1°N, 81.3°W) is a 780-km^2 former nuclear weapons production and current research facility operated by the DOE since the early 1950s (Fig. 5.2). The DOE used the river as a source of cooling water for nuclear reactors, and water was discharged to artificial thermal cooling reservoirs. Streams from the SRS flow directly into the Savannah River, and fish can move freely between the on-site streams and the river (Workman and McLeod 1990). Although there is controversy about the upstream sources of mercury to the Savannah River, the on-site SRS activities themselves also resulted in contamination by heavy metals (including mercury) and radionuclides, and there is some discharge of mercury from the coal-burning power plant on DOE site (Kvartek et al. 1994; Sugg et al. 1995). Further, the use of water for cooling the reactors redistributed the mercury on SRS lands, and in the marshes and streams on site, and off-site. These sources could result in higher mercury levels in fish adjacent to the SRS. Atmospheric deposition of mercury is relatively high in this region (>10 μg/m²/yr, EPA 1980; Downs et al. 1998). If mercury exposure was coming from the SRS, then mercury levels should be lower above the site than along or below it (and they generally were not). Regardless of the source of mercury, DOE is mindful of its economic and social role in the region (Greenberg et al. 1998).

5.3.1.2 Consumption Advisories for the Savannah River

At the time (1999), the South Carolina Department of Health and Environmental Control (SCDHEC) had issued fish consumption advisories that included the Savannah River (SCDHEC 1999). While Georgia did not issue an advisory for its side of the Savannah River, it did issue "Guidelines for eating fish from Georgia waters" (GDNR 1999). Thus, important discrepancies existed in the advice given by their states to fishers and consumers. Since EPA, through its regional office in Atlanta, Georgia, provides CERCLA (Comprehensive Environmental Response Compensation and Liability Act) oversight for all federal facilities, it had some interest in the mercury contamination in fish issue. EPA wanted each state to maintain its jurisdictional primacy with respect to human health, while insisting that a tangible form of risk communication resulted. EPA was at the time particularly attentive to its mandate (Federal Order 12898) to address environmental justice issues (EPA 2002, 2009b; DOE 2009).

There was resistance to the need for fish advisories (by Georgia), confusion about the likely pattern of public response to the consumption advisories, and important differences of opinion among the several public authorities as to whether there was significant fishing in the Savannah River along the SRS. The Consortium for Risk Evaluation with Stakeholder Participation (CRESP) was encouraged by DOE (and then also volunteered) to help resolve the issues by developing and providing new data to address these conflicts, and by facilitating discussions and collaborations among all parties.

5.3.2 Identification of Stakeholders

One of the most important steps to formulating issues and solving environmental problems is the level and timing of involvement of stakeholders. A mantra among most public agencies is "early and often," a mantra more often stated than enacted. While there were a number of obvious stakeholders (fisherfolk, state agencies responsible for issuing advisories) in situations such as this, there is a fuller range of stakeholders that should be included (Table 5.1). Additionally, although knowledge and expertise may vary among stakeholders, they all have something worthwhile to contribute to the discussion. In particular, often long-time residents, fisherfolk, and others familiar with the site will have information not available in the literature.

Sufficient time and attention needs to be devoted to identifying the relevant stakeholders, and for the SRS, this process involved talking to the relevant federal and state agencies, attending meetings (particularly of the Citizens' Advisory Board (CAB)), and talking with the fishers themselves. That is, CRESP talked to the regulatory agencies (GDNR, SCDHEC, EPA), as well as DOE, and they each suggested others (such as CDC, CAB) that should also be included. Then the CDC and CAB were asked who else should be included. This resulted in a list of the key stakeholders that needed to be involved throughout the process. In the case of mercury in fish from the Savannah River, the key stakeholders involved federal and state agencies, advisory boards, a scientific research laboratory, and scientists (from the CRESP), as well as the people fishing and consuming the fish. Further, information was put on the CRESP web site, making it possible for the general public to access information and provide comments to CRESP (http://www.CRESP.org).

5.3.3 *Formulating the Problem*

Formulating the problem and identification of stakeholders is an iterative process that involves initial problem formulation, followed by meetings with stakeholders, and then a reformulation of the problem in collaboration with the key stakeholders. Although the individuals (or agencies) who initiated the process may initially define the problem, once it is identified, it is critical to involve all stakeholders to refine the problem, the hypotheses to be tested, the temporal and spatial features of the issue, and the possible methodologies to employ. Problem formulation should involve both determination of assessment endpoints (what is important) and measurement endpoints (what can be measured to reflect the important questions, Norton et al. 1992).

The general problem faced by DOE and the other stakeholders was whether mercury levels in fish actually did pose a risk to humans (and to other biota) who consumed them at the levels they were, in fact, consuming the fish. The state and federal agencies did not think there were very many people who fished above, along, or just below the SRS (Fig. 5.3). Thus, the problem identified was to determine whether any people were at risk from eating fish from the Savannah River, which involved examining both fishing and consumption patterns, and contaminants in fish.

While there was general agreement on the problem itself, there were priority and other nuanced differences among the diverse different stakeholders. In particular, the primary questions of concern to the several kinds of stakeholders differed (Table 5.2). The regulatory agencies were primarily interested in whether they needed to modify their consumption advisories, and if so, how. EPA, on the one hand, was interested in having uniformity between the states, but on the other hand respected the two states' rights to develop the advisories. Additionally, EPA was also responsible for overseeing cleanup on SRS, and both EPA and DOE were interested in arriving at an agreeable solution to deal with the mercury contamination in fish that would accommodate the different interests and perspectives of the two adjoining states. Of course, one potential outcome was a requirement for DOE to

Fig. 5.3 Photographs of fishermen along the Savannah River who were interviewed for the study. Another factor necessary to consider is that some fishermen give inferior fish (often with higher mercury levels) to others

Table 5.2 The problem as seen through the eyes of different stakeholders, including self-interest and management goals

Stakeholder	Problem formulation and questions of interest
Fisherfolk and the General Public	What fish are safe to eat, and in what quantities?
Department of Energy	Are remediation actions necessary to prevent mercury contamination in the fish or other biota?
Citizens' Advisory Board	What advice should we give to DOE regarding cleanup or risk from mercury in fish (because of human exposure through consumption)?
Environmental Protection Agency	Should we intervene in the issuance of fish consumption advisories? How can advisories be similar for the two states? What are the implications for cleanup at SRS?
Savannah River Ecology Laboratory	What levels of mercury are there in fish, and what is the import to functioning ecosystems?
Center for Disease Control	What is the extent of exposure that could prove harmful to human health?
South Carolina Department of Health and Environmental Control	Are there any regulatory controls or actions required? Are any fish at risk? Should there be any changes in fish consumption advisories?
Georgia department of natural resources	Are any fish at risk? Should there be any changes in fish consumption advisories?
Consortium for Risk Evaluation with Stakeholder Participation	What data need to be collected to address the major concerns of the different stakeholders? How can the data be used to formulate a solution for the fisherfolk?

conduct further remediation to remove mercury from on-site source (that could move off-site via streams and creeks).

Discussions were held among the above parties in open meetings with each group individually, and in meetings (and conference calls) among groups (e.g., GDNR, SCDHEC, EPA, DOE). The discussions among the range of stakeholders resulted in agreement that data were needed to address the basic questions, which included the following: (1) What are the levels of mercury in fish from the Savannah River? (2) Do the levels of mercury vary above, along, and below the Savannah Rive Site? (3) Is there fishing along this stretch of the Savannah River, and (4) What are the consumption rates (and do they vary by gender or ethnicity)? All stakeholders agreed that site-specific data were required to determine if there was a risk to people from consuming fish from the Savannah River. And all agreed that if there was a risk, they had to address what could or should be done about it. Some agencies were willing to issue advisories on existing information, others not until the site specific data became available.

As a result of the discussions, two studies were designed to assess (1) fishing behavior and consumption patterns in anglers along the Savannah River (Fig. 5.4) and (2) mercury levels in fish as a function of species, size, and location along the river. In the latter case, fish were collected from above SRS, along SRS, and below SRS (Fig. 5.1).

5.3.4 Fishing Behavior and Consumption Patterns

In response to stakeholder requests, these two separate studies were initiated by CRESP to provide the information needed to move forward on the advisory process: (1) fishing behavior and consumption patterns and (2) mercury levels in fish as a function of species, size of fish, trophic level, and location of collection. The major steps for examining fishing behavior and consumption patterns involved establishing the study site (above, along, and below the SRS), designing a survey instrument to ascertain fishing behavior and consumption patterns, conducting a pilot study to make sure that the survey instrument was appropriate, conducting the study, analyzing the data, and presenting the results to all the stakeholders in a series of meetings (Fig. 5.4, see Burger et al. 2001a for more details).

The process of designing the survey instrument involved the full range of stakeholders, both in the initial design and in the redesign of the instrument. The stakeholders provided information not only on what questions to ask but on the overall protocol. While the scientists were responsible for the initial design of the survey form, the information gathered in the iterative process assured that information gathered would satisfy the regulatory agencies' needs as well as inform the fishing public.

A pilot study of 40 fishers was essential to make sure that the study questions were relevant to people fishing along the Savannah River, and could be administered easily. People fishing along the river provided input during the pilot study. In subsequent meetings with all the stakeholders, considerable refinements were made to the survey instrument to make it relevant to the fishers and to the agencies involved. As a result of this process, we added questions about cooking practices, specific fish, and the age at which children first consumed fish. The pilot study also allowed CRESP researchers to determine the sample size necessary for appropriate statistical analysis. All other components of the study were similarly taken to the stakeholders for comments and suggestions in open meetings or in conference calls.

Two aspects of the stakeholder involvement process bear comment: (1) inclusion of local scientists, and (2) use of local interviewers familiar with the culture and locations along the river (and who spoke the local language). The involvement of local scientists from the Savannah River Ecology Laboratory (SREL) provided CRESP with advice on the form of questions, advice on how to approach local fishers,

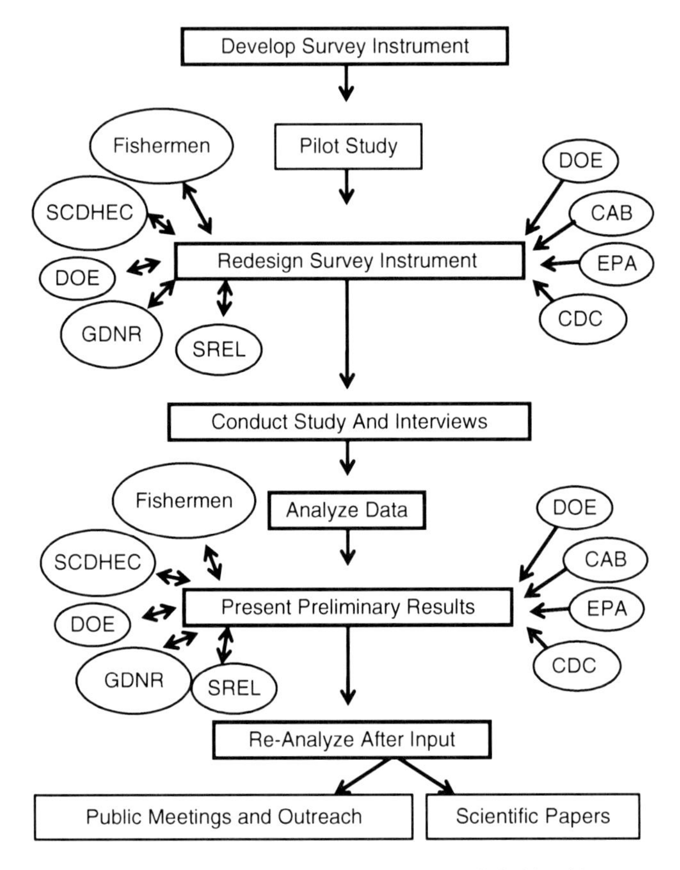

Fig. 5.4 Schematic of the process involved in conducting a stakeholder-driven survey of fishing behavior and consumption patterns of fishers along the Savannah River. *EPA* environmental protection agency; *DOE* department of energy; *CAB* citizens' advisory board for SRS; *SREL* university of Georgia's Savannah River ecology laboratory located adjacent to SRS; *CDC* centers for disease control; *GDNR* Georgia department of natural resources; *SCDHEC* South Carolina department of health and environmental control. Note: "Fisherman" includes all fishers of all ages and genders. CRESP was involved in all phases and conducted the study

advice on the protocol, and logistical information. The training of local people (fisherfolk themselves) as interviewers ensured that the interviews were conducted in an appropriate manner (with sufficient time for conversing amiably). Both were essential to the process.

The results of the study of fishing behavior and consumption patterns indicated that (1) portion size increased with the number of times people ate self-caught fish, (2) a substantial number of people ate more fish than that used by state agencies to compute risk to recreational fishers (19 kg/year for SC), (3) some people consume more fish than the default assumption (50 kg/yr), and (4) blacks consumed more locally caught fish per year than did whites, putting them at greater risk from mercury exposure (Fig. 5.5; Burger et al. 1999b) Further, people with less education, and

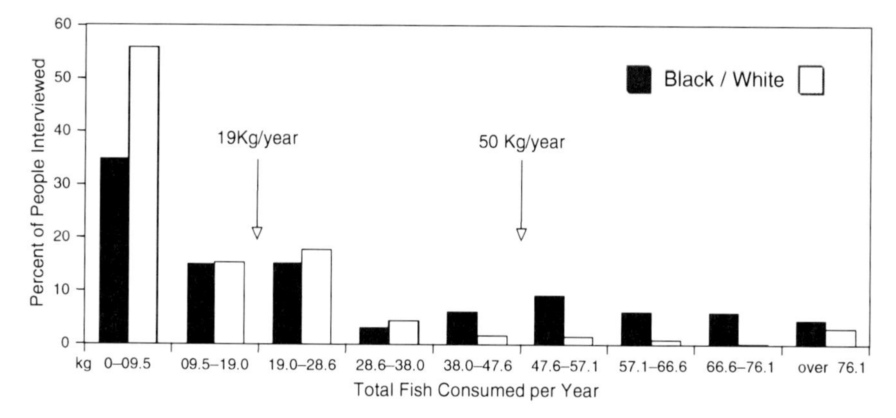

Fig. 5.5 Relative consumption patterns of black and white people interviewed who were fishing along the Savannah River. 19 kg/year is the fish consumption rate used to design the fish consumption advisories used at the time, and 50 kg/year was the maximum default assumption

those with lower incomes, ate more fish than others. Thus, the data clearly indicated that there was a problem with people eating more fish than agencies thought (Fig. 5.5), and it was an environmental justice issue as well (Bullard and Wright 1987; Bullard 1990, 1994; EPA 1994, 2002, 2009b)

5.3.5 *Mercury in Fish and Risk to Fish Consumers*

Determining mercury levels in fish requires designing a protocol, collecting the fish, and analyzing the fish according to the protocol. A key aspect is to select fish species that are of interest to the fishing public, regulators, managers, DOE, and other stakeholders. A smaller group of stakeholders was involved in the process, although all the stakeholders named above made comments about the initial study design, locations for collecting, and fish species to be examined (Fig. 5.6). Fish were collected by electroshocking the fish to force them to the surface; all fish were dissected at SREL and shipped to Rutgers University for analysis.

Not surprisingly, we found that there were significant differences in mercury levels among fish, with bowfin (*Amia calva*, a fish largely eaten by black people) and largemouth bass (*Micropterus salmoides*, a preferred fish) having the highest mercury concentrations (Burger et al. 2001b, c). There were few differences in mercury levels in fish collected above, along, and below the Savannah River. Where there were differences, they were small and contradictory. Mercury levels were highest in bowfin above the SRS, while mercury levels in largemouth bass were highest below the SRS (Burger et al. 2001b). The lack of a strong difference among the three study locations may relate to the migratory patterns of the fish, moving up and down river. Thus, the data did not, in fact, provide strong support for SRS being a major contributor to mercury levels in fish. On the other hand, those same migratory patterns also did not

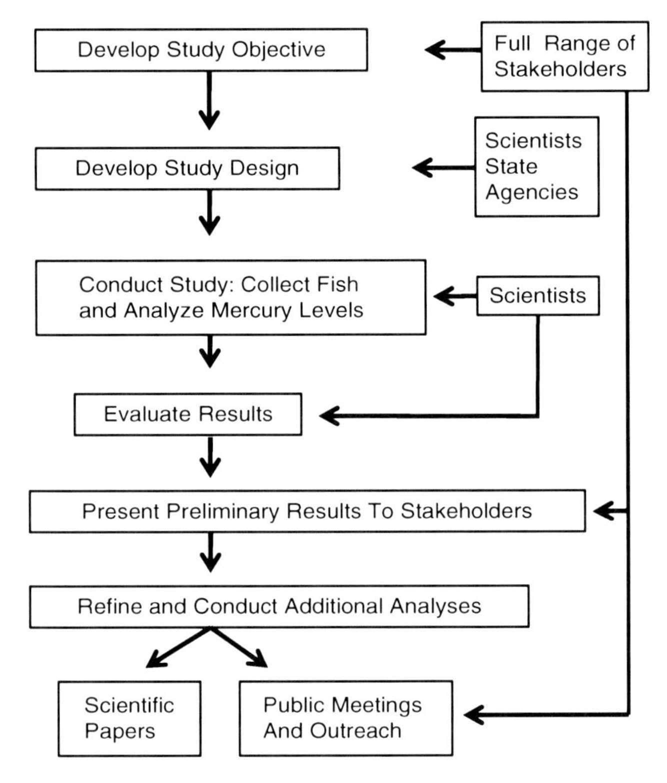

Fig. 5.6 Schematic showing the overall process involved in examining the levels of mercury found in fish from the Savannah River. The full ranges of stakeholders are those described in Fig. 5.4

exonerate the site as being a contributing source to the mercury levels in the fish. Data from the study were presented to the full range of stakeholders, many of whom suggested other analyses or questions to address with the data. In all cases, these analyses were performed and returned to the appropriate stakeholders. In general, this improved the analyses by making them more relevant to the users (Figs. 5.4 and 5.6).

The ultimate goal of the study was to determine if people who eat fish from the Savannah River are at risk. The consumption study indicated that many people were eating fish at higher levels than those used to develop the fish consumption advisories (Burger et al. 1999b), and the mercury study in fish indicated that levels varied by species and size of the fish (Burger et al. 2001b). Although several methods were used to examine risk from consuming fish from the Savannah River (Burger et al. 2001b), one of the easiest to understand is to use the action levels developed by various states or other agencies, such as the FDA. This action level is the mercury level used by the FDA to condemn a shipment of commercial fish between states or prevent importation. While these levels were initially designed to prevent commerce between states, they can serve as a guide for fish that should generally be avoided.

Table 5.3 Mercury concentrations (arithmetic mean and standard error) (μg/g wet weight=ppm) and percent of each species exceeding 0.5 and 1.0 ppm (after Burger et al. 2001b, c)

		Mean (μg/g) ± SE	% >0.5 ppm	% >1 ppm
Bowfin	Amia calva	0.94 ± 0.05	81	45
Largemouth bass	Micropterus salmoides	0.46 ± 0.04	38	4
Chain pickerel	Esox niger	0.36 ± 0.03	21	0
Yellow perch	Perca flavescens	0.28 ± 0.02	10	0
Spotted sucker	Minytrema melanops	0.27 ± 0.04	14	3
Black crappie	Pomoxis nigromaculatus	0.24 ± 0.02	9	0
Channel catfish	Ictalurus punctatus	0.20 ± 0.02	1	0
American eel	Anguilla rostrata	0.15 ± 0.03	8	0
Bluegill sunfish	Lepomis macrochirus	0.14 ± 0.02	3	0
Red-breast sunfish	Lepomis auritus	0.13 ± 0.02	6	0

The U.S. FDA (FDA 2001, 2005) has an action level of 1.0 μg/g (=ppm), and has issued a series of consumption advisories based on methylmercury, advising that pregnant women and women of childbearing age who may become pregnant should limit their fish consumption, should avoid eating four types of marine fish (shark, swordfish, king mackerel, tilefish), and should also limit their consumption of all other fish to just 12 oz per week (FDA 2003).

The FDA Action Level is not a health-based level. Some states have developed action levels, and many states use 0.5 ppm or even 0.3 ppm. Since information on levels of mercury is reported in ppm, providing this information can aid in making informed decisions. Table 5.3 shows the mean mercury levels for the fish, along with the percent of samples greater than 0.5 and 1.0 ppm. This indicates that many of the fish caught in the river, both by recreationists and by subsistence fishermen, have mercury levels that are of concern. At a minimum, this analysis allows people to decide which fish they should generally avoid (bowfin, largemouth bass), and which are safe to eat in nearly unlimited quantities (sunfish).

5.3.6 Development of a Fish Fact Sheet

As noted, in many regions, including along the Savannah River, adherence to the advice given in [compliance with] state-issued fish consumption advisories is low (Burger et al. 1999b; Burger and Gochfeld 2006; Wentholt et al. 2009). Low compliance can be due to a number of factors, such as confusing messages, conflicting advisories (as was the case with the Savannah River), controversies concerning the health benefits and risks of fish consumption, personal preferences, and an unwillingness to comply because of personal beliefs or preferences. In short, people often ignore advisories because they distrust the government, don't believe it applies to them, assume they can detect tainted fish, or just love fish.

After numerous meetings and conferences with the full complement of stakeholders, it was agreed that there were people (particularly blacks) who were consuming more fish than the advisories advocated, and that they ate significantly more than the amount used to devise the advisories. Information on consumption and mercury levels indicated that some people (not only pregnant women), are at risk from consuming fish from the Savannah River. This led to agreement among the several responsible public entities that advice in a form likely to be understood by the affected community (i.e., a brochure or other communication device) was needed that gave more information on species to avoid (and to eat), preferred cooking methods, and the adverse effects of mercury (particularly on fetuses and young children). The device (brochure) and the method by which it was delivered needed to be clear, simple, readable, attractive, and responsive to the information needs of the fishing public. Since advisories were generally being ignored, there needed to be a plan for bringing it to people's attention. One recommendation was that the brochures should be distributed to people fishing on the Savannah River (not just left at clinics and fishing tackle shops, as is the case with the state-generated consumption advice). The concept emerged that the brochure should be distributed by local student interns during the summer, an approach which had the added advantage that it involved yet another group of stakeholders.

This process also involved other steps that included all the major stakeholders. It involved even more conference calls and meetings than the science-based studies because it involved direct actions, as well as an agreed-upon Fish Fact Sheet (Fig. 5.7). Given the diverse interests that we have identified among the public entities (see Table 5.1), not only did it take time for the several public entities to agree on the message, it required many meetings to agree upon the exact wording even though everyone agreed there was a need for a directed message to consumers of fish (particularly pregnant women) from this area of the Savannah River.

Having data on consumption patterns (including what species of fish were preferred by local anglers) and on contaminant levels was a crucial aid to bringing all stakeholders together. Since this information could have resulted in disagreements between the states, and with the EPA about the issuance of consumption advisories, we sought instead to find common ground. By finding a communications mechanism that effectively removed the jurisdictional issues (between the states and with EPA), it was possible to work on a design and wording of the brochure that was acceptable to everyone, yet provided the public with much-needed information.

Before embarking on creating a brochure, the group agreed on several principles: (1) fish are a healthy source of protein, (2) there was a population of the fisher public at risk because they ate more fish with higher mercury levels than previously suspected, (3) information on demographics was useful in helping outreach and communication specialists, (4) pregnant women and children should be the focus, (5) complete site characterization and extensive knowledge of the pathways was not necessary to develop a brochure for fish consumption, and (6) it should be clear to everyone, including the fishers, that the brochure represented a consensus among the two state agencies and EPA (Burger et al. 2001d). This was accomplished by adding logos of each agency on the brochure, and giving contact phone numbers of individuals from each of the three relevant agencies.

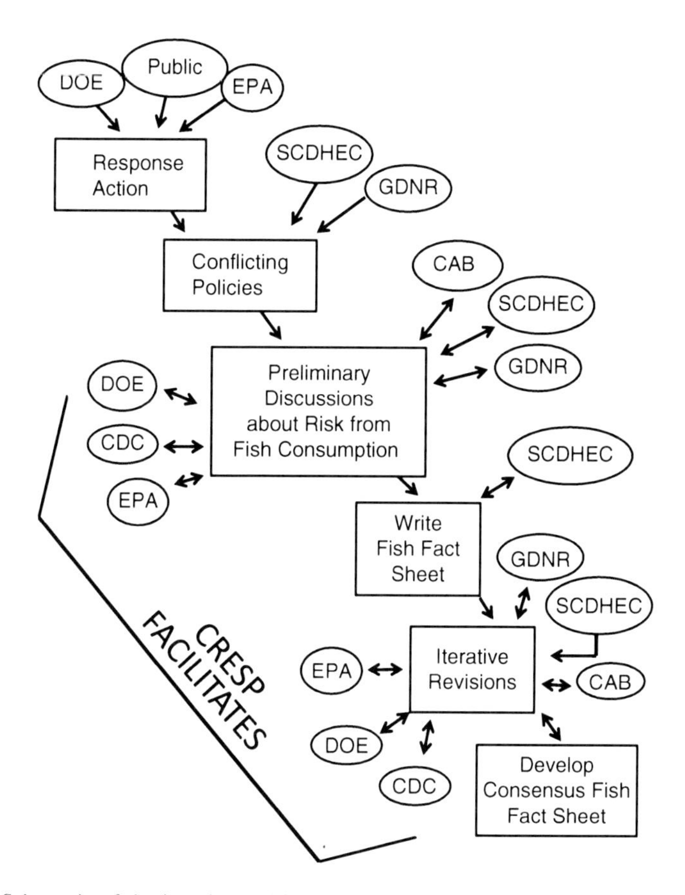

Fig. 5.7 Schematic of the iterative and interactive process of developing a fish fact sheet for consumers of fish from the Savannah River. *EPA* environmental protection agency; *DOE* department of energy; *CAB* citizens' advisory board for SRS; *SREL* university of Georgia's Savannah River ecology laboratory located adjacent to SRS; *CDC* centers for disease control; *GDNR* Georgia department of natural resources; *SCDHEC* South Carolina department of health and environmental control. CRESP was involved in all phases and conducted the study

In informal meetings it was agreed that EPA would provide the initial content, and SCDHEC would write the initial draft. This was followed by numerous communications concerning the intent and working of the draft, including conference calls where representatives of all agencies were involved. The draft was also reviewed by the CAB, SRS–CDC health effects committees, and others for content and presentation (Fig. 5.7). While the process took more time than a top–down approach by any agency, it resulted in a product that all parties were pleased with. This was the first such information brochure about fish consumption that involved two states and the EPA. Further, this agreement contributed to the willingness of DOE to fund an intern to distribute the brochure during each fishing season.

Taken as a whole, the processes described here are examples of how credible scientific data gathered with stakeholder involvement can in turn be used to implement important risk-reduction management and policies, and to provide a basis for consensus building on difficult risk communication issues. It further suggests that consensus building and risk communication are continuing processes that involve assimilation of new information on contaminants in the food chain, variations in mercury levels among fish species (and sizes), state and federal laws, public policy, and public responses.

5.3.7 *Efficacy of the Fish Fact Sheet: Evaluating the Stakeholder Involvement Process Itself*

Having reached consensus about a path forward, and developed a Fish Fact Sheet to distribute to fishers above, along, and below the SRS, we wanted to determine whether the brochure was clear and the meaning received. We designed a questionnaire to determine if we were reaching the target audience, using the same overall methods as described for the other stakeholder-driven process described above. Our objectives were to determine (1) whether people fishing had previously read the Fish Fact Sheet or had heard about consumption advisories, (2) what major message they obtained from the Sheet, (3) who they felt the Fact Sheet was aimed at, and who should get it, and (4) who should be concerned about health risks from fish consumption. We also gathered the same demographic information as obtained in the fish consumption study. The same interviewers (local fishers who had lived there all their lives) conducted the interviews by first giving the Fish Fact Sheet to them, asking them to read it, and then interviewing them. Nearly everyone approached participated, and carefully read the brochure before answering the questions.

Over half of those interviewed felt that everyone should get the Fish Fact Sheet, even if they did not fish along the river, and the messages they obtained were accurate (Burger and Waishwell 2001). The major message, that people should limit fish intake in some way (e.g., by eating fish lower in mercury), was recognized by 86% of the people, a very high comprehension rate for this kind of information. Further, many people asked for additional copies for family members and friends.

This last step in the stakeholder-driven process to address potential mercury exposure of people fishing on the Savannah River is one usually not taken for environmental actions. That is, remediation, restoration, and educational and behavioral intervention actions taken around contaminated sites are not evaluated to see if they are effective in reducing risk to people fishing and to other community members living around the site. Even large-scale remediation/restoration projects seldom conduct in-depth follow-up studies to determine efficacy and effectiveness. Merely continuing to monitor levels of contaminants in media (soil, sediment), the usual method for determining success, may not effectively determine whether contaminants have reached the food chain, and ultimately humans.

This step in the process was again time-consuming, since it involved developing a questionnaire specifically for this project, pilot testing it, conducting the interviews, analyzing the results, and having stakeholders involved in the process. However, the end results convinced the agencies that the Fish Fact Sheet was effective, and that the distribution was worth the effort. It further indicated that the primary stakeholders, the fishers and their family/friends who ate the fish, were interested in the message, wanted more brochures to distribute, and obtained the correct message. That the message was designed specifically for this target audience improved the willingness to read the brochure, distribute it to friends and family, and act on the advice. Further, the interviewers found that those interviewed were very interested in more information about the source of contamination, ecological pathways, exposure, risk levels, and why the river had not been cleaned up (Burger and Waishwell 2001).

5.4 The Role of Science in the Process

The role of science was to provide site-specific information that was credible and addressed the questions as identified in the problem formulation phase. It was the presence of the site-specific data on consumption patterns and behavior that convinced the agencies (SCDHEC, GDNR, EPA) that there was a problem, DOE and CDC that something needed to be done, and the whole group that it was necessary to stop worrying about jurisdiction issues and past differences in the issuing of advisories. By changing the game from disagreements over consumption advisories to consensus over the message that needed to reach stakeholders, the science could form the basis for moving forward. Without site-specific data dealing with the key aspects of the issue, the parties had not agreed on any solution. The data provided a consensus for moving forward.

The two main advantages of collaborative science are that there can be some disagreements among the scientists about the approaches and methods to be used in the studies (which result in better approaches), and the scientists have to be willing to make changes in the design and protocol as a result of collaborations. That is, scientists usually like to design their own studies individually, and conduct them as they wish. Collaborative science requires that scientists listen to different viewpoints, and fully intend to implement changes in protocols.

5.5 The Role of Stakeholders in the Process

Jardine, Burger, and others suggested that public participation in the establishment of fish consumption advisories would greatly improve not only the advisories themselves, but compliance (Jardine 2003; Jardine et al. 2003; Burger and Gochfeld 2006). Jardine established 14 guiding principles for public participation that included the inclusion of community needs and values into the advisory process, in a timely fashion, in a collegial fashion, and with transparency. This happened at Savannah River.

The full range of stakeholders were involved in every step of the process, from problem formulation, identification of the key questions, design of the studies (one on fishing behavior and consumption patterns, and one on mercury levels in fish), examining the results of the pilot study with the survey questionnaire, redesigning the survey form, examining the final analyses, suggesting additional analyses, participating in public meetings, deciding on a path forward, creating a brochure, and testing the efficacy of the brochure. While the interviews themselves and the collection of fish for analysis were conducted by scientists, this involved using local research assistants who were also fishers as interviewers, and scientists from both CRESP and SREL (refer to Table 5.4). The beginning-to-end stakeholder involvement process was not only good public policy, rather it was fundamental to actually improving the quality of the scientific results and laid the groundwork for successful risk reduction once the risk management strategy had been defined.

The collaborative process of inclusion of regulators, other state and federal agencies, scientists, citizens' groups, fishers, and others requires the recognition that there will be disagreements. And in this case, there were times when phone calls were aborted, meetings got heated or canceled until issues could be resolved among two or more parties, and everyone felt the process might fall apart. But the bottom line was that everyone wanted to address the issue because there was a clear need to determine if the fish had high mercury levels and if there were people at risk. The agencies all recognized that the fishers were confused about the safety of fish from this stretch of the Savannah River, and that a clear, united message had to be presented that had the blessing of the state (SCDHEC, GDNR) and federal agencies (EPA), and the cooperation and agreement of DOE. More importantly, the message (form or content) had to be on target for the people fishing along the Savannah River, as well as clear, concise, and accurate.

5.6 Lessons Learned and Paths Forward

Until we succeed in removing or at least containing environmental contaminants and preventing further contamination, we must deal with reducing risk to humans and the environment, often by education or by blocking the exposure pathway. Both education and consumption advisories are short-term measures to reduce exposure and to deal with high levels of mercury contamination until such time as source reduction has significantly reduced the risk. These are solutions until our nation's waters are sufficiently clean so that all fish can be eaten as often as desired. The decline in PCB concentrations in the environment over the past 25 years provides a basis for optimism (EPA 2009a).

5.6.1 Conflict Resolution

During the process there were obviously contentious issues that revolved around (1) jurisdictional autonomy (state's rights), (2) federal authority to obtain uniformity

Table 5.4 Role of different stakeholders in refinement of the fishing behavior and consumption patterns study along the Savannah River

Activity	Anglers	EPA	CDC	DOE	CAB	SREL	SCDJEC	GDMR
Formulating the problem	X	X	X	X	X	X	X	X
Identifying additional stakeholders		X		X		X	X	X
Designing of survey instrument		X	X	X		X	X	X
Modification of survey instrument	X	X	X	X	X	X	X	X
Conducting study	X					X		
Analyzing results						X		
Presenting and commenting on preliminary results	X	X	X	X	X	X	X	X
Suggesting additional analyses		X		X		X	X	X
Reanalyzing data after input						X		
Meetings with stakeholders	X	X	X	X	X	X	X	X
Modifying analyses	X	X		X		X	X	X
Writing scientific papers		X				X		
Discussing path forward		X	X		X	X	X	X
Involvement in developing fish fact sheet	X	X	X	X	X	X	X	X
Efficacy survey of fish fact sheet	X			X			X	X
Meetings with stakeholders	X	X	X	X	X	X	X	X
Writing scientific paper								
Distributing fish fact sheet to fishing public along the river (each summer)	X			X			X	X

Anglers = fisherfolk; other abbreviations found in Table 5.1. An X indicates that the agency or group contributed to this aspect of the study. CRESP is not included because they participated in all phases. Fisherfolk and the General Public are combined for this table as anglers

for advisories pertaining to the same waters, (3) different health objectives (long-term goals of CDC vs. immediate mandate of the states to issue advisories), and (4) source reduction, cleanup standards (for DOE) vs. behavioral modifications in the form of issuance of specific fish consumption advisories or information brochures for this section of the Savannah River. These were major and important concerns, and they were always present as undercurrents to discussions. However, these old and contentious issues were mitigated by shifting the discussion from consumption advisories to information brochures that everyone could agree were needed.

There were several times during the nearly 3-year process that one or more parties became annoyed with progress or the views of others, disgusted with delays, or impatient with jurisdictional issues. What held the process together was the desire for site-specific data gathered from fishers and their families on fishing and consumption, site-specific data on mercury levels in fish gathered to answer very specific questions, and a desire to move forward to provide the public with a unified message. The role of CRESP was to facilitate the process and to design the initial protocol (for others to collaborate on modifying), conduct the study with help from local interviewers who were long-time residents and fishers and from SREL scientists. Since all parties participated in both the design of the survey form and the design of the mercury in fish study, there was uniform buy-in of the resultant data.

5.6.2 Lessons Learned

All participants in the process learned the importance of inclusion, iteration, and involvement of the state and federal agencies, local scientists, citizen's boards, and the fishers themselves. While the process took longer than some participants

Table 5.5 The main lessons learned from considering the risks from mercury in fish in the Savannah River, that flows between Georgia and South Carolina

1. Define the problem initially
2. Identify the stakeholders early and be inclusive
3. Involve stakeholders early and often
4. Include all the relevant state and federal agencies
5. Include a range of scientists, including local agency or university scientists
6. Modify the survey instrument to fit local biological and social conditions
7. Use local fisherfolk to conduct the interviews (with scientific training)
8. Conduct a pilot study to determine appropriate sample sizes, and refine survey forms
9. Be prepared for the process to take longer than a top–down approach
10. Be prepared (the scientists) to change the objectives, protocols, methods, and data collection, and to reanalyze data at the request of a range of stakeholders
11. Recognition by all participants that iteration and inclusion of others is essential, and that all viewpoints are valid
12. Be prepared for disagreements, and find ways to change the definition of the immediate issue to one that everyone can work with

The stakeholder-driven and collaborative process involved a wide range of state and federal agencies, citizens groups, research scientists and laboratories, and the fishing public had expected or desired at the outset, it did result in resolution and in coming up with a communication plan (and a brochure) that all parties could agree met the needs of the fishers and their families. The inclusion of the wide range of stakeholders early, and at every point in the process, meant that everyone was aware of progress and had an opportunity to participate. While not everyone agreed with every decision about protocols, methods, approaches, or data analysis, everyone had a chance to comment, make suggestions, and to have their concerns addressed and questions clarified.

Some of the main lessons learned are described in Table 5.5. In the final analysis, the issue of risk from mercury that might have been increased by SRS had been festering among the stakeholders for some time without any resolution. It was only the stakeholder-driven process that moved it forward.

Acknowledgments I particularly thank Michael Gochfeld, Charles W. Powers, David S. Kosson, Bernard D. Goldstein, Lynne Waishwell, and Camilla Warren for valuable discussions about SRS, science, stakeholders, and fish consumption; Shane Boring, I.Lehr Brisbin Jr., Patricia A. Cunningham, Karen F. Gaines, J.Whitfield Gibbons, and Joel Snodgrass for advice and logistical help while on SRS; Caron Chess, James Clarke, MichaelGreenberg, , and Lisa Bliss for helpful discussions about science over the years; Tracey Shelly and Robert Marino (SCDHEC), Randy Manning (GDNR), John Stockwell (USEPA), Thomas Johnson and Wade Whittaker (USDOE), Christian Jeitner, Taryn Pittfield, and Mark Donio (Rutgers University) for technical help. This research was funded mainly by the Consortium for Stakeholder Participation (CRESP) through a contract from the Department of Energy (AI # DE-FFG-26-00NT 40938 and DE-FC01-06EW07053) to Vanderbilt University and Rutgers University, as well as the Nuclear Regulatory Commission (NRC 38-07-502M02), NIEHS (P30ES005022), and the New Jersey Department of Environmental Protection. The conclusions and interpretations reported herein are the sole responsibility of the author, and should not in any way be interpreted as representing the views of the funding agencies.

References

Albert CM, Campos H, et al (2002) Blood levels of long-chain n-3 fatty acids and the risk of sudden death. New Engl J Med 346:1113–1118

Anderson PD, Wiener JB (1995) Eating fish. In: Graham JD, Wiener JB (ed) Risk versus risk: tradeoffs in protecting health and the environment. Harvard University Press, Cambridge

Agency for Toxic Substances and Disease Registry (ATSDR) (1996) Mercury awareness for the public. Hazard Subst Public Health 6:4–5

Buck GM, Vena JE, Schisterman, EF, et al (2000) Parental consumption of contaminated sport fish from Lake Ontario and predicted fecundability. Epidem 11:388–383

Bullard RD (1990) Dumping in Dixie: race, class, and environmental quality. Westview Press, Colorado

Bullard RD (1994) Overcoming racism in environmental decision making. Environment 30:39–44

Bullard RD, Wright BH (1987) Environmentalism and the politics of equity: emergent trends in the black community. Mid-Am Rev Sociol 12: 21–32

Burger J (2000) Consumption advisories and compliance: The fishing public and the deamplification of risk. J Environ Plan Manage 43:471–488

Burger J (2002) Consumption patterns and why people fish. Environ Res 90:125–135

Burger J (2005) Fishing, fish consumption and knowledge about advisories in college students and others in central New Jersey. Environ Res 98:268–275

Burger J, Gochfeld M (2006) A framework and information needs for the management of the risks from consumption of self-caught fish. Environ Res 101:275–285
Burger J, Waishwell L (2001) Are we reaching the target audience? Evaluation of a Fish Fact Sheet. Sci Total Environ 277:77–86
Burger J, Gaines KF, Gochfeld M (2001a) Ethnic differences in risk from mercury among Savannah River fishermen. Risk Anal 21:533–544
Burger J, Cooper K, Gochfeld, M (1992) Exposure assessment for heavy metal ingestion from a sport fish in Puerto Rico: estimating risk for local fishermen. J Toxicol Environ Health 36:355–365
Burger J, Staine K, Gochfeld M (1993) Fishing in contaminated waters: knowledge and risk perception of hazards by fishermen in New York City. J Toxicol Environ Health 3:95–105
Burger J, Pflugh KK, Lurig L, von Hagen LA, von Hagen SA (1999a) Fishing in urban New Jersey: Ethnicity affects information sources, perception, and compliance. Risk Anal 19:217–229
Burger JW, Stephens CS, Boring CS, Kuklinski M, Gibbons JW, Gochfeld M (1999b) Factors in exposure assessment: Ethnic and socioeconomic differences in Fishing and Consumption of fish caught along the Savannah River. Risk Anal 19:427–438
Burger J, Gaines KF, Gochfeld M (2001b) Ethnic differences in risk from mercury among Savannah River fishermen. Risk Anal 21:533–544
Burger J, Gaines KF, Boring CS et al (2001c) Mercury and selenium in fish from the Savannah River: species, trophic level, and locational differences. Environ Res 87:108–118
Burger J, Gochfeld M, et al (2001d) Science, policy, stakeholders, and fish consumption advisories: developing a fish fact sheet for the Savannah River. Environ Manage 27:501–514
Chess C, Burger J, McDermott MH (2005) Speaking like a state: environmental justice and fish consumption advisories. Soc Nat Res 18:267–278
Connelly NA, Knuth BA (1998) Evaluating risk communication: examining target audience perceptions about four presentation formats for fish consumption health advisory information. Risk Anal 18:649–659
Davis S Ogden JC (1994) Everglades: the ecosystem and its restoration. CRC Press, Florida
Department of Energy (DOE) (2009) Environmental justice at the U.S. Department of Energy. http://www.1m.doe.gov/spotlight/ej3.html. Accessed 9 Feb 2010
Downs SG, Macleod CL, Lester JN (1998) Mercury precipitation and its relation to bioaccumulation in fish: a literature review. Water Air Soil Pollut 108:149–187
Driscoll CT, Abbot M, et al (2006) Airsheds and watersheds. In: Harris R, Krabbenhoft DP, Mason R, Murray MW, Reash R, Saltman T, (ed) Ecosystem responses to mercury contamination. CRC Press, Florida
EPA (Environmental Protection Agency) (1980) Ambient water quality criteria for mercury. US Environmental Protection Agency Report 440/5-80-058. Natl Tech Inform Serv, 5285 Port Royal Road, Springfield, VA
Environmental Protection Agency (EPA) (1994) Environmental justice initiatives 1993. (EPA 2000-R-93-001) US Environmental Protection Agency Washington
Environmental Protection Agency (EPA) (2002) National Environmental Justice Advisory Council: fish consumption and environmental justice http://www.epa.gov/compliance/resources/publications/ej/fish_consump_report_1102.pdf. Accessed April 2009
EPA (Environmental Protection Agency) (2009a) National listing of fish advisories: technical fact sheet: 2008 biennial national listing. http://www.epa.gov/fishadvisories/tech2008.html. Accessed 9 Nov 2009
Environmental Protection Agency (EPA) (2009b) Environmental justice: compliance and enviorcement. http://www.epa.gov/environmentaljustice. Accesed 1 Feb 2010
Food and Drug Administration (FDA) (2001) FDA consumer advisory. http://www.fda.gov/bbs/topics/ANSWERS/2000/advisory.html. Accessed 1 Dec 2005
Food and Drug Administration (FDA) (2003) FDA consumer advisory. http://www.fda.gov/bbs/topics/ANSWERS/2000/advisory.html. Accessed 1 Jan 2008
Food and Drug Administration (FDA) (2005) What you need to know about mercury in fish and shellfish – March 2004: 2004 EPA and FDA Advice For: Women Who Might Become Pregnant Women Who are Pregnant Nursing Mothers Young Children. http://www.fda.gov/Food/

FoodSafety/Product-SpecificInformation/Seafood/FoodbornePathogensContaminants/Methylmercury/ucm115662.html. Accessed 8 June 2009

Fitzgerald WF, Mason RP (1996) The global mercury cycle: oceanic and atmospheric aspects. In: Baeyens W, Ebinghaus R, Vasiliev O (ed) Global and regional mercury cycles: sources, forces and mass balances. Kluwer Academic Publishers, Dordrecht

Fleming LE, Watkins S et al (1995) Mercury exposure in humans through food consumption from the Everglades of Florida. Water Air Soil Pollut 80:41–48

FOA (1998) Seafood safety – Economics of hazard analysis and critical control point (HACCP) programmes. United Nations Food and Agricultural Organization, Fisheries Technical Paper # 381, Rome, Italy. http://www.fao.org/documents/show_cdr.asp?url_file=/DOCREP/003/XO465E12.html. Accessed 10 March 2008

GDNR (Georgia Department of Natural Resources) (1999) Guidelines for eating fish from Georgia waters. GDNR, Atlanta

Ginsberg GL, Toal BF (2000) Development of a single-meal fish consumption advisory for methylmercury. Risk Anal 20:41–47

Gochfeld M, Burger J (2005) Good fish/bad fish: a composite dose and benefit-risk curve. NeuroToxicol 26:511–520

Greenberg M, Krueckeberg D et al (1998) Socioeconomic impacts of U.S. nuclear weapons facilities: a local scale analysis of Savannah River, 1950-1993. Appl Geograph 18:101–116

Guallar E, Sanz-Gallardo MI, et al (2002) Heavy metals and Myocardial Infarction Study Group: Mercury, fish oils, and the risk of myocardial infarction. New Eng J Med 347:1747–1754

Halkier B (1999) Consequences of the politicization of consumption: the example of environmentally friendly consumption practices. J Environ Policy Plan 1:25–41

Hightower JM, Moore D (2003) Mercury levels in high-end consumers of fish. Environ Health Perspect 111:604–608

Hites RA, Foran JA et al (2004) Global assessment of organic contaminants in farmed salmon. Science 303:226–229

Institute of Medicine (IOM) (1991) Seafood Safety. National Academy Press. Washington

Iribarren C, Markovitz JH et al (2004) Dietary intake of n-3, n-6 fatty acids and fish: relationship with hostility in young adults – the CARDIA study. Eur J Clin Nutr 58:24–31

Iso H, Rexrode KM (2001) Intake of fish and omega-3 fatty acids and risk of stroke in women. J Am Med Assoc 285:304–312

Jakus PM, Downing M, Bevelhimer MS, Fly JM (1997) Do sportfish consumption advisories affect reservoir anglers' site choice? Agricult Res Econom Rev 26:196–204

Jardine CG (2003) Development of a public participation and communication protocol for establishing fish consumption advisories. Risk Anal 23:461–471

Jardine CG, Hrudey SE et al (2003) Risk management frameworks for human health and environmental risks. J Toxicol Environ Health Part B 6:569–641

Jewett SC, Zhang X et al (2003) Comparison of mercury and methylmercury in northern pike and Arctic grayling from western Alaskan rivers. Chemosphere 50:386–392

Johnsson C, Schutz A, Sallsten G (2005) Impact of consumption of freshwater fish on mercury levels in hair, blood, urine, and alveolar air. J Toxicol Environ Health (part A). 68:129–141

Knobeloch L, Anderson HA et al (2005) Fish consumption, advisory awareness, and hair mercury levels among women of childbearing age. Environ Res 97:220–227

Kvartek EJ, Carlton WH et al (1994) Assessment of mercury in the Savannah River Site environment. Westinghouse Savannah River Co (WSRC-TR-94-0218ET). Aiken

Lange TR, Royals HE, Connor LL (1994) Mercury accumulation in largemouth bass (*Micropterus salmoides*) in a Florida lake. Environ Contam Toxico 127:466–471

Montiero LR, Costa V, Furness RW, Santos RS (1996) Mercury concentrations in prey fish indicate enhanced bioaccumulation in mesopelagic environments. Mar Ecol Prog Ser 141:21–25

Moya J (2004) Overview of fish consumption rates in the United States. Human Ecol Risk Assess 10:1195–1211

Mueezzinoglu, A (2003) A review of environmental consideration on gold mining and production. Critical Reviews in Environmental Science and Technology 33:45–71

National Research Council (NRC) (2000) Toxicological effects of methylmercury. National Academy Press. Washington
Neuringer M, Reisbick S, Janowsky J (1994) The role of n-3 fatty acids in visual and cognitive development: Current evidence and methods of assessment. J Pediatr 125:39–47
Nichols JW (2001) Use of indicators in ecological risk assessment for persistent, bioaccumulative toxicants. Human Ecol Risk Assess 7:1043–1057
NOAA. (2004) Seafood consumption rose again in 2003. NOAA Magazine, U.S. Commerce Dept., National Oceanographic and Atmospheric Administration. http://www.noaanews.noaa.gov/stories2004/s2322.html. Accessed 10 March 2005
Norton SB, Rodier DR et al (1992) A framework for ecological risk assessment at the EPA. Environ Toxicol Chem 11:1663–1672
Oken E, Kleinman KP et al (2003) Decline in fish consumption among pregnant women after a national mercury advisory. Obstet Gynecol 102:346–351
Olsen SF, Secher NJ (2002) Low consumption of seafood in early pregnancy as a risk factor for preterm delivery: prospective cohort study. Brit Med J 324:447–450
Pacyna EG, Pacyna JM, Steenhuisen F, Wilson, S (2006) Global anthropogenic mercury emissions inventory for 2000. Atmosph Environ 40:4048–4063
Patterson J (2002) Introduction – comparative dietary risk: balance the risks and benefits of fish consumption. Comments Toxicol 8:337–344
Pflugh KK, Lurig L, vonHagen LA, vonHagen S, Burger J (1999) Urban angler's perception of risk from contaminated fish. Sci Total Environ 228:203–218
Ramos AM, Crain EF (2001) Potential health risks of recreational fishing in New York City. Ambulat Pediatr 1:252–255
SCDHEC (South Carolina Department of Health and Environmental Control) (1999) Public health evaluation: Cesium-137 and strontium-90 in fish. Attachment to the fish consumption advisory for the Savannah River, Columbia
Schantz SL (1996) Developmental neurotoxicity of PCBs in humans: What do we know and where do we go from here? Neurotox Teratol 18:217–227
Schantz SL, Widholm JJ, Rice DC (2003) Effects of PCB exposure on neuropsychological function in children. Environ Health Perspect 111:357–376
Scudder BC, Chaser LC et al (2009) Mercury in fish, bed sediment, and water from streams across the United States, 1998–2005. U.S. Dept of Interior, Report 2009–5109. Reston, Virginia. 74 pp
SFWMD (South Florida Water Management District) (2007) South Florida Environmental Report: 2007 (G. Redfield, ed). West Palm Beach. SFWMD
Sparks P, Shepherd R (1994) Public perceptions of the potential hazards associated with food production: an empirical study. Risk Anal 14:799–808
Stern AH (1993) Re-evaluation of the reference dose for methylmercury and assessment of current exposure levels. Risk Anal 13:355–364
Stern AH, Jacobson JL, Ryan L, Burke TA (2004) Do recent data from the Seychelles Islands alter the conclusions of the NRC Report on the toxicological effects of methylmercury? Environ Health 3:2
Sugg DW, Chesser RK, Brooks JA, Grasman BT (1995) The association of DNA damage to concentrations of mercury and radiocesium in largemouth bass. Environ Toxicol Chem 14:661–668
Toth Jr JF, Brown RB (1997) Racial and gender meanings of why people participate in recreational fishing. Leisure Sci 19:129–146
Vupputuri S, Longnecker MP, Daniels JL, Xuguang G, Sandler DP (2005) Blood mercury level and blood pressure among US women: results from the national Health and Nutrition Examination Survey 1999–2000. Environ Res 97:195–200
Wentholt MTA, Rose G, Konig A, Marvin HJP, Frewer LJ (2009) The views of key stakeholders on an evolving food risk governance framework: results from a Delphi study. Food Policy 34:539–548
Workman SW, McLeod KW (1990) Vegetation of the Savannah River Site: major community types. Savannah River Ecology Laboratory (SRO-NERP-19). Aiken

Chapter 6
Helping Mother Earth Heal: Diné College and Enhanced Natural Attenuation Research at U. S. Department of Energy Uranium Processing Sites on Navajo Land

William J. Waugh, Edward P. Glenn, Perry H. Charley, Marnie K. Carroll, Beverly Maxwell, and Michael K. O'Neill

Contents

Abstract Diné College is a key stakeholder and partner with the U.S. Department of Energy in efforts to develop and implement sustainable and culturally acceptable remedies for soil and groundwater contamination at uranium mill tailings processing and disposal sites on Navajo Nation land. Through an educational philosophy grounded in the Navajo traditional living system which places human life in harmony

W.J. Waugh (✉)
DOE Environmental Sciences Laboratory, S.M. Stoller Corporation,
Grand Junction, CO 81503, USA
e-mail: Jody.Waugh@lm.doe.gov

J. Burger (ed.), *Stakeholders and Scientists: Achieving Implementable Solutions to Energy and Environmental Issues*, DOI 10.1007/978-1-4419-8813-3_6,

with the natural world, the College has helped guide researchers to look beyond traditional engineering approaches and seek more sustainable remedies for soil and groundwater contamination at former uranium mill sites near Monument Valley, Arizona, and Shiprock, New Mexico. Students and researchers are asking first, what is Mother Earth already doing to heal a land injured by uranium mill tailings, and second, what can we do to help her? This guidance has led researchers to investigate applications of natural and enhanced attenuation remedies involving native plants – phytoremediation, and indigenous microorganisms – bioremediation. College faculty, student interns, and local residents have contributed to several aspects of the pilot studies including site characterization, sampling designs, installation and maintenance of plantings and irrigation systems, monitoring, and data interpretation. Research results look promising.

6.1 Introduction

With bright yellow cottonwood canopies illuminating the San Juan River floodplain below, students from Diné College, a Navajo-owned community college, gather in a fenced plot on an ancient river terrace in Shiprock, New Mexico, armed with tape measures and pruners to record the growth of native phreatophytes and clip stems and leaves for chemical analysis. Phreatophytes – literally, "well plants" – survive in this desert environment by extending their roots down like straws to suck groundwater. As students work on an autumn day in 2008, a film crew from the National Science Foundation interviews some of their peers and instructors for the documentary, "Weaving STEM (science, technology, engineering, and mathematics) Education and Culture: The Faces, Places, and Projects of the Tribal Colleges and Universities Program." Curriculum in the Diné Environmental Institute (DEI) at the College is designed to weave environmental science methods with Navajo cultural traditions – Navajo Science.

The students, their instructors, and scientists from the U.S. Department of Energy (DOE), University of Arizona, and New Mexico State University have teamed up on a phytoremediation research project. Phytoremediation is the name given to the science and practice of using plants as part of the remedy for contaminated soil and groundwater. An objective of the research is to determine if the native phreatophytes can be grown to withdraw groundwater and slow the spread of contamination away from a nearby uranium mill tailings disposal site, and do so without contaminating the plants.

Phytoremediation, a type of enhanced natural attenuation, fits well with the College's approach for weaving Navajo culture into environmental science education. Navajo tradition teaches us that we are connected to the land, that we should live in harmony with Mother Earth. By teaming with the DEI and incorporating the goals of Navajo Science, DOE is learning to take a more holistic approach in developing remedies for contamination related to past uranium milling on Navajo land. Phytoremediation, bioremediation, and enhanced natural attenuation are scientific approaches that ask, allegorically, "What is Mother Earth already doing to heal a land injured by uranium mill tailings, and what can we do to help her?"

As future scientists and community leaders, today's Diné College students are living a new chapter in a Navajo story about uranium mining, milling, and the U.S. Government. Many of these students' grandfathers and community elders, when they were young, moved their families to far corners of the Navajo Nation to work in mines and mills that had sprung up across the Colorado Plateau to supply uranium to fuel the weapons of the Cold War. These families were not forewarned that the colorless, odorless radon gas, which emanates from the uranium ore and mill tailings, would eventually cause lung cancer among the Navajo miners at a rate 20–30 times higher than that of nonminers in the region, a tragedy that continues to impact their families and communities even today.

This chapter serves as an example of how Native American students and their way of life can be incorporated into an ongoing remediation and research project to better understand how to restore Mother Earth. It provides an overview of the environmental legacy of Cold War uranium mining and milling, of efforts by the U.S. Congress and federal agencies to repair and provide long-term care of land and water contaminated by uranium milling, and of the role Diné College students and faculty are playing as stakeholders and researchers, in collaboration with students and researchers from the University of Arizona, New Mexico State University, and the U.S. DOE, to discover and enhance natural remedies for cleaning up soil and groundwater at uranium mill tailings sites on Navajo land near Monument Valley, Arizona, and Shiprock, New Mexico.

6.2 Uranium Mill Tailings: A Cold War Legacy

Uranium mill tailings are residues of crushed ore following extraction of uranium oxide, commonly called yellowcake. After a sequence of grinding, separating, and concentrating uranium oxide during the milling process, most of the original ore is discarded as tailings and other residual wastes. Almost all of the uranium oxide processed in the United States during the mid 1900s, including that unearthed on Navajo Land, was purchased by the U.S. Government to fuel the massive Cold War weapons production effort. Federal purchasing of uranium triggered a mining boom in the Four Corners states of Utah, Colorado, Arizona, and New Mexico, as prospectors equipped with radiation detectors combed the sandstone outcrops in the region. By 1955, there were hundreds of mines producing high-grade ore and several AEC-funded milling facilities and buying stations operating on the Colorado Plateau. The uranium mining and milling boom abated in 1962 with a drop in AEC purchases of uranium, and then nearly ceased altogether in 1970 when the AEC stopped purchasing uranium. Many mines and mills were abandoned, leaving a legacy of tailings and processing residues that can adversely affect human and environmental health.

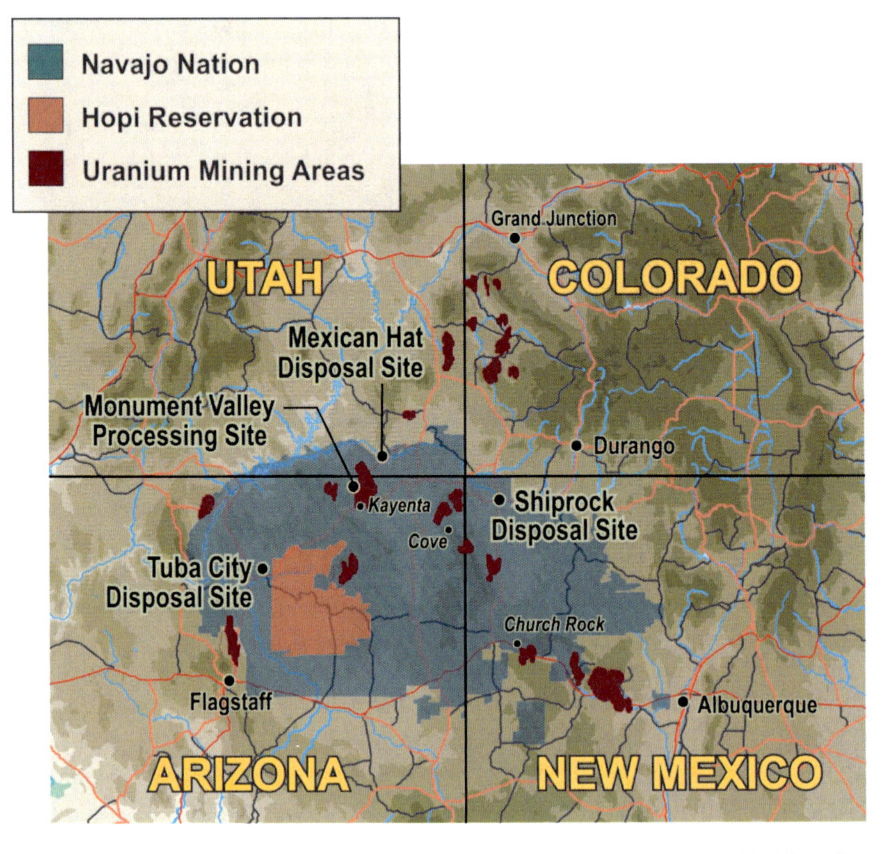

Fig. 6.1 Map of the Navajo Nation showing uranium mining areas and uranium mill tailings sites managed by the U.S. Department of Energy under the Uranium Mill Tailings Radiation Control Act of 1978

6.2.1 Navajo Uranium Mining, Milling and Health Effects

The Navajo Nation, covering about seven million hectares (over 27,000 square miles) of the Colorado Plateau in northeastern Arizona, northwestern New Mexico, and southeastern Utah, played a major role in the Cold War uranium mining boom. Between 1943 and 1945, the Manhattan Project, a secret U.S. military project instituted during World War II to develop the first atomic bomb, recovered approximately 34,000 kg (about 76,000 pounds) of uranium oxide from vanadium mill tailings on Navajo land (Chenoweth 1997). Navajos and other Native Americans guided prospectors and miners to large deposits of uranium-bearing ore on reservation land (Johnston et al. 2007). Discoveries of uranium, first near Cove, Arizona (Brugge and Goble 2006), and then elsewhere on Navajo land, eventually led to the development of mining and milling regions near Shiprock, New Mexico; Church Rock, New Mexico; Monument Valley, Arizona; and Kayenta, Arizona (Fig. 6.1).

Unaware of the health risks posed to miners, millers, and local residents, the Navajo Tribal Council endorsed these private contracts because of the employment opportunities they created for Tribal members (Johnston et al. 2007).

After many years of epidemiology studies showing much higher rates of lung cancer and other diseases in Navajo miners than in the general population of the region (Gilliland et al. 2000; Brugge and Goble 2002) and after years of thwarted efforts by Navajos and their advocates to seek restitution from the federal government, the Radiation Exposure Compensation Act (RECA) was passed by the U.S. Congress and signed into law in 1990. The RECA states that the government "offers an apology and monetary compensation to individuals who contracted certain cancers and other serious diseases ... following their occupational exposure to radiation while employed in the uranium industry during the build-up to the Cold War." The RECA Amendment of 2000 "broadened the scope of eligibility for benefits ... to include uranium mill workers and uranium ore transporters and adding compensable diseases, thus allowing more individuals to be eligible to qualify."

Given this history, Navajo people often perceive uranium as a monster, as described here by Yazzie-Lewis (2006). "The Navajo word for monster is *nayee*. The literal translation is 'that which gets in the way of a successful life.' Navajo people also believe that one of the best ways to start to overcome or weaken a monster as a barrier to life is to name it. Every evil—each monster—has a name. Uranium has a name in Navajo. It is *leetso*, which means 'yellow brown' or 'yellow dirt.' Aside from its literal translation, the word carries a powerful connotation. Sometimes when we translate a Navajo word into English, we say it 'sounds like' something. We think *leetso* sounds like a reptile, a monster."

6.2.2 Uranium Mill Tailings Radiation Control Act

The United States Congress enacted the regulatory framework for cleanup, containment, and long-term care of uranium mill tailings with passage of the Uranium Mill Tailings Radiation Control Act (UMTRCA) in 1978. The fundamental purpose of this legislation was to mitigate health risks to the public, in perpetuity and in an environmentally sound manner, from residual radioactive materials related to processing of uranium ore. The act authorized the U.S. Environmental Protection Agency (EPA) to issue standards for cleanup and long-term management of uranium mill tailings including standards for remedial action, groundwater quality, and performance of tailings containment systems called disposal cells. The EPA standards were promulgated in 40 Code of Federal Regulations (CFR) 192, "Health and Environmental Protection Standards for Uranium and Thorium Mill Tailings."

UMTRCA designated the U.S. Nuclear Regulatory Commission (NRC) as the agency responsible for enforcing the EPA standards and assigned the responsibility for cleanup, remediation, and long-term care of tailings sites to DOE. It required the NRC to evaluate the design and implementation of remedies by DOE and, after remediation, to concur that remedies satisfy the standards developed by EPA. DOE, as licensee to the NRC for long-term care, is responsible for site inspections, monitoring, reporting, and record keeping. After receipt of an NRC license, sites located on Tribal land revert to Tribal control. Tribes generally allow DOE to fulfill its custody and long-term care responsibilities through a site access agreement.

6.2.3 *U.S. Department of Energy and Long-Term Stewardship*

DOE created the Office of Legacy Management (DOE-LM) in 2003 to function as the Federal licensee to NRC for long-term care of UMTRCA sites (http://www.lm.doe.gov). The accepted remedy was to contain tailings and other residual contamination from the milling operation in an engineered, near-surface disposal cell designed to limit radon escape into the atmosphere, limit percolation of rainwater into tailings – and subsequent leaching of contaminants into groundwater – and continue limiting radon and water flux for 200 to 1,000 years. If former uranium processing activities contaminated groundwater at a particular site, the NRC general license pertains only to the surface remediation; NRC will not fully license the site until groundwater quality satisfies applicable EPA standards.

UMTRCA sites are generally located in the vicinity of uranium ore deposits in arid and semiarid regions of Western states. Four sites are located on Navajo land: the Mexican Hat Disposal Site in southeastern Utah, the Shiprock Disposal Site in northwestern New Mexico, and the Monument Valley Processing Site and Tuba City Disposal Site, both in northeastern Arizona (Fig. 6.1).

6.3 Helping Mother Earth Heal: Navajo Tradition and Science

Navajo cultural tradition teaches the Diné (Navajo for 'child of the Holy People') to fulfill their duty as caretakers of Mother Earth; to reciprocate her nurturing by helping her restore and maintain the health of a desert land. Putting this into practice, Diné College students, together with students and researchers from the University of Arizona, New Mexico State University, and funded by DOE-LM, are exploring natural remedies for groundwater contamination at the Monument Valley Processing Site and the Shiprock Disposal Site. This section provides a brief history of Diné College, a Navajo-owned college on the Navajo Nation, and describes how an educational policy instituted by the College has helped to shape DOE's research approaches at Monument Valley and Shiprock, and to focus on natural and enhanced attenuation (EA) remedies.

6.3.1 *Diné College and Sá'ah Naagháí Bik'eh Hózhóón*

Diné College was founded in 1968 as the first nonprofit public institution of higher learning established by Native Americans for Native Americans (http://www.dinecollege.edu). Formerly Navajo Community College, Diné College is the

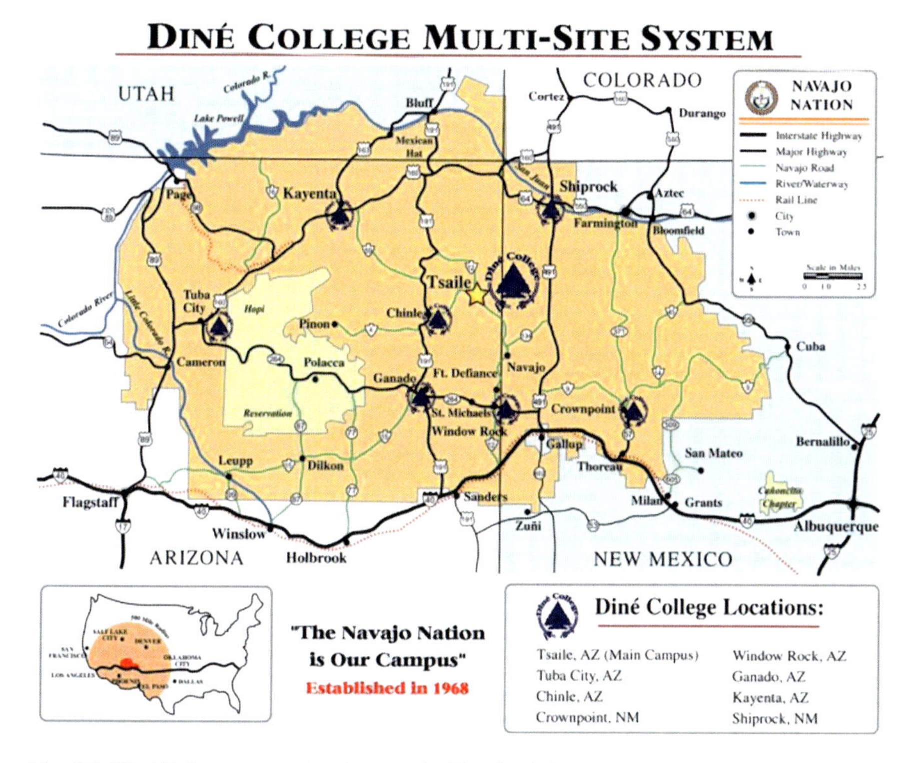

Fig. 6.2 Diné College campus locations on the Navajo Nation

oldest Tribally controlled college in the United States, with eight campuses and centers in Arizona and New Mexico (Fig. 6.2). The main residential campus at Tsaile, Arizona, with buildings designed in the tradition of the Navajo Hogan, adorns the flanks of the Chuska Mountains near the eastern, upper end of Canyon De Chelly National Monument.

Diné College was chartered by the Navajo Nation as the "Higher Education Institution of the Navajo." The College currently enrolls over 2,000 students each semester. The College is accredited by the Higher Learning Commission and is a member of the North Central Association of Colleges and Schools. The landmark decision by the Navajo Nation to create a Tribally owned, postsecondary institution set the precedent for subsequent establishment of several community colleges owned and operated by Native Americans on or near other reservations in the United States. Diné College is one of the founding members of the American Indian Higher Education Consortium, which now represents 33 Tribal colleges and universities.

In creating an institution of higher learning, the Navajo Nation sought to encourage Navajo youth to become contributing members of both the Navajo traditional community and the greater world community. To help fulfill this two-fold mission, Diné College actively fosters a unique combination of traditional and Western learning. The College's educational philosophy, in all academic departments, is grounded in the Navajo traditional living system called Sá'ah

Naagháí Bik'eh Hózhóón, roughly interpreted as "walking or being in the pattern of beauty that surrounds you," or, in other words, placing human life in harmony with the natural world. Sá'ah Naagháí Bik'eh Hózhóón philosophy evolved from Hajiinai Bahané (ancient creation story) which explains the life, mistakes, and struggles of hardship, chaos, and harmony of Diyin Diné (Holy People). They corrected their own mistakes through prayers, songs, and ceremonies to heal themselves. The corrective measures taken by Diyin Diné became the basic teachings of Sá'ah Naagháí Bik'eh Hózhóón philosophy which consists of four inseparable parts:

- Nitsáhákees – consciousness or creative thought.
- Nahatá – planning, actions, and implementation of our thoughts and ideas.
- Iiná – living by achieving quality outcomes of thoughts and actions as a community.
- Siih Hasin – having the assurance of personal stability and satisfaction with life's achievements.

These key principles bring recognition and understanding of disturbances in the natural world caused by human misconduct and of ways to seek restoration. As such, all life forms, the natural world, and all of creation must be treated with utmost reverence and with understanding of natural order and properties. Therefore, within these ancient teachings the principles of Sá'ah Naagháí Bik'eh Hózhóón are relevant and applicable to modern environmental science, law, and policy issues.

Instruction at DEI, located at the Shiprock, New Mexico branch campus of Diné College, endeavors to unite the traditional Sá'ah Naagháí Bik'eh Hózhóón philosophy with Western environmental science methods. DEI environmental science teaching and research programs have been developed around the four Navajo sacred elements of natural systems – fire/light, air, water, and earth. DEI links objectives for curriculum, research, and community outreach to the four sacred elements as follows:

- Fire/Light – Assess different forms and increase the applications of *renewable energy* technology on the Navajo Nation, including solar and wind resources for families not connected to the grid, in a culturally sensitive manner.
- Air – Evaluate and improve both outdoor and indoor *air quality* affecting the health of Navajo people, including indoor radon and proper fuel use, and provide community education and outreach programs.
- Water – Evaluate and improve water quality including research and community outreach efforts with respect to groundwater quality, watershed management, erosion control, drought mitigation, rainwater catchment, and waterborne diseases.
- Earth – Conduct research and provide policy recommendations with respect to environmental health, sustainable and traditional agriculture, solid and hazardous waste management, reclamation of coal mines and abandoned mines, and soil and groundwater remediation, including natural and EA approaches.

Thus, this Navajo conception of and relationship to the environment – the natural elements of life – is strongly linked to the use of core ancient Navajo principles and values that guide environmental education and research. DEI students and interns are introduced and held to the principles and values that are framed and integrated with the process.

6.3.2 *Diné Environmental Institute as a Stakeholder*

DEI is currently expanding its role as the higher education center of the Navajo Nation for instruction, research, and community outreach addressing environmental and energy issues of importance to the Navajo people including sustainable land management, improvement of air and water quality, and development of clean energy, in addition to remediation of lands impacted by uranium mining and milling. DEI faculty and students administer and contribute directly to research and community outreach programs that are linked to classroom, field, and laboratory instruction. DEI faculty also have many years of experience addressing the environmental and energy issues of importance to the people of the Navajo Nation, including firsthand knowledge of the human health and environmental issues associated with the history of uranium mining and milling on Navajo land. Hence, in addition to its role as a contributor to the philosophy and science of remedies, DEI has emerged as a critical bridge among the larger group of stakeholders including federal regulators, research scientists, Navajo Nation agencies, and the Navajo people.

DEI's Uranium Education Program, in particular, plays a key role as an education and outreach umbrella for public health and environmental risks associated with the former uranium mining and milling industries on Navajo land. With comprehensive institutional knowledge of human health and environmental issues associated with uranium mining, milling, and remediation, coupled with lifelong practice of Navajo traditions and culture, DEI faculty have been instrumental in fostering stakeholder interaction and communication including DOE/Navajo agency meetings, Navajo Nation environmental conferences, and public gatherings. For example, at a DOE/Navajo Nation quarterly meeting, a DEI faculty member successfully argued the value of including a traditional healing ceremony for a Navajo resident employed as a DOE subcontractor at Monument Valley.

Although this chapter focuses on DOE's teaming with Diné College to help ask the right questions and then find answers for groundwater contamination at Monument Valley and Shiprock, many other stakeholders are involved. Table 6.1 presents an overview of all stakeholders and their roles in the remediation and long-term stewardship of uranium mill tailings sites on Navajo land administered under UMTRCA.

6.3.3 *Science of Natural and Enhanced Attenuation*

DOE and stakeholders involved in the remediation and long-term stewardship of uranium mill tailings sites on Navajo land are focused on natural and EA remedies in part because the approach compliments both Navajo tradition and Navajo Science.

Table 6.1 Stakeholders affected or involved in evaluating and mitigating human health and environmental risks from uranium mill tailings contamination on Navajo Nation land

Agency or group	Role and interests
General public	People interested in implementation of environmental laws, human health and environmental risks associated with uranium mill tailings, the design and long-term performance of remedies for uranium mill tailings contamination, and the science and implementation of natural attenuation remedies
Navajo people	Navajo people who share the interests of the general public and who are also concerned about the legacy of uranium mining and milling contamination on Navajo homelands, U.S. Government efforts to address this legacy, and the future environmental condition of legacy sites once remedies are in place
People in the vicinity of uranium mill sites and disposal sites	People who live near uranium mill sites on Navajo land, pump groundwater for domestic use and livestock, graze livestock and harvest game, and collect plants and soil for healing ceremonies, medicinal uses, and dyes
U.S. Environmental Protection Agency (EPA)	Federal agency responsible for developing standards, as authorized under the Uranium Mill Tailings Radiation Control Act (UMTRCA) of 1978, for mitigating health risks to the public, and for cleanup and long-term management of uranium mill tailings including standards for remedial action, groundwater quality, and long-term performance of tailings disposal cells; interested in the state of the science of enhanced natural attenuation strategies
U.S. Nuclear Regulatory Commission (NRC)	Federal agency responsible for enforcing U.S. EPA standards as authorized under the UMTRCA of 1978; also interested in the state of the science of enhanced natural attenuation strategies
Office of Legacy Management, U.S. Department of Energy	Federal agency responsible for compliance with U.S. EPA standards, as authorized under the UMTRCA of 1978 and as licensee to the U.S. NRC for long-term surveillance and maintenance of licensed Title I sites on Navajo land. Interested also, as a best management practice, in advancing the science of enhanced natural attenuation remedies
Navajo Abandoned Mine Land Reclamation Office, Navajo Nation	Navajo Nation agency responsible to the Navajo people for ensuring the remediation and long-term stewardship by the U.S. Government of all uranium mill tailings sites and disposal sites on Navajo land, and responsible for communicating to the Navajo people the current understanding of human health and environmental risks associated with these sites

(continued)

Table 6.1 (continued)

Agency or group	Role and interests
Diné Environmental Institute (DEI), Diné College	Community college institution of the Navajo Nation, responsible to the Navajo people for educating youth in the methods of Western environmental science in concert with Navajo traditions, language, and customs. As a research partner, DEI is an advocate of sustainable natural attenuation remedies. As an institution that blends science and culture, DEI has become a critical bridge between federal regulators, scientists, Navajo Nation agencies, and the Navajo people
Environmental Research Laboratory, University of Arizona	As a leader of environmental research and education in arid regions, the overall goal of the environmental research laboratory is to improve the health, welfare, and living standards of communities in desert areas through the application of appropriate and sustainable technologies, and more specifically, to advance the science of natural and enhanced attenuation in arid environments
Farmington Agricultural Science Center, New Mexico State University	Research and extension service campus near Shiprock, New Mexico with overall goals of advancing sustainable agriculture, landscaping, energy, and soil and groundwater remediation, specifically phytoremediation with native phreatophytes, in northwestern New Mexico and the Navajo Nation

The connection between enhanced natural attenuation science and Sá'ah Naagháí Bik'eh Hózhóón tradition became apparent in early meetings between DOE and DEI and helped shape subsequent research and student involvement.

Although natural attenuation has been accepted elsewhere by regulatory agencies for many years, EA has only recently been forwarded by the scientific community as a distinct strategy. Before and into the early 1990s, most large-scale attempts to clean up contaminated soil and groundwater focused on engineering strategies. Engineering approaches included excavating and hauling large volumes of soil to landfills, and drilling wells and pumping large volumes of water to the surface for treatment (NRC 2000). By the mid 1990s, studies and experience had revealed several shortcomings. Excavating and hauling contaminated soil can damage natural ecosystems and potentially expose workers or nearby residents. Also, many conventional pump-and-treat remedies for groundwater contamination had not achieved cleanup goals (NRC 2000). Overall, engineered remedies have not always been successful in restoring contaminated soil and groundwater.

As awareness of the limitations of engineering approaches grew, research began revealing more fully how naturally occurring processes in soils and groundwater can transform or prevent the migration of contaminants (NRC 2000). Reliance on natural attenuation has increased as a consequence. Natural attenuation is now considered a tool for supplementing or even replacing engineered treatment systems. In some cases, including sites with uranium mill tailings contamination, natural attenuation can be used to manage groundwater contamination remaining after engineering approaches have removed or isolated the source of contamination (DOE 1996). The term "monitored natural attenuation" (MNA), as an alternative to active engineering approaches "...refers to the reliance on natural attenuation processes to achieve site-specific remedial objectives within a time frame that is reasonable compared to that offered by other more active methods. The 'natural attenuation processes' that are at work in such a remediation approach include a variety of physical, chemical, or biological processes that, under favorable conditions, act without human intervention to reduce mass, toxicity, mobility, volume, or concentration of contaminants in soil or groundwater." (EPA 1999)

The natural physical, chemical, or biological processes most often referenced that can degrade or dissipate contaminants in soil and groundwater include aerobic and anaerobic biodegradation, dispersion, volatilization, and sorption (e.g., see Ford et al. 2008). Phytoremediation is another attenuation process that is often categorized separate from microbiological, physical, and chemical processes. Phytoremediation and microbial denitrification are the natural attenuation processes DOE, Diné College, University of Arizona, and New Mexico State University students and researchers are jointly investigating at Navajo uranium mill tailings processing and disposal sites.

Although the basic idea is quite old, the concept of using plants for natural attenuation didn't take root until the 1970s, and since then has been studied and applied primarily in wetland and humid upland settings. EPA defines phytoremediation as a set of technologies that use different types of plants for containment, destruction, or extraction of contaminants (EPA 2000). Some general categories of phytoremediation include degradation, the breakdown of contaminants in the root zone or through plant metabolism; extraction, the accumulation of contaminants in shoots and leaves and subsequent harvesting of the crop to remove the contaminant from the site; and immobilization, sequestration of contaminants in soil or hydraulic control of groundwater via evapotranspiration. A review of literature suggests that research using native, desert, phreatophytic shrubs for phytoremediation in the Monument Valley, Arizona and Shiprock, New Mexico deserts is new and innovative.

Microbial denitrification, as discussed here, is a technology that encourages growth and reproduction of indigenous microorganisms to enhance denitrification in both soil and the saturated zone. Denitrification ultimately produces molecular nitrogen (N_2) through a multistep process that results first in the intermediate gaseous nitric oxide (NO), then nitrous oxide (N_2O) (Tiedje 1994). Denitrification completes the cycle by returning molecular N_2 to the atmosphere. The process is performed primarily by heterotrophic bacteria and several species of bacteria that may be involved in the complete reduction of nitrate to nitrogen gas. Denitrification requires electron donors such as organic matter or another carbon source to reduce oxidized forms of nitrogen.

In 2003, DOE introduced the concept of EA and developed the technical basis and documentation to use EA as a transition between active engineered remedies and sustainable remedies that rely solely on natural processes (SRNL 2006). The EA concept is a departure from the classical definition of MNA (EPA 1999).

An *enhancement* is any type of human intervention that might be implemented in a source-plume system that increases the magnitude of or accelerates attenuation by natural processes beyond what occurs without intervention. EA is a strategy that bridges the gap between active, engineered solutions, and passive MNA. A successful enhancement is also a sustainable manipulation – it does not require continuous, long-term intervention. Hence, EA requires a short-term, sustainable manipulation of a natural attenuation process leading to a reduction in mass flux of contaminants. In many cases, sustainable enhancements of natural processes are needed to achieve a favorable balance between the release of contaminants from a source (source loading) and attenuation processes that degrade or retard migration of contaminants in resultant plumes. For DOE and Diné College research purposes at Navajo UMTRCA sites, EA refers to sustainable interventions that enhance phytodegradation of nitrate, evapotranspiration for hydraulic control, and microbial denitrification.

6.4 Monument Valley, Arizona Pilot Studies

Students and faculty from Diné College and University of Arizona researchers are working with DOE on pilot studies of EA remedies for contaminated groundwater at a former uranium-ore processing site near Monument Valley, Arizona (Waugh et al. 2010). Nitrate and ammonium levels are elevated in an alluvial aquifer, and the contaminant plume is spreading away from a source area where a uranium mill tailings pile once stood. Pilot studies were designed to answer two questions: (1) what is the capacity of natural processes to remove nitrogen and slow plume dispersion, and (2) can we efficiently enhance natural attenuation if necessary? In other words, what is Mother Earth already doing to cleanse desert soil and groundwater, and how can we help her? Phytoremediation is also in harmony with the Navajo Nation's revegetation and range management goals for the site.

This section highlights several aspects of the Monument Valley pilot studies that have been supported by Diné College students and local residents, including site characterization, planting, monitoring, and data interpretation. Researchers and students developed the following objectives for the pilot studies:

1. Manage soil water balance and deep percolation, much like an evaporation cover for landfills, to control loading of nitrate and ammonia from the soil source into the alluvial aquifer.
2. Remove nitrate and ammonium from the soil source by enhancing natural phytoremediation and microbial denitrification.
3. Reduce nitrate and ammonia concentrations in the alluvial aquifer to less than 44 mg/L, the EPA standard, and slow the spread of the plume, again by enhancing natural phytoremediation and microbial denitrification.
4. Create a beneficial use of nitrate and ammonium by growing plants that produce seed for use in rangeland improvement plantings and mine land reclamation on the Navajo Nation.

5. Restore the ecology of land disturbed by the milling operation and by site remediation to native plant communities with the goal of improving management of land for both wildlife habitat and sustainable livestock grazing.

6.4.1 Background Information

The DOE Monument Valley Processing Site is located in Cane Valley in northeastern Arizona, 26 km south of Mexican Hat, Utah (Fig. 6.1). Uranium was first discovered in 1942 approximately 1 km west of the site by Luke Yazzie, a local resident. An estimated 696,000 metric tons of uranium and vanadium ore were mined from the deposit between 1943 and 1968. From 1955 until 1964, ore was processed by mechanical milling followed by chemical flocculation. The finer-grained material, higher in uranium content, was shipped to other mills such as the one at Shiprock, New Mexico, for chemical processing. Coarser-grained materials were stored on site.

From 1964 until 1968 an estimated 998,000 metric tons of tailings and low-grade ore were processed using batch and heap leaching. Uranium and vanadium were batch-leached by flowing sulfuric acid solution through sandy tailings placed in lined steel tanks. Heap leaching consisted of percolating a sulfuric acid solution through crushed, low-grade ore spread on polyethylene sheeting. Both operations used ammonia, ammonium nitrate, and quicklime (calcium oxide) to produce a bulk precipitate of concentrated uranium and vanadium. The tailings and processing solutions were discharged to a tailings pile and evaporation pond downslope from the processing area. The mill closed in 1968, and most of the mill buildings were removed shortly thereafter. Surface remediation of the site, from 1992 to 1994, included excavation and hauling of tailings and other site-related contamination to the Mexican Hat Disposal Cell. Analysis of soil within the footprint of the tailings piles at the time of tailings remediation indicated that residual ammonium and nitrate may be contributing to nitrogen contamination in the shallow, alluvial groundwater. Nitrate is the constituent of greatest concern in alluvial groundwater because concentrations exceeded the EPA groundwater standard of 44 mg/L for nitrate.

6.4.2 Plume Source Containment and Removal

Phytoremediation and enhanced microbial denitrification are the natural processes DOE and stakeholders are investigating to contain and remove the nitrate plume source. The evaluation of phytoremediation began with characterization of the ecology of the site in part to determine if native plant species could be used for phytoremediation.

Two native phreatophyte populations occur at the site: *Atriplex canescens* and *Sarcobatus vermiculatus* (díwózhii_beii and díwózhiishzhiin in Navajo, and four-wing saltbush and black greasewood in English). Phreatophytes at the Monument

Fig. 6.3 Diné College students Alverae Laughter and Westin Lee measuring fourwing saltbush plants as part of a greenhouse study of stunted growth and micronutrient supplements at the Tsaile, Arizona campus

Valley site may act, in essence, as passive, solar-powered, pump-and-treat systems for nitrate and ammonium in the source area and alluvial aquifer. *S. vermiculatus* is considered an obligate phreatophyte requiring a permanent groundwater supply and can transpire water from aquifers as deep as 18 m below the land surface (Nichols 1994). *A. canescens*, a facultative phreatophyte, takes advantage of groundwater when present but can tolerate periods of low water availability. The rooting depth of *A. canescens* may exceed 12 m (Foxx et al. 1984).

With assistance from local residents, about 1.7 ha of the source area for the nitrate plume, where tailings had been removed, was planted in 1999, mainly with the native desert shrub *A. canescens*. Another 1.6 ha of the source area was planted in 2006. The purposes of this phytoremediation cover were (1) to control the soil water balance through evapotranspiration, limiting deep percolation and contaminant seepage and (2) to extract and convert ammonium and nitrate into plant tissue. A rectangular irrigated plot was planted with approximately 4,000 *A. canescens* seedlings raised in a greenhouse at the University of Arizona. Navajo Department of Agriculture had confiscated the seed from an illegal harvesting operation on Navajo land. A drip irrigation system was installed to accelerate growth and enhance denitrification. In the future, replacement plants will be grown by students in a Diné College greenhouse near Tuba City, Arizona.

Growth of *A. canescens* transplants in portions of the 1999 planting remained stunted for several years. A greenhouse study conducted by Diné College students (Fig. 6.3) and faculty at the Tsaile, Arizona campus helped identify the cause of

Fig. 6.4 Diné College students Garry Jay and Rita White sampling soil for nitrogen content as part of soil and groundwater phytoremediation pilot studies at Monument Valley, Arizona

stunted growth. Results suggest that stunted growth occurs where iron and manganese coprecipitated as soft concretions during the milling process, possibly reducing the plant-availability of certain micronutrients. The greenhouse study suggested that micronutrient supplements could restore healthy growth rates (DOE 2007).

A. canescens shrub growth and nitrogen uptake have been monitored since 2000 using field sampling and QuickBird satellite images (DOE 2008). Plant canopy cover and growth rates have steady increased but varied across the plantings in response to the age of the planting, irrigation rates, and soil fertility. Annual nitrogen uptake, estimated from plant canopy volume, plant biomass, and nitrogen content on the basis of biomass, has been over 200 kg in mature plantings (DOE 2008).

Soil cores are collected annually with help from DEI students and local residents in the source area soils and analyzed for nitrate, ammonium, and sulfate (Fig. 6.4). Total soil nitrogen had been reduced from 350 to 200 mg/kg. These data show that planting and irrigating the source area has been exceptionally effective in removing nitrate from the soil. However, the data also show that nitrogen removal far exceeds what can be attributed to plant uptake. A salt-balance evaluation and a study of ^{15}N enrichment in the residual nitrate show that the nitrate loss can be attributed primarily to microbiological processes and not leaching (DOE 2009). A soil microcosm study and field observations support a hypothesis that nitrification occurs when source area soils are drier, and denitrification occurs at higher moisture contents (Jordan et al. 2008).

One objective of planting phreatophytes in the source area was to control the soil water balance and limit percolation and leaching of nitrate, much like an evapotranspiration disposal cell cover (Albright et al. 2010). Plantings were purposefully underirrigated to prevent recharge. Soil moisture profiles are monitored with help from local residents using neutron hydroprobes and time-domain reflectometry, and percolation flux is monitored with water fluxmeters (Gee et al. 2009) to evaluate the dynamic soil water balance. Results indicated yearly declines in water content at all depths, a likely response to increasing leaf area and transpiration, and zero percolation in all locations (DOE 2009). These results are evidence that the phytoremediation planting has cut off the plume from its source. Precipitation and irrigation are stored in the fine sand until seasonally removed by evapotranspiration and are not percolating and leaching nitrate.

6.4.3 Enhanced Attenuation of the Plume

DOE, Diné College, and University of Arizona researchers are also evaluating natural and EA remedies for groundwater contamination in the alluvial aquifer at Monument Valley, with a focus on two attenuation processes: phytoremediation to remove nitrate and ammonia and to slow plume dispersion, and microbial denitrification.

A. canescens and *S. vermiculatus*, if rooted into the nitrate plume, could be contributing to natural attenuation in two ways: (1) transpiration of water from the plume, slowing its dispersion from the site and (2) uptake of nitrate from the plume. Stable isotope methods, used to evaluate plant extraction of water and nitrate (DOE 2009), support the hypothesis that *S. vermiculatus* is an obligate phreatophyte rooted into the plume, whereas *A. canescens* is a facultative phreatophyte that uses both plume water and vadose zone water, and that both plant species are extracting nitrate from the plume.

Preliminary studies found that protecting existing stands of *A. canescens* and *S. vermiculatus* from grazing could double biomass production, transpiration rates (water extraction from the aquifer by plants), and nitrogen uptake rates (McKeon et al. 2006). These studies also demonstrated how, on a small scale, greenhouse-grown transplants of native shrubs could be established in denuded areas of the plume, and with managed irrigation, send roots 9 m and deeper into the alluvial aquifer. With managed grazing, phreatophytic shrubs growing over the nitrate plume could extract enough water to slow the spread of the plume during the time it takes for denitrification to reduce nitrate to safe levels. Transpiration rates of individual *A. canescens* and *S. vermiculatus* plants, measured both inside and outside grazing exclosure plots using sap-flow instrumentation, coupled with landscape-scale monitoring using QuickBird and Moderate Resolution Imaging Spectrometer (MODIS) satellite estimates of shrub cover (Glenn et al. 2008), suggest that an increase of 30 mm/year in annual evapotranspiration over the plume through enhanced vegetation abundance could tip the water balance of the aquifer from recharge to discharge.

Early pilot studies suggested that natural denitrification is occurring in the plume (McKeon et al. 2005, 2006). Nitrate levels in the alluvial aquifer decrease with distance from the source area and have also decreased over time. Part of the decrease is likely due to dilution, but part of the nitrate may have been lost to microbial denitrification. An investigation of natural process called ^{15}N enrichment in the plume suggested that up to 60% of a drop in nitrate from the source out to the leading edge of the plume can be attributed to denitrification.

Results of a feasibility study of enhancing natural groundwater denitrification processes (DOE 2008; Carroll et al. 2009) confirmed that the natural attenuation of nitrate is occurring at the site, and that although natural attenuation is occurring, it may take more than 150 years to achieve cleanup standards without enhancements. However, the feasibility study also suggested that the injection of ethanol as a substrate for denitrification could substantially increase groundwater denitrification rates and shorten the cleanup time by more than 100 years. A field-scale ethanol injection study is underway.

With assistance from local residents, University of Arizona developed an unobtrusive approach for evaluating changes in phreatophytic shrub populations based on remote sensing technologies at Monument Valley (Glenn et al. 2008). The research used a combination of field measurements and remote sensing to measure transpiration by *S. vermiculatus* and *A. canescens* growing over the nitrate plume at the site. Heat balance sap flow sensors were used to measure transpiration by the two phreatophytes, and results were scaled to larger landscape units and longer time scales using leaf area index (LAI), fractional vegetation cover, meteorological data, and the enhanced vegetation index from the MODIS sensors on the Terra satellite (Fig. 6.5). *S. vermiculatus* tended to have higher transpiration rates than *A. canescens*. The results support the premise that managing grazing could slow or halt the movement of the contamination plume by allowing the shrub community to extract more water than is recharged in the aquifer.

6.4.4 *Summary: Article in the Gallup Independent, a Local Newspaper*

In 2006, a journalist with the Diné Bureau of the Gallup Independent, after hearing about the research at Monument Valley, journeyed to the remote site to see for herself. The following quotes from her May 1, 2006 article in the Gallup Independent tell what she learned, show how she communicated the benefits of the Diné College collaboration to her Navajo readers (stakeholders), and provide a fitting summary of the Monument Valley project.

> "When people in this area think of plants, they may naturally think of dyes for weaving, or medicinal herbs, but they don't usually think of plants as remedies to remove contamination from places like uranium mills. However, that is exactly what some plants in the area are doing. There is currently a pilot study in Monument Valley involving plants and soil microbes that will help remove nitrate from groundwater."

Fig. 6.5 Cane Valley residents Mary and Ben Stanley sampling grazed black greasewood plants to determine leaf area index as part of a study of remote sensing methods for long-term, landscape-scale monitoring of phreatophyte transpiration

> "The project is a collaborative effort between Navajo Nation officials, scientists with the U.S. Department of Energy, University of Arizona researchers, and students from Diné College. The study is jointly funded by DOE and the University of Arizona. The DOE is also providing funds to Diné College so that interns can participate in the research."
>
> "The group strongly believes that caring for Mother Earth and restoring the health of the desert land are part of the Navajo way of life, and have implemented these same concepts into their project. A conventional cleanup strategy would be to drill wells and pump the groundwater to an aboveground treatment facility, but the group is looking into alternative remedies that would be more sustainable and would require less intervention."
>
> "Instead of a more Westernized cleanup method, the group began looking into more natural methods. They observed that two native plants in particular were withdrawing nitrate from both the soil source and the plume, and are actually converting nitrate into healthy plant tissue. Along with restoring and putting things back into balance, which is a primary concept in Navajo culture, the project is helping prevent deep seepage of nitrate into groundwater."

6.5 Shiprock, New Mexico Phytoremediation Pilot Studies

DEI students from the Shiprock, New Mexico campus, researchers from the New Mexico State University Agricultural Science Center south of Farmington, New Mexico, and researchers from the University of Arizona Environmental

Research Laboratory are collaborating with the DOE on phytoremediation pilot studies near the Shiprock UMTRCA Disposal Site. Groundwater in the vicinity of the site was contaminated with uranium, selenium, nitrate, sulfate, and associated constituents as a result of uranium milling operations in the 1950s and 1960s. The goal of phytoremediation in these areas is hydraulic control, to limit the spread of groundwater contaminants.

6.5.1 Shiprock Uranium Processing and Remedial Action

The Shiprock Disposal Site sits on a terrace above the San Juan River within the Navajo Nation town of Shiprock, about 28 miles west of Farmington, New Mexico (Fig. 6.1). The 93 ha (230 acres) of land occupied by the mill, ore storage area, tailings piles, and raffinate ponds (ponds that contain spent liquids from the milling process), were leased from the Navajo Nation starting in 1954 until the lease expired in 1973, when control of the land reverted to the Navajo Nation.

The Shiprock mill processed uranium–vanadium ore hauled primarily from mines located on Navajo land in the Carrizo Mountains, Lukachukai Mountains, and Sanostee Wash areas in northeastern Arizona and adjacent San Juan County, New Mexico. The Shiprock mill also processed uranium–vanadium products from the Monument Valley mill, uranium ore from the Lisbon Valley area in Utah and the Grants area of New Mexico, and, after the Durango, Colorado, mill was closed in 1963, uranium ore from mining districts in southwestern Colorado.

At the Shiprock mill, ore was crushed and then leached in a bath of sulfuric acid and oxidant to solubilize uranium and vanadium. Precipitation of uranium from the solution was accomplished by increasing the acidity and boiling to expel carbonate, followed by neutralization with magnesia.

When milling operations ceased in 1968, contaminated materials including the mill and other buildings, the raffinate pond area, and about 1.5 metric tons of mill tailings contained in two piles remained at the mill site above the San Juan River floodplain. After the facility reverted to Navajo Nation control in 1973, Navajo Engineering and Construction Authority (NECA) used the tailings as a training ground for heavy equipment operators. Between 1974 and 1978, NECA worked to consolidate and stabilize the tailings piles with guidance from the U.S. EPA.

In 1983, after passage of UMTRCA, the DOE entered an agreement with the Navajo Nation for cleanup, and afterwards, long-term care of the Shiprock mill site. Contaminated materials were consolidated in a disposal cell and covered. In 1996, NRC issued a general license to DOE for custody and long-term care of the disposal cell; however, contaminated groundwater remained in shallow alluvial material and in weathered and fractured shale bedrock beneath the former mill site. In 2003, DOE began pumping groundwater from the terrace and floodplain areas into an evaporation pond, but by 2004, pumping had produced only about half the expected amount. In 2004, DOE developed recommendations for improving the groundwater treatment system including an evaluation of phytoremediation, in this case the use of deep-rooted native plants to enhance evapotranspiration of terrace water and thus limit spread of the plumes.

6.5.2 Phytoremediation Pilot Study Objectives and Progress

In 2006, DOE, Diné College students and their collaborators from University of Arizona and New Mexico State University began to evaluate the feasibility of phytoremediation at the site. The concept was to use deep-rooted native plants to enhance evapotranspiration in an area south of the disposal cell where nitrate levels are elevated in alluvial sediments, and on a terrace between the disposal cell and an escarpment above the San Juan River floodplain to the north of the disposal cell where a uranium plume enters the floodplain. The goal of phytoremediation in these areas is hydraulic control, to limit the spread of contaminants in groundwater.

At Shiprock, Diné College students and faculty are helping to evaluate transplanting methods for native phreatophytes, plant water extraction rates, and contaminant uptake risks in phytoremediation test plots established overlying the nitrate plume and above the floodplain escarpment. Students participate in field activities and data analysis in the classroom.

The objectives of the pilot studies follow:

1. Establish native phreatophytes by transplanting seedlings started in a greenhouse and then irrigating transplants until roots have accessed plume groundwater.
2. Once plant roots have accessed groundwater, evaluate the human health and ecological risks associated with uptake of groundwater constituents and accumulation in aboveground plant tissue.
3. Evaluate the potential beneficial effects of phytoremediation on plume water volume, plume migration, and flow in existing contaminated seeps at the base of the escarpment and in floodplain groundwater.

Hydraulic control, in the context of phytoremediation, can be defined as the use of plants to transpire groundwater in order to contain or control the migration of contaminants (EPA 2000). An increase in water extraction rates may occur naturally over time as populations of phreatophytes establish above the nitrate plume and on the terrace above the San Juan River floodplain. However, if feasible, manipulation or enhancement of the plant ecology with the goal of accelerating water extraction by plants may be an economical addition to the current groundwater remedy.

Passive phytoremediation (no human intervention) and hydraulic control are already ongoing at Shiprock above the nitrate plume and on the river terrace. Volunteer plants of black greasewood (*Sarcobatus vermiculatus*), four-wing saltbush (*Atriplex canescens*), and rubber rabbitbrush (*Ericameria nauseosa*) currently growing above the nitrate plume are likely extracting water, nitrate, and possibly other groundwater constituents. A few scattered black greasewood plants that have "volunteered" on the terrace above the floodplain are likely removing water that

Fig. 6.6 Diné College students Thoer Peterman and Beverly Maxwell sampling soils for physical and chemical properties, and transplanting native phreatophytes to establish phytoremediation test plots at Shiprock, New Mexico

might otherwise surface in contaminated seeps at the base of the escarpment. Higher rates of water extraction by woody plants in both locations may improve hydraulic control.

Planting these areas – enhanced phytoremediation – may be an economical addition to the current groundwater compliance strategy. The success of enhanced phytoremediation would depend on several factors: depth to groundwater, phytotoxicity of groundwater constituents, site preparation methods, plant species selection, planting methods, soil amendments, and natural disturbances. The purpose of this pilot study is to begin evaluating the feasibility of phytoremediation at Shiprock.

Diné College students and collaborators set up two test plots in 2006 in a soil borrow pit overlying the nitrate plume, and two test plots in 2007 on the terrace between the disposal cell and the escarpment above the San Juan River floodplain. Students and faculty planted the four test plots using native *A. canescens* and *S. vermiculatus* transplants grown in greenhouses at the University of Arizona from seed acquired on Navajo land (Fig. 6.6). Students assembled the irrigation system with assistance from New Mexico State University. An employee of the Navajo Nation Abandoned Mine Lands Reclamation Department regularly filled the irrigation holding tanks with San Juan River water. Diné College students irrigated plants on a regular schedule and maintained the plantings, plot fences, and the irrigation system.

Students measured plant canopy dimensions in all plots in October 2007 and again in October 2008. Overall, plants in the terrace plots had grown considerably more than plants in the nitrate plume plots even though the terrace plots were planted a year later. At each location, differences in plant growth between plots were not significant. Values for the different growth parameters were more dispersed (greater variability) in 2008 than in 2007; some plants grew rapidly between 2007 and 2008 while others grew very little. The inconsistent growth patterns may be attributable to insufficient irrigation in 2008. At this stage of the study, soil type and depth to groundwater do not appear to have influenced canopy size. In 2008, based on canopy cover and canopy volume measurements, *A. canescens* appeared to be a better candidate than *S. vermiculatus* for phytoremediation at Shiprock.

Water isotope signatures can provide evidence of volunteer and translanted phreatophytes rooting into the shallow groundwater plumes and, therefore, the feasibility of enhancing phytoremediation and hydraulic control. Oxygen and hydrogen isotope signatures were determined for plants growing naturally overlying the nitrate plume and on the terrace above the San Juan River floodplain, and for water from groundwater monitoring wells in these locations. Salt cedar (*Tamarix ramosissima*) and *A. canescens* plants were sampled from the nitrate plume area, *S. vermiculatus* plants were sampled on the terrace above the escarpment, and *T. ramosissima* were sampled in the San Juan River floodplain.

Enrichment of water in heavy isotopes is expressed as δD and $\delta^{18}O$, in units of per mile (‰) compared to a seawater standard, with positive numbers representing enrichment and negative numbers representing depletion of heavy isotopes relative to the standard (Coplen et al. 2000). Water samples extracted from stem sections of plants generally have isotope signatures similar to the source of water tapped by plant roots. This makes it possible to infer the source of water used by a plant by comparing isotope signatures in the plant to those of potential sources of water in the environment accessible to the roots.

These principles were used to infer water sources of plants and well samples in this study. For this study, we used δD and $\delta^{18}O$ values reported for summer and winter rains at Page, Arizona, to plot the local meteoric water line (Lin et al. 1996). Diné College students and University of Arizona researchers sampled plant water and groundwater near the phytoremediation test plots in 2006 and again in 2007. Water isotope signatures for water in groundwater monitoring wells in the San Juan River floodplain, near the *T. ramosissima* plants sampled on the floodplain, were similar to river water, indicating that these wells are intercepting the floodplain aquifer recharged by the river. Water isotope signatures for the *T. ramosissima* plants sampled near the wells indicated that these plants are rooted into and using aquifer water for transpiration. An interpretation of water isotope signatures for volunteer *S. vermiculatus* plants growing on the escarpment suggests that these obligate phreatophytes are extracting plume water that rises by capillary action up into the escarpment. Water isotope signatures for *T. ramosissima* and *A. canescens* plants growing over the nitrate plume generally indicate that they are likely using locally recharged rainwater to support growth.

The primary purpose of the phytoremediation test plots located on the terrace at Shiprock is hydraulic control, to reduce the source of water for uranium-contaminated seeps at the base of the escarpment and for the uranium plume in the San Juan River floodplain below the escarpment. If plants are also accumulating toxic levels

Fig. 6.7 Diné College students (*left* to *right*: Stephanie Garcia with hood, unidentified student, Michelle John, Rita White, and Vanessa Todacheeny) sampling fourwing saltbush plants for analysis of heavy metal uptake in phytoremediation test plots at Shiprock, New Mexico

of uranium or other heavy metals in above-ground tissues, then the risks of bioaccumulation would be greater than the benefits of hydraulic control. Diné College students are assisting DOE to determine if native phreatophytic shrubs, both those that have naturally "volunteered" on the terrace and those planted to influence hydraulic gradients, are taking up uranium and other metals at levels high enough to be harmful. In 2009, students and DOE scientists designed and carried out a sampling plan to determine uranium and metal levels in stems, leaves, and seeds of planted and volunteer shrubs in and near test plots located on the terrace (Fig. 6.7). In 2010, students will apply standard statistical procedures to analyze data and evaluate human health and the environmental risks of uranium and heavy metal bioaccumulation.

6.5.3 Summary: National Science Foundation Documentary

Collaboration between DOE and the DEI of Diné College on phytoremediation research at the Shiprock Disposal Site received national recognition in the National Science Foundation documentary film, " Weaving STEM Education and Culture: The Faces, Places, and Projects of the Tribal Colleges and Universities Program." The documentary highlights high-quality STEM instructional and outreach programs within the National Science Foundation's Tribal Colleges and Universities Program (TCUP) (http://www.nsftcup.org/).

The following quotes, from interviews with Diné College instructors and students in the documentary, are a fitting summary of the Shiprock collaboration. The quotes tell a story of collaboration among scientists from different cultures, respect for tradition, the value of diversity in learning, and hope for a new generation.

> "When I went to school, we only learned about Western science. When I started working here at this school, we came to learn about science from the Navajo perspective. Our teaching is 'honor thy mother; honor thy father.' That's the Indian people's teaching. That means your mother, the one that gave birth to you, and all the way back to First Woman, and then back to Mother Earth—Mother Earth is your mother. So if you mistreat your mother, there will be consequences."

Jack C. Jackson, a 75-year old instructor at the Tsaile campus of Diné College, explaining the importance of linking Western environmental science education with Navajo cultural traditions. As a young man, Professor Jackson was part of a Navajo delegation that convinced the U.S Congress to appropriate funds to build Diné College.

> "A lot of students we have here at Diné College, the majority of them, have not really heard the stories based on the old traditional teachings, and how to relate the teachings to science. That's what I do. I like to teach what I call Navajo Science. I learned a lot of it from my elders. In the past, many of our students were going out to other universities to study off of the reservation. They don't realize that right here on the reservation, there are many things that need to be studied. Look at all of this area, this natural laboratory."

Arnold Clifford, Navajo botanist and part-time Diné College instructor at the Shiprock campus, explaining the importance of incorporating traditional teachings in science curriculum and the need for continued research on the Navajo reservation.

> "What we're doing with the students at Diné College is a phytoremediation research project: phyto meaning 'plants', and remediation meaning 'a remedy to fix a problem.' So plants are being used to control the hydrology so the groundwater plume doesn't flow down into the San Juan River floodplain."
>
> "The Department of Energy is very interested in having students involved, to be aware of the risks and what the Department of Energy is doing to alleviate the risks in the long term. This is their land; they are the prime stakeholders. I think it's important for people who live here to understand and help come up with the remedies themselves. Fortunately, Diné College provides that opportunity."

Dr. Jody Waugh, an Environmental Scientist with a DOE contractor, Project Lead for phytoremediation research at Shirock, and part-time instructor at Diné College, explaining the need for DOE to collaborate with Diné College.

> "I think it's very important, especially from my own Tribe, getting involved and becoming a scientist, and being able to make an impact in our future."

Vanessa Todacheeny, Diné College student working on the phytoremediation research project at Shiprock.

> "I know that as a student, you can make changes. In the past, we couldn't speak out, but now we can become educated and learn more about things through research. Our ancestors, our elders, they tell you to respect Mother Nature, that you're not supposed to put it out of balance. Well, there is a way to put things back in balance, and you, as a student, have to find out that way. I want to someday be that student."

Rita White, Diné College student working on the phytoremediation research project at Shiprock.

6.6 Summary and Conclusions

DEI has become a key stakeholder and partner with the U.S. DOE in efforts to develop and implement sustainable and culturally acceptable remedies for soil and groundwater contamination at uranium mill tailings processing and disposal sites on Navajo Nation land. DEI is a center for environmental education, research, and community outreach located on the Shiprock, New Mexico campus of Diné College, the Navajo Nation institution of higher education. As a stakeholder, DEI plays a key role in shaping the philosophy of remedial actions, advancing the science of sustainable remedies, bridging communication and interaction among other stakeholders, listening to and responding to the concerns of the Navajo people, and training a new generation of scientists to address the uranium mining legacy and other environmental and energy issues on the Navajo homeland.

Through an educational philosophy grounded in the Navajo traditional living system called Sá'ah Naagháí Bik'eh Hózhóón, which places human life in harmony with the natural world, DEI has helped guide researchers to look beyond traditional engineering approaches and seek more sustainable remedies for contaminated soil and groundwater at former uranium mill sites near Monument Valley, Arizona, and Shiprock, New Mexico. Following this philosophy, researchers are asking first, what is Mother Earth already doing to heal a land injured by uranium mill tailings, and second, what can we do to help her? This has led researchers to investigate applications of first, natural, and then, EA remedies involving native plants – phytoremediation, and indigenous microorganisms – bioremediation. Although such applications are fairly common in wetland and humid environments, EA in the desert is new and innovative.

DEI faculty and students are working side by side with university and DOE scientists on pilot studies aimed at developing sustainable remedies for contaminated soil and groundwater at Monument Valley and Shiprock. Diné College faculty, student interns, and local residents have contributed to several aspects of the pilot studies including site characterization, sampling designs, installation and maintenance of plantings and irrigation systems, monitoring, and data interpretation. Research results look promising.

At Monument valley, DOE removed radioactive tailings from the site in 1994. Nitrate and ammonium, waste products of the milling process, remain in an alluvial groundwater plume spreading from the soil source where tailings were removed. Planting and irrigating two native phreatophytic shrubs, fourwing saltbush and black greasewood, has markedly reduced both nitrate and ammonium in the source area over an 8-year period. Most of the reduction is attributable to irrigation-enhanced microbial denitrification rather than plant uptake. However, soil moisture and percolation flux monitoring show that the plantings control the soil water balance in the source area, preventing additional leaching of nitrogen compounds. Enhanced denitrification and phytoremediation also look promising for plume remediation. Microcosm experiments, nitrogen isotopic fractionation analysis, and solute transport modeling results suggest that most of the plume nitrate has been lost through natural denitrification since the mill was closed in 1968. Injection of ethanol may accelerate microbial denitrification in plume hot spots. Finally, landscape-scale remote sensing methods developed for the project suggest that transpiration from restored native phreatophyte populations rooted in the aquifer could limit further expansion of the plume.

At Shiprock, DOE contained mill tailings in an engineered disposal cell in 1986. Groundwater is contaminated by uranium, nitrate, and other constituents as a result of milling operations. Passive phytoremediation and hydraulic control are ongoing at Shiprock. Native phreatophytes are extracting water and possibly other groundwater constituents. Phytoremediation test plots were set up in 2006 with assistance from DEI students and faculty to evaluate the feasibility of enhancing hydraulic control. Researchers are evaluating several factors that will influence the success of enhanced phytoremediation including site preparation methods, establishment and growth of different plant species, root access of plume groundwater, and uptake and toxicity of groundwater constituents.

DEI's insight and experience implementing an educational policy that fosters diversity of thought, the joining of tradition and science, and the importance of community has been instrumental in building stakeholder relations. With firsthand knowledge of human health and environmental issues associated with the Navajo uranium legacy, lifelong practice of Navajo way of life, and experience directing community outreach programs, DEI faculty have been influential in helping mediate communication and interaction among stakeholders including federal regulators and administrators, research scientists, Navajo Nation agencies, and the Navajo people.

Finally, DEI and Diné College are training a new generation of scientists and community leaders who will write the next chapter in the Navajo story about uranium mining, milling, and environmental stewardship. They will know the history, they will continue the traditions, they will advance the science, they will facilitate the needed partnerships, they will inform the people, they will protect human health, and they will fulfill their duty as caretakers of Mother Earth, helping her restore and sustain the health of the land.

Acknowledgments Funding for the preparation of this chapter was provided by the U.S. Department of Energy (USDOE) Office of Legacy Management (Contract No. DE-AM01-071M00060). Its contents are solely the responsibility of the authors and do not necessarily represent the official views of the USDOE.

References

Albright WH, Benson CH, Waugh WJ (2010) Water Balance covers for waste containment. ASCE Press, Reston

Brugge D, Goble R (2002) The history of uranium mining and Navajo people. Amer J Public Health 92:1410–1419

Brugge D, Goble R (2006) A documentary history of uranium mining and the Navajo people. In: Brugge D, Benally T, Yazzie-Lewis (ed) The Navajo people and uranium mining. University of New Mexico Press, New Mexico

Carroll KC, Jordan FL, Glenn FP, Waugh WJ, Brusseau ML (2009) Comparison of nitrate attenuation characterization methods at the uranium mill tailing site in Monument Valley, Arizona. J Hydrol 378:72–81

Chenoweth WL (1997) Summary of uranium-vanadium mining in the Carrizo Mountains, Arizona and New Mexico, 1920–1968. In: Anderson OJ, Kues BS, Lucas SG (eds) Mesozoic geology and paleontology of the Four Corners region: New Mexico Geological Society, Guidebook 48, pp 267–268

Coplen T, Herczeg A, Barnes C (2000) Isotope engineering using stable isotopes of the water molecule to solve practical problems. In: Cook PG, Herczeg AL (eds) Environmental tracers in subsurface hydrology. Kluwer Academic Publishers, Massachusetts

DOE (U.S. Department of Energy) (2008) Natural and enhanced attenuation of soil and ground water at Monument Valley, Arizona, and Shiprock, New Mexico: 2007 status report. LMS/MON/S04243, U.S. Department of Energy Office of Legacy Management, Grand Junction, Colorado

DOE (U.S. Department of Energy) (2009) Natural and enhanced attenuation of soil and ground water at Monument Valley, Arizona, DOE legacy waste site: 2008 pilot study status report. LMS/MON/S05418, U.S. Department of Energy Office of Legacy Management, Grand Junction, Colorado

DOE (U.S. Department of Energy) (1996) Final programmatic environmental impact statement for the Uranium Mill Tailings Remedial Action Ground Water Project. DOE/EIS-0198, Grand Junction Projects Office, Grand Junction, Colorado

DOE (U.S. Department of Energy) (2007) Natural and enhanced attenuation of soil and ground water at Monument Valley, Arizona, and Shiprock, New Mexico: 2006 status report. DOE-LM/1428, U.S. Department of Energy Office of Legacy Management, Grand Junction, Colorado

EPA (Environmental Protection Agency) (1999) Use of monitored natural attenuation at Superfund, RCRA Corrective Action, and Underground Storage Tank Sites. Directive 9200.4–17P, EPA Office of Solid Waste and Emergency Response, Washington, DC

EPA (U.S. Environmental Protection Agency) (2000) Introduction to phytoremediation. Publication no. 600/R-99/107. U.S. Environmental Protection Agency, Cincinnati, Ohio

Ford RG, Wilkin RT, Acree S (2008) Site characterization to support use of monitored natural attenuation for remediation of inorganic contaminants in ground water. EPA/600/R-08–114, National Risk Management Research Laboratory, Environmental Protection Agency, Cincinnati, Ohio

Foxx TS, Tierney GT, Williams JM (1984) Rooting depths of pants on low-level waste disposal sites. LA-10253-MS, Los Alamos National Laboratory, Los Alamos, New Mexico

Gee GW, Newman BD, Green SR et al (2009) Passive wick fluxmeters: Design considerations and field applications. Water Resour Res: 45, W04420, doi:10.1029/2008WR007088

Gilliland FD, Hunt WC, Pardilla M et al (2000) Uranium mining and lung cancer among Navajo men in New Mexico and Arizona, 1969 to 1993. J Occup Environ Med 42:278–283

Glenn EP, Morino K, Didan K et al (2008) Scaling sap flux measurements of grazed and ungrazed shrub communities with fine and coarse-resolution remote sensing. Ecohydrol 1:316–329

Johnston BR, Dawson SE, Madsen GE (2007) Uranium mining and milling: Navajo experiences in the American Southwest. In: Johnston BR (ed) Half-lives and half-truths: Confronting the radioactive legacies of the Cold War. School for Advanced Research Press, New Mexico

Jordan F, Waugh WJ, Glenn EP et al (2008) Natural bioremediation of a nitrate-contaminated soil-and-aquifer system in a desert environment. J Arid Environ 72:748–763

Lin G, Phillips S, Ehleringer J (1996) Monsoonal precipitation responses of shrubs in a cold desert community on the Colorado Plateau. Oecologia 106:8–17

McKeon, C, Glenn E, Waugh WJ et al (2006) Growth and water and nitrate uptake patterns of grazed and ungrazed desert shrubs growing over a nitrate contamination plume. J Arid Environ 64:1–21

McKeon C, Jordan FL, Glenn EP, Waugh WJ et al (2005) Rapid nitrate and ammonium loss from a contaminated desert soil. J Arid Environ 61:119–136

Nichols W (1994) Groundwater discharge by phreatophyte shrubs in the Great Basin as related to depth to groundwater. Water Resour Res 30:3265–3274

NRC (National Research Council) (2000) Natural attenuation for ground water remediation. National Academy Press, Washington

SRNL (Savannah River National Laboratory) (2006) Enhanced attenuation: A reference guide on approaches to increase natural treatment capacity. WSRC-TR-2005-00198, Westinghouse Savannah River Company, Aiken, South Carolina, 160pp

Tiedje J (1994) Denitrifiers. In: Weaver RM, Angle S, Bottomley P et al (ed) Methods of soil analysis. Part 2: Microbiological and biochemical properties. Soil Science Society of America Inc, Wisconsin

Waugh WJ, Miller DE, Glenn EP et al (2010) Natural and enhanced attenuation of soil and groundwater at the Monument Valley, Arizona, DOE Legacy Waste Site. Proceedings of Waste Management 2010 Symposium, Phoenix, Arizona

Yazzie-Lewis E (2006) Leetso, the powerful yellow monster: A Navajo cultural interpretation of uranium mining. In: Brugge D, Benally T, Yazzie-Lewis E. The Navajo people and uranium mining. University of New Mexico Press, New Mexico

Chapter 7
Nez Perce Involvement with Solving Environmental Problems: History, Perspectives, Treaty Rights, and Obligations

Gabriel Bohnee, Jonathan Paul Matthews, Josiah Pinkham, Anthony Smith, and John Stanfill

Contents

J.P. Matthews (✉)
Environmental Restoration and Waste Management (ERWM), Nez Perce Tribe, P.O. Box 365, Lapwai, ID 83540, USA
e-mail: jonathanm@nezperce.org

J. Burger (ed.), *Stakeholders and Scientists: Achieving Implementable Solutions to Energy and Environmental Issues*, DOI 10.1007/978-1-4419-8813-3_7,

Abstract The Nez Perce, like other federally recognized Tribes, is a sovereign Nation, and the United States is required to consult on a government-to-government basis with the Tribe on action that stand to effect the Tribal resources, such as the cleanup of nuclear wastes at the Hanford Facility near Richland, WA. This chapter examined the Nez Perce perspective on treaty rights and the U.S. government's obligations, using the case study of the handling of Greater-than-Class C (GTTC) Low-Level Radioactive Waste, with an emphasis on the Department of Energy's Hanford Site. It also provides an overview of how the Nez Perce view the environmental features and values that effect their lifeways, including seasonal rounds, gathering times, Tribal values, and Tribal perspectives. While the chapter focuses on Hanford, the history, perspectives, treaty rights, and obligations are common to other Tribes and other environmental situations.

7.1 Introduction

The Nez Perce Tribe has powers and authorities derived from its inherent sovereignty, from its status as the owner of land, and from legislative delegations from the Federal government. The Tribe is also a cultural entity charged with the responsibility of protecting and transmitting that culture which is uniquely Nez Perce. The Tribe is a beneficiary within the context of federal trust relationship, and a trustee responsible for the protection and betterment of its members and the protection of their rights and privileges.

The department of energy (DOE) – Nez Perce Tribe relationship at Hanford is defined by the trust relationship between the Federal government and the Tribe by treaty, federal statute, executive orders, administrative rules, case law, DOE's American Indian Policy, and by the mutual and generally convergent interests of the efficient and expeditious cleanup of the DOE weapons complex, which is expressed in a Cooperative Agreement between the Nez Perce Tribe and DOE Hanford. The Cooperative Agreement is grounded in the site-specific cleanup of Hanford and extends to all trust-related activities by DOE.

The Tribe sees itself not only as a trustee of resources at Hanford, but also as technical and cultural advisors to DOE decision-making. The continuation of the Cooperative Agreement contemplates an approach that will integrate these and other roles into a comprehensive Nez Perce-DOE program. The Tribe is asked to review and comment on documents and activities by DOE as a means to uphold their trust responsibilities and comply with other federal statutes, laws, regulations, executive orders, and memoranda governing the United States' relationship with Native Americans and the Nez Perce people. Several Tribal departments lend their respective technical expertise to DOE Hanford issues and present recommendations to the Nez Perce Tribal Executive Committee (NPTEC) for consideration and guidance. The NPTEC also may requests formal consultation with the federal agency to discuss a proposal or issue further.

There are limitations of the National Environmental Policy Act (NEPA). Federal regulations like the NEPA define a set of rules for generating alternatives, evaluating the natural and human environment, and engaging the public. The NEPA process does not consider Native Americans as part of the natural environment nor does it adequately provide a framework where Tribal values or traditional lifeways are given equal weight in comparison to those of a modern society. It has been difficult to adequately communicate Native American culture and spiritualism, which culminates into a holistic environmental ethic that encapsulates long-term stewardship before the term became popular. The NEPA process legally allows for affected Tribes to participate during scoping, alternatives development, and impacts analysis, but where has our perspective been invited to the process or better yet, where has our participation influenced federal decision-making? Resource values from a Tribal perspective are just as valid as those articulated by a government entity or the general public.

Key questions are: How can DOE's trust responsibilities to Tribes be met, if not by allowing equal input into their federal decision-making? How can the Nez Perce Tribe fully carry out their culture and preserve elements of the lower Columbia at Hanford that supports it?

7.2 Background on Nez Perce Lifeways

How can the Nez Perce Tribe provide DOE staff a fuller meaning of our connection to the land and our concern for the lower reach of the Columbia River at Hanford? After all, the Nez Perce homeland is nearly 200 miles from the DOE Hanford site. For DOE decision-makers to fully understand our perspective, they must understand our past at Hanford, its historical value to us as a people, and accept its present and future value towards preserving our culture. In the past, the Nez Perce traditional lifestyle was often mislabeled as nomadic. We were a people that relied on the salmon, but more importantly, we followed a seasonal round.

7.2.1 Seasonal Rounds

The seasonal round is best described as a *return to a specific area* for the purpose of gathering resources: food, medicinal, or otherwise. The seasonal round advanced in

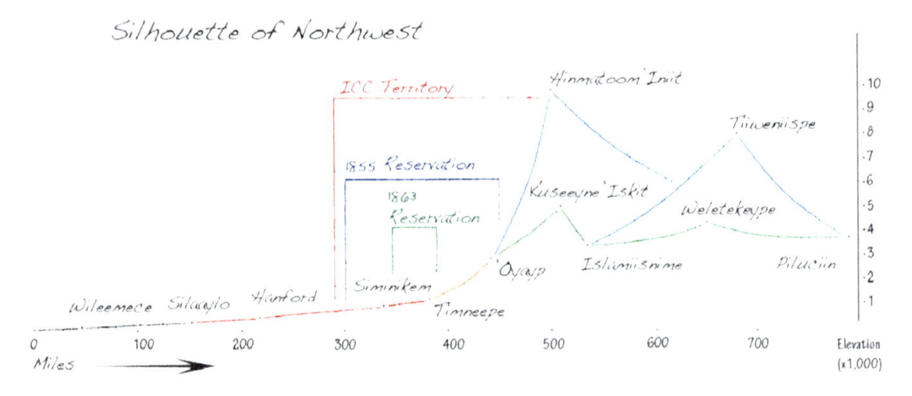

Fig. 7.1 Elevational profile illustrating the extent of travels by the Nez Perce

area and elevation simultaneously. It is not the act of following resources wherever they occur but rather a return to an area to gather resources based on prior knowledge or experience. It is also marked by the availability as warming seasonal temperatures foster development of the resource. Examples are the return to root digging areas as spring or summer temperatures have warmed plants to the point of opening the opportunity to harvest, or a return to a hunting area in the fall before temperatures drop to low. Figure 7.2 shows how the Hanford area fits into the area used by the Nez Perce over time. The time for gathering resources is marked by lunar changes. Since there were more foods than there were moons during the year, some resource gathering times were simultaneous. The diagram below shows how the seasons for gathering various foods correspond to the commonly used 12-month calendar and four seasons. The Nez Perce changed elevations depending on the warming weather and this is shown through another diagram showing the names of the gathering seasons and the elevations.

The seasonal round also covered an elevation from sea level up to ten thousand feet. Figures 7.1 and 7.2 show the elevation difference in the usual and accustomed areas used by the Nez Perce. The beginning of the seasonal round is marked with a Ke'uyit or first foods ceremony in the spring. Ke'uyit translates to "first bite" and is an annual ritual of prayer immersed in song for the first foods of the year. Traditional foods are laid out on the floor in the order in which they are gathered throughout the year beginning with Salmon. This annual ritual is an expression of gratitude to the foods for their return and for those gathered during the seasonal round. Other Tribes have more than one feast such as a root feast and a huckleberry feast but the Nez Perce only have one and it is held toward the latter part of the spring (Fig. 7.3).

7.2.2 *Gathering Times*

Gathering times are extremely important to the Nez Perce. Examples of resource gathering times are discussed herein.

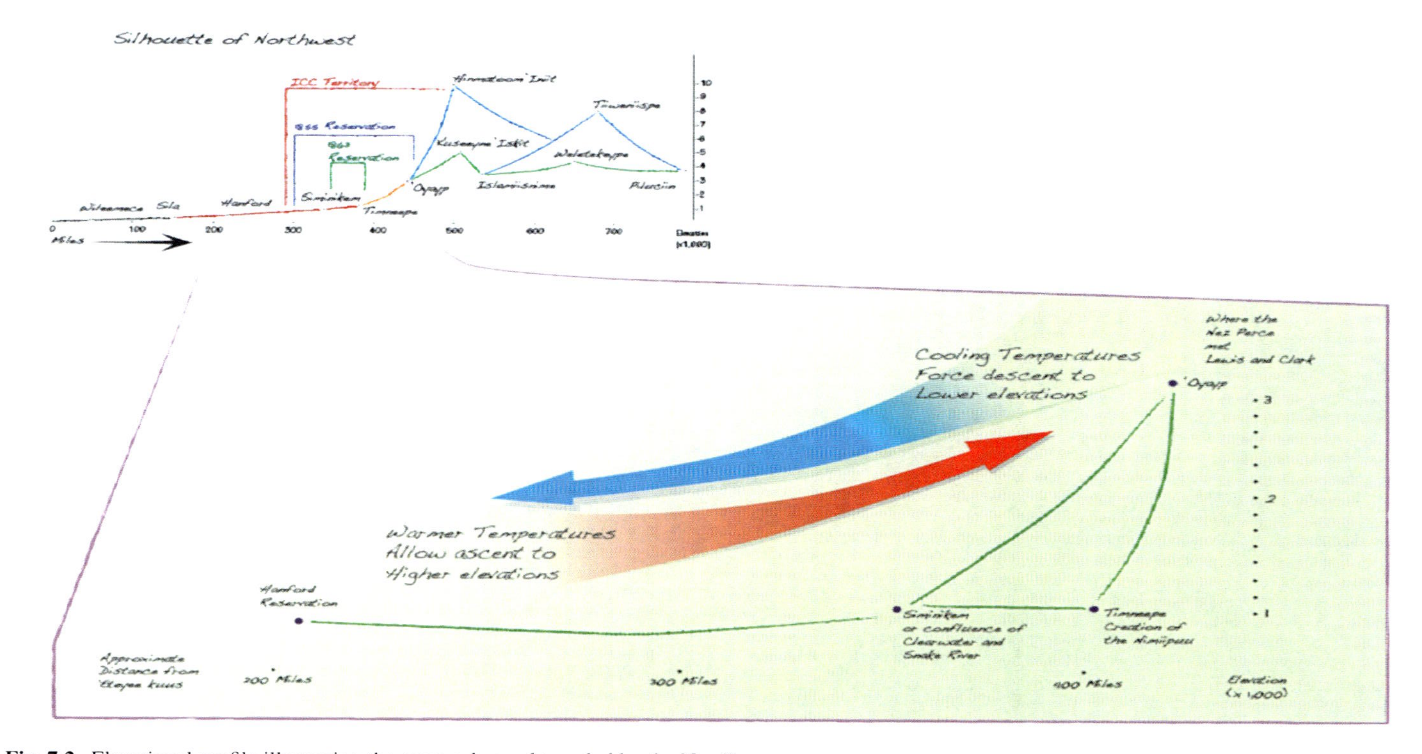

Fig. 7.2 Elevational profile illustrating the seasonal round traveled by the Nez Perce

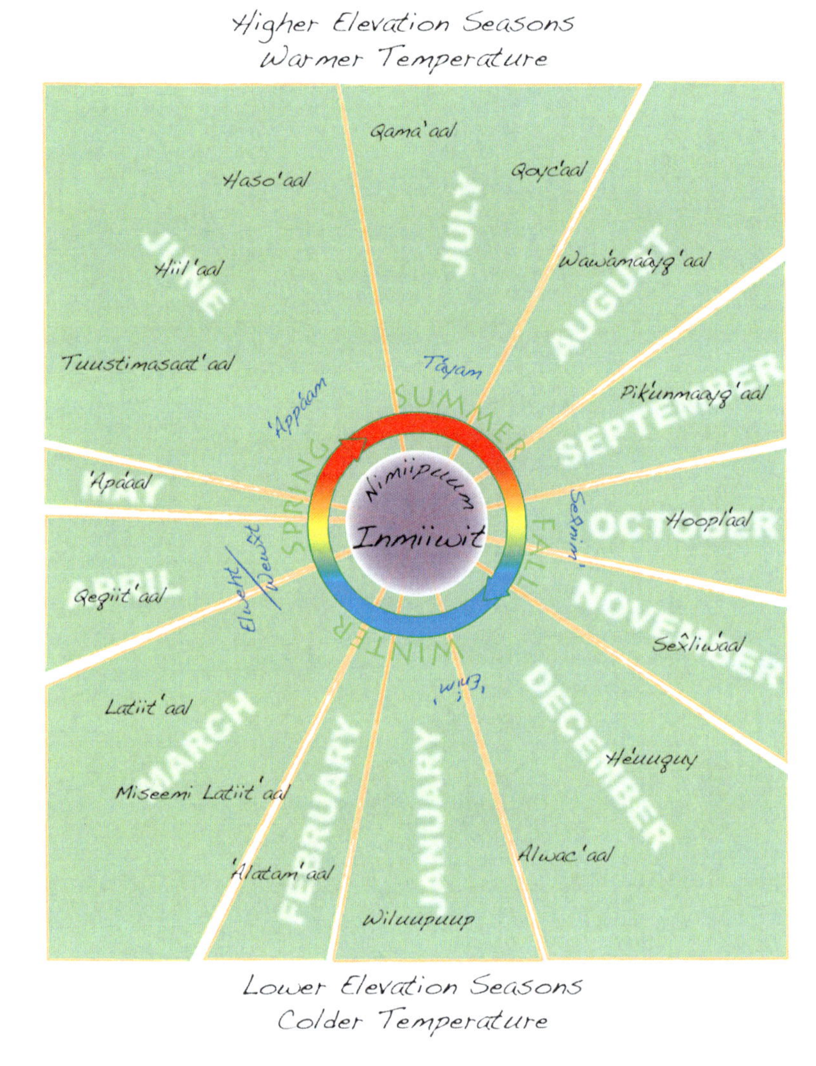

Fig. 7.3 Seasonal periods recognized by the Nez Perce and their correspondence to the 12-month calendar

Wiluupup: Time when cold air travels. Often corresponds to the month of January.

'Alatam'aal: Time between winter and spring or the time for fires (often corresponds to the month of February) 'Alafire.

Miseemi latiit'al: Time of false blossoms roughly corresponding to early March. Miseemito lie or speak falsely, Latiito bloom or blossom.

Latiit'al or Latiit'aal: Time when flowers bloom. Roughly corresponds to the month of March. Latiito bloom or blossom.

Qeqiit'aal or qaqiit'aal: Time of gathering qeqiit roots. Roughly corresponds to April.

'Apa'aal: Time for digging roots and making them into small cakes called 'Apa. Roughly corresponds to the month of May or June.

Tustimasaatal: Ascend to higher mountain areas. Roughly corresponds to the month of June. Tustihigher/above.

'Il'aal: The time of the first run of Salmon. Roughly corresponds to the month of June.

Haso'al': The time to gather eels or Pacific Lamprey. Roughly corresponds to the month of June. Heesueel.

Qama'aal: Time for digging and roasting qem'es bulbs. Often corresponds to the month of July. Qem'escamas bulbs.

Q'oyxc'aal: Time of gathering Blueback Salmon. Often around the month of July. Q'oyxcBlueback Salmon.

Waw'ama'aq'aal: Season when salmon swim to the headwaters of streams (often corresponds to August) Waaw'amheadwaters.

Pik'unma'ayq'al or pik'onma'ayq'aal: Time when Chinook Salmon return to the main river and steelhead begin their ascent. Roughly corresponds to September. Piik'unriver.

Hoopl'al: Time when Tamarack needles begin to fall. Huupto fall (as Pine needles do). Roughly corresponds to October.

Sexliw'aal: Autumn or the time roughly corresponding to November.

He'uquy: Time of elk fetus gestation roughly corresponding with winter and the month of December.

'Alwac'aal: Time of Bison Yearling roughly corresponding to December. 'Alawabison yearling.

7.3 Nez Perce Tribal Values and Environmental/Tribal Health

7.3.1 Oral Histories

Oral histories impart basic beliefs, taught moral values, and explained the creation of the world, the origin of rituals and customs, the location of food, and the meaning of natural phenomena. The oral tradition provides accounts and descriptions of the region's flora, fauna, and geology. Fish and other animals are characters in many of these stories. Coyote is the main character in many of the stories because it exhibits all the good and bad traits of human beings. Although some of the characters and themes may differ slightly, many of these same stories are held in common by Columbia Basin Tribes.

7.3.2 Tribal Values

Tribal values lie imbedded within the rich cultural context of oral tradition and are conveyed to the next generation by the depth of the Nez Perce language. How to properly perceive life and land are among the core tenets of which the stories speak. The numerous landmarks that season the landscape are reminders to the events, stories, and cultural practices of our people. The values are what must and they can only be properly conveyed by the oral traditions and language. Overall the values are intent on protecting, preserving, and perpetuating resources for the sake of survival. The Nez Perce taught these values for generations to our children just as we still teach them today. The most appropriate way to understand these cultural values is to view our cultural practices still conducted today on our landscape. They reflect a complex tradition of high regard for the land by utilizing the resources, but not using so much that the resource cannot propagate to preserve their continued existence.

Land was managed by cultural practices so that resources would not be jeopardized by the actions of one generation. The Nez Perce Tribe utilized resource areas with several other Tribes that carried similar resource values. We value the landscape for the rich resources it offers our children for survival. The landscape is full of powerful reminders in the form of rock features associated with oral traditions that relate exploits of the animal people. The Nez Perce elders recall hunting and fishing areas taught to them when they were young. These are the same places they learned about in the same way from their elder kinsmen. The women dig roots and harvest berries in the same places that they learned from their grandmothers. Each place utilized for resources was maintained to sustain children and future generations.

Each plant had a window of harvest in which it could be gathered. The window of harvest was always honored because gathering at another time would either affect its strength or viability. When women were gathering *qem'es* bulbs, they would evaluate the field to ensure that others had not already gathered past the threshold of the resource's stability. If the field looked as though others had already been there and the resource needed to be left so it could continue on, then they would simply go to another place. When a place was found which could be used for harvest, the digging would begin with prayer songs and it was common for many of the women to sing as they continued to dig. When the work was finished for the day it was closed with a prayer song just as it had began. They were cautious about the way in which they gathered the roots as well. Arguing and fighting did not occur while gathering foods, even among the young, because it was strictly forbidden. Root diggers were reminded by the elderly to be prayerful and concentrate on good thoughts as they conducted their work avoiding negative feelings that might be carried by the foods to those that would consume them. Peelings from the roots always were to be returned to the original grounds from which they came or buried in the earth. They are never to be simply thrown in the garbage. There are traditional stories that communicate values that regardless of where the oral tradition originated, it applies during times that native Tribes are on site and practicing usual and accustomed rights. These are teachings tied to the landscape and the land ethic that is our culture.

Fishing and hunting were conducted in the same way. Young boys were raised with the guidance of elder kinsmen. A group of hunters or fishermen would depart for areas that were, on occasion, previously scouted for the presence of fish and/or game. Young hunters and fishermen would observe the actions of those that were responsible for imparting knowledge of how to conduct oneself appropriately as game was stalked or fish were caught. Expectations were similar to those of the young women; concentrate on good thoughts and feelings, prohibited acts included fighting and arguing. Excessive pride and boasting were frowned upon by elder kinfolk since the hunt was to be conducted with the utmost humility. Hunters and fisherman learned to avoid catching the largest fish or killing the largest animal they could find because it preserved the gene pool that replaced that size animal. Upon return, the hunters were not questioned as to the number each hunter killed and it was never announced because it was deemed as a group activity. One exception was when a young hunter killed an animal for the first time or caught his first fish. At this time the family recognized the young hunter or fisherman as a provider with a ceremonial feast. The elder fisherman and hunters sat around the meat which was to be boiled, baked, or prepared in some traditional fashion as stories were told conveying more teachings and proper conduct. As the elder hunters and fishermen consumed the meat the newly recognized hunter or fisherman was not allowed to partake of even a morsel of the meal. Everyone else was to eat before the hunter or fisherman could consume a meal. This reinforced their role as a provider rather than someone that merely killed game or caught fish for recreational purposes. Young hunters were taught proper shot placement, as it was crucial to the hunting experience. Young hunters were taught to shoot an animal so that it would be killed as quickly and limit the animal's suffering as much as possible. Shooting an animal or catching a fish was only part of the overall commitment to the animal's sacrifice. It had to be cleaned and taken care of with the same regard as the roots and berries. The utmost gratitude and respect was offered to the animal's spirit for imparting a tremendous gift of life to the people.

Spiritual or religious aspects of natural resources are the heart of Indian culture. There is a connection to the daily activities of a traditional lifestyle communicated through the oral traditions that tell how to take care of the land. Even landmarks have oral traditions associated with them. These landmarks are tangible cultural reminders.

7.3.3 *Value of Uncontaminated Resources*

For natural resources to be uncontaminated as part of Niimiipuu physical and spiritual well-being, the land and waters and air from which they come should be uncontaminated otherwise the risk to human health increases the potential for illness and other ailments. For Tribal use of natural resources to be fully utilized, the example of manufacturing and using a *wistiitam'o* or sweat lodge is presented. One purpose of a sweat lodge is for purification. It is for cleansing and a time for meditation, spiritual reflection, healing, sharing oral history, and teaching. The *wistiitam'o* is often a place where the Nez Perce return to have spiritual well-being restored after family losses. It is a place of contemplation and an opportunity to relieve stress and anxiety built up from the day's activities. It is a place for centering your soul through prayer and meditation. It is also a place where many socialize with family and friends and learn what is happening in the community.

For these reasons, it is imperative that the materials used in making a sweat lodge come from the natural environment. The structure is made of willows gathered from the immediate vicinity of where the sweat lodge will stand. The covering is to be of animal hides, or other natural materials. The water for the bathing after sweating is to be from a natural spring or stream. Herbs are collected in their proper season with prayers and gratitude offered for their service.

Sitting in a sweat bath is a rigorous activity. While outwardly relaxed, your inner organs are as active as though you were exercising. The skin is the largest organ of the body and through the pores it plays a major role in the detoxifying process along with the lungs, kidneys, bowels, liver, and the lymphatic and immune systems. Capillaries dilate permitting increased flow of blood to the skin in an attempt to draw heat from the surface and disperse it inside the body. The heart is accelerated to keep up with the additional demands for circulation. Impurities in the liver, stomach, muscles, brain, and most other organs are flushed from the body. It is in this way that purification occurs.

7.4 NEPA and DOE Fiduciary Responsibility

The following sections of the CEQ (Council of Environmental Quality) regulations afford Tribes the right to participate throughout the NEPA process and provide comment to the lead agency. As a result, DOE's request of Tribal involvement provides the opportunity to communicate a Nez Perce perspective of Hanford resources.

Section 1501.1.6(a) and 1508.5 states that affected Tribes have the right to be invited as a cooperating agency. A cooperating agency would participate throughout the entire NEPA process as a partner to the lead agency and can request the role as lead agency. Section 1501.7(a)(1) states that affected Tribes are afforded the right to be a participant in the scoping process. Scoping is the term for the early meetings that define the purpose and need of the project and develops the initial range of preliminary alternatives that defines the area of potential effect (APE). Section 1503.1(a)(2ii) recognizes that Tribal governments have the right to comment on NEPA proposals. An important regulation is Section 1507.2(b) that states that "presently unquantified environmental entities and values may be given appropriate consideration." In other words, Tribal perspectives, traditional values, and spiritual significance can be considered as part of the NEPA evaluation process.

In essence, Tribal values are intent on protecting, preserving, and perpetuating resources for the sake of perpetuating our culture. While completing NEPA, DOE must invite us early to the process and allow us to determine the extent of our involvement. DOE can meet trust obligations by incorporating Tribal views on resource protection while moving forward with their proposed action. When Tribal views conflict with the proposed actions, then consultation becomes an important resolution exercise for the benefit of both DOE and Tribes.

Often times federal trust obligations are not clearly articulated during the NEPA process or in their document. When there are foreseen conflicts between the agency's proposed action and their fiduciary responsibility of trust resources, DOE personnel sometimes will avoid Tribal involvement to the point of exclusion, except for commenting opportunities with the general public. Tribes are kept uninformed, and with limited resources may not know the full extent of the impacts to treaty reserved rights until after implementation.

The Nez Perce Tribe's approach is to fully engage DOE early when making important decisions about cleanup strategies and long-term stewardship of Hanford trust resources. By participating early and communicating through government consultation, we believe better decisions will be made for both DOE and the Nez Perce for future generations.

7.5 Programmatic EIS Greater-Than-Class C (GTCC) Low-Level Radioactive Waste and GTCC-Like Waste

Recently, DOE invited Affected Tribes to participate in the development of a Programmatic EIS that would look at several locations around the country to place Greater-Than Class C nuclear waste (waste that must be disposed of in a geological facility). Even though Tribe's have a legal right to be invited as cooperating Agencies to such a process, we were only invited to participate in drafting a Tribal perspective of the Hanford Affected Environment Section. We chose to participate and develop this limited narrative for the benefit of the grander scheme of communicating our perspective and fostering more open dialog with DOE. With coordination with technical staff of the Confederated Tribes of the Umatilla Indian Reservation (CTUIR) and the Wanapum people, we created a list of specific issues that are uniquely a Tribal perspective with regard to the GTCC Programmatic EIS. Partly this narrative can serve as a template to build upon for future decision-making at Hanford.

The Nez Perce Tribe expects that DOE will incorporate the following *Tribal Perspective* in their decision-making process. But more importantly, we had an opportunity to begin a more detailed discussion of Tribal values and their need to be included in the NEPA process. The following is a brief summary of the issues we identified and follow the general outline of a NEPA document.

7.5.1 Climate, Air Quality, and Noise

7.5.1.1 Climate

Climate is one of the dominate issues of our time. Any programmatic EIS that makes decisions about radioactive waste storage for thousands of years must give serious consideration to the likelihood of climate change on a storage facility. The false assumption that the climate is a constant when considering long-term storage decisions could lead to inadequate design. The reality is that nuclear waste storage will last for thousands of years and climate will likely be different with potential to reach similar condition of history. For instance, the last glacial period ended approximately 11,000 years ago. The maximum extent of glaciation was approximately 18,000 years ago. This is a brief time period considering the half-life of many radioactive isotopes.

Columbia Plateau Tribes have stories about the world being transformed from a time considered prehistoric to what is known today. The Nez Perce remember volcanoes, great floods, and animals now extinct. Oral histories also indicate a time when the climate was much wetter and supported vast forests in the region.

These distinct climatic periods have occurred during which Tribal life had to adapt for our people to survive. Our oral history tells of our struggles against volcanic activity where our world seemed on fire, of great floods, and of the previous ice age. Scientific and historic knowledge validates our oral history for many thousands of years.

Oral histories describe a time when Gable Mountain or *Nookshia* (Relander 1986), a major landscape feature on the Hanford Reservation, rose out of the Missoula floods. There is a story about Indian people who fought severe winds that were common a long time ago. One story tells of how a family trained their son by having him fight with the ice in the river until he became strong enough to fight the cold winds.

Holocene (Roberts 1998) is the term used to describe the climate during the last glaciers (110,000–11,700 years ago), covering much of the northwestern North America. Arctic foxes found at Marmes Rock Shelter provide some of this archeological record (Browman and Munsell 1969; Hicks 2004). The Palynological data would be a good source for recreating climates that supported ecosystems of the past 10,000 years. This information should be a minimal basis for climate analysis relative to decision-making on long-term storage of radioactive waste.

7.5.1.2 Air Quality

Air quality monitoring results of past and present monitoring of the Hanford site should be summarized and presented in a NEPA document. This should include measures of radioactive dust at locations like the environmental restoration disposal facility (ERDF), various plant emission stacks, venting systems, and power generation sites. Also, fugitive dust needs to be described relative to inversions and health risks. Also, this section should describe seasonal and daily wind patterns where fugitive dust could impact visibility and the Hanford viewshed.

The Nez Perce believe that radioactivity is brought into the air and distributed by the high winds that commonly occur at Hanford. Past Hanford NEPA documents provided little if any information about radioactive soil/dust dispersal capabilities of wind. ERDF Site managers occasionally send workers home and close down the facility due to blowing dust impairing worker visibility, thus creating an unsafe work environment. These situations are part of the existing environment and yet are not described.

There is no mention in the Hanford portion of the GTCC EIS where high winds could pick up contaminated soils from demolition areas or those placed at ERDF. Do ERDF or demolition sites operate with wind thresholds where work stops if wind speeds exceed some level? Do excavation or demolition sites that create radioactive debris operate under temporary structures to prevent wind dispersals? None of this information is presented.

Winds commonly blow 40–45 miles per hour and intermittently much stronger at Hanford (http://www.bces.wa.gov/windstorms.pdf). High winds over 150-mile per hour were recorded in 1972 on Rattlesnake Mountain; and in 1990, winds on the mountain were recorded at 90 miles per hour. Dust devils can be massive in size, spin up to 60 miles per hour, and frequently occur at the site. Tornadoes have been observed in Benton County which is regionally famous for receiving strong winds (Benton County 2009). It is important to understand how wind has the potential to distribute radioactive and chemical waste at Hanford during excavation, transport, handling, and storage of these contaminates.

7.5.1.3 Noise

Nonnatural noise can be offensive to native people during traditional ceremonies. Noise generating projects can interrupt the thoughts and focus and thus the spiritual balance and harmony of the Tribal community at a ceremony (Greider 1993). The general values or attributes from a Tribal perspective is for the natural environment to provide solitude, quietness, darkness, and an uncontaminated environment. These attributes provide unquantifiable value that allows for spiritual connection to mother earth. These attributes of nature are fragile.

The noise generated by the Hanford facility may have the potential to interfere with ceremonies held at sites like Gable Mountain and Rattlesnake Mountain. The disruption of natural harmony at ceremonial sites has not been surveyed or even discussed.

The Nez Perce Tribe recommends that quiet zones and time periods be identified for known Native American ceremonial locations on and near the Hanford site. Not all ceremonial sites have been shared with DOE or the non-Indian public. For this reason, Tribal values of the Hanford environment that already supports solitude should be documented. These values are also discussed in our new recommended section that we titled "Viewshed."

7.5.1.4 Light Pollution

Light pollution is a broad term that refers to multiple problems, all of which are caused by inefficient, unappealing, or (arguably) unnecessary use of artificial light.

Artificial light can create measurable harm to the environment by affecting nocturnal and diurnal animals. It can affect reproduction, migration, feeding, and other aspects of animal survival. Artificial light can also reduce the quality of experience during Tribal cultural and ceremonial activities. Presently, there is no discussion in an EIS about how artificial light may cause harm to the Hanford environment, especially those areas regularly visited by Tribal members.

7.5.2 Geology and Soils

7.5.2.1 Geology

Physiography

The Yakima Fold Belt and the Palouse Slope play potentially very significant roles at Hanford both culturally and geologically. Rattlesnake and Gable Mountains are examples of folded basalt structures within the Yakima Fold Belt. These geological features have direct bearing on the groundwater and its flow direction. There are oral history accounts of these basalt features above the floodwaters of Lake Missoula. Many other topography features have oral history explanations such as the Mooli Mooli (ground undulations found along the river terrace) and the sand dunes.

Site Geology and Stratigraphy

The 200 West area location for the proposed repository has underlying sediments of either the Hanford Formation and/or the Cold Creek formation. There is uncertainly about the geology and hydraulic conductivity in this area. The vadose zone needs to be discussed as part of the Stratigraphy Section of the EIS and is probably one of the most important elements for evaluating a potential Hanford waste repository. It should be noted that within the sediments exists a major subsurface trough feature (an eroded channel at the surface of the Ringold Formation) that can be traced in the stratigraphy from Gable Gap across the eastern part of 200 East and on to the southeast. This trough contains the Cold Creek sedimentary unit. Geologists do not understand the effects this subsurface feature has on contaminant transport.

Clastic dikes are networks of vertical features like cracks that developed in the vadose zone and thought to be related to seismic activity. Their sediments either upwell from a deeper layer or by filling in from above, or a combination of both. How clastic dikes may influence contaminant transport is not understood. There is a question as to whether or not the DOE has looked for them at the proposed site. They were noted to be present in the 200 Areas during the tank farm construction.

Regional Seismicity – The Pacific Northwest has been geologically active and this needs to be discussed if there is interest in putting more contaminants in the ground at Hanford. Geologic structure of the Pacific Northwest includes a feature called the Olympic-Wallowa Lineament (the OWL). Surface and depth data have identified a structural "line" within the earth's crust that can be traced roughly from southeast of the Wallowa Mountains, under Hanford, through the Cascades and under Seattle and the Sound. Such lineaments are signals of crustal structure that are not yet well identified. Emerging research being reported through the USGS is highlighting the importance of Seattle area faults connecting under the Cascades into the Yakima Fold Belt and on along the OWL. The geologic stress on the surface of the earth in the local region have a north–south compressional force that has caused the surface to wrinkle in folds that trend approximately east–west, thus creating the Yakima Fold Belt. Fault movement along these folds occurs all the time and studies have shown these to be considered active fault zones (Repasky et al. 1998; Campbell et al. 1995).

The 1936 earthquake and the 1973 earthquakes at Hanford need to be mentioned as recent activities and justification for requiring an earthquake-resistant structure. Any storage structure of highly contaminated nuclear waste should also have backup safety systems as a line of defense against earthquakes.

7.5.2.2 Soils

Soil is part of mother earth that supports plant and animal life which Native people rely for our traditional lifeways. We understand the importance of soils and minerals through our traditional use of them. Clays were used as a building material, for creating mud baths, and for making pottery. One of the best known attributes of soils is its ability to filter water. Hanford has delineated contamination areas called operable units (OUs) for surface contamination. It is essential for the soils section of the Affected Environment Chapter to graphically illustrate and describe the surface contamination OUs. The influence of past tank leaks on changing soil chemistry and properties are not understood. Sandy soils at Hanford already have high transmissivity. Such changes could increase water and contaminant transport.

Oral histories document medicinal properties of soil for healing wounds. Soils from the White bluffs were used for cleaning hides, making paints, and whitewashing villages.

7.5.3 Minerals and Energy Resources

The extent and value of mineral resources displaced by the present contamination in the 200 East and 200 West areas has not been documented. DOE has designated this area as industrial use according to the comprehensive land use plan (CLUP). It appears that DOE's present vision is to allow temporary and long-term waste storage at the uncontaminated surface in this area while continuing pump and treat technology and natural attenuation as a means of managing vadose and groundwater contamination that is under much of this site. This may seem like a reasonable strategy by DOE from a technical standpoint but this strategy will likely prevent Tribal use of the area for thousands of years. As a result, there is a loss of resource use to Tribes.

7.5.4 Water Resources

7.5.4.1 Groundwater

Purity of water is very important to the Nez Perce, and thus DOE should be managing for an optimum condition considering Tribal cultural connection and direct use of water, rather than managing for a minimum water quality threshold. Hanford has delineated OUs for subsurface contamination. It is essential for the soils and groundwater sections of the Affected Environment Chapter to graphically illustrate and describe existing groundwater contamination. From the perspective of the Nez Perce Tribe, the greatest long-term threat at the Hanford site lies in the contaminated groundwater. There is insufficient characterization of the vadose zone and groundwater. There is a tremendous volume of radioactive and chemical contamination in the groundwater. The mechanisms of flow and transport of contaminants through the soil to the groundwater are still largely unknown. The volumes of contamination within the groundwater and direction of flow are still only speculative. Due to lack of knowledge and limited technical ability to remediate the vadose zone and groundwater puts the Columbia River at continual risk.

7.5.4.2 Water Use

The Columbia River is the lifeblood of the Nez Perce people. It supports the salmon and every food or material that they rely on for subsistence. It is an essential human right to have clean water. If water is contaminated it then contaminates all living things. Tribal members that exercise a traditional lifestyle would also become contaminated. A perfect example is making a sweat lodge and sweating. It is a process of cleansing and purification. If water is contaminated then the sweat lodge materials and process of cleansing would actually contaminate the individual.

Tribal people are well known for adopting technology if it were instituted wisely and did not sacrifice or threaten the survival of the group as a whole. This approach applies to Tribal use of groundwater. Even though groundwater was not used except at springs, Tribes would have potentially used technology for developing wells and would have used groundwater if seen to be an appropriate action. The existing contamination is considered an impact to Tribal rights to utilize this valuable resource.

The hyporheic zone in the Columbia River needs to be more fully characterized to understand the location and potential of groundwater contaminants discharging to the Columbia River. Contaminated groundwater plumes at Hanford are moving towards the Columbia River and some contaminants like chromium are already recharging to the river. It is the philosophy of the Columbia River Tribes that groundwater restoration and protection be paramount to DOE's management of Hanford. Institutional controls, such as preventing use of groundwater, should only be a temporary measure for the safety of people and animals. Tribe's prefer a proactive corrective cleanup strategy over DOE's inference to use surface barriers, natural attenuation, and institutional controls as a viable long-term management option. In our opinion, monitoring natural attenuation is not a cleanup strategy. By not actively pursuing cleanup of vadose and groundwater contamination, DOE is open to placing additional waste like GTCC or Mercury in the 200 areas since the site will have limitations for any other use.

7.5.5 Human Health

Nez Perce health involves access to traditional foods and places. Both of these are located on the Hanford facility and can be impacted by placement of the GTCC waste in the 200 area.

Definition of Tribal health: Native American ties to the environment are much more complex and intense than is generally understood by risk assessors (Harris and Harper 1997, Oren Lyons; http://www.ratical.org/many_worlds/6Nations/OLatUNin92.html; http://www.youtube.com/watch?v=hDF7ia23hVg.). All of the foods and implements gathered and manufactured by the traditional American Indian are interconnected in at least one way, but more often in many ways. Therefore, if the link between a person and his/her environment is severed through the introduction of contamination or physical or administrative disruption, the person's health suffers, and the well-being of the entire community is affected.

To many American Indians, individual and collective well-being is derived from membership in a healthy community that has access to, and utilization of, ancestral lands and traditional resources. This wellness stems from and is enhanced by having the opportunity and ability to live within traditional community activities and values. If the links between a Tribal person and his or her environment were severed through contamination or DOE administrative controls, the well-being of the entire community is affected.

7.5.6 Risk Assessments

Risk assessments should take a public health approach to defining community and individual health. Public health naturally integrates human, ecological, and cultural health into an overall definition of community health and well-being. This broader approach used with risk assessments is adaptable to indigenous communities that, unlike westernized communities, turn to the local ecology for food, medicine, education, religion, occupation, income, and all aspects of a good life (Harris and Harper 1997; Harper and Harris 2000).

"Subsistence" in the narrow sense refers to the hunting, fishing, and gathering activities that are fundamental to the way of life and health of many indigenous peoples. The more concrete aspects of a subsistence lifestyle are important to understanding the degree of environmental contact and how subsistence is performed in contemporary times. Also, traditional knowledge can be learned directly from nature. Through observation this knowledge is recognized and a spiritual connection is often attained as a result. Subsistence utilizes traditional and modern technologies for harvesting and preserving foods as well as for distributing the produce through communal networks of sharing and bartering. The following is a useful explanation of "subsistence," slightly modified from the National Park Service:

> "While non-native people tend to define subsistence in terms of poverty or the minimum amount of food necessary to support life, native people equate subsistence with their culture. It defines who they are as a people. Among many Tribes, maintaining a subsistence lifestyle has become the symbol of their survival in the face of mounting political and economic pressures. To Native Americans who continue to depend on natural resources, subsistence is more than eking out a living. The subsistence lifestyle is a communal activity that is the basis of cultural existence and survival. It unifies communities as cohesive functioning units through collective production and distribution of the harvest. Some groups have formalized patterns of sharing, while others do so in more informal ways. Entire families participate, including elders, who assist with less physically demanding tasks. Parents teach the young to hunt, fish, and farm. Food and goods are also distributed through native cultural institutions. Nez Perce young hunters and fisherman are required to distribute their first catch throughout the community at a first feast (first bite) ceremony. It is a ceremony that illustrates the young hunter is now a man and a provider for his community. Subsistence embodies cultural values that recognize both the social obligation to share as well as the special spiritual relationship to the land and resources." (National Park Service: http://www.cr.nps.gov/aad/cg.fa_1999/subsist.html)

The following four categories of an undisturbed environment contribute to individual and community health. Impacts to any of these functions can adversely affect health. Metrics associated with impacts within each of these categories are presented in Harper and Harris (1999).

7.5.6.1 Human Health-Related Goods and Services

This category includes the provision of water, air, food, and native medicines. In a Tribal subsistence situation, the land provided all the food and medicine that was necessary to enjoy long and healthy lives. From a risk perspective, those goods and services can also be exposure pathways.

7.5.6.2 Environmental Functions and Services

This category includes environmental functions such as soil stabilization and the human services that this provides, such as erosion control or dust reduction. Dust control in turn would provide a human health service related to asthma reduction. Environmental functions such as nutrient production and plant cover would provide wildlife services such as shelter, nesting areas, and food, which in turn might contribute to the health of a species important to ecotourism. Ecological risk assessment includes narrow examination of exposure pathways to biota as well as examination of impacts to the quality of ecosystems and the services provided by individual biota, ecosystems, and ecology.

7.5.6.3 Social and Cultural Goods, Functions, Services, and Uses

This category includes many things valued by suburban and Tribal communities about particular places or resources associated with intact ecosystems and landscapes. Some values are common to all communities, such as the esthetics of undeveloped areas, intrinsic existence value, environmental education, and so on.

7.5.6.4 Economic Goods and Services

This category includes conventional dollar-based items such as jobs, education, health care, housing, and so on. There is also a parallel nondollar indigenous economy that provides the same types of services, including employment (i.e., the functional role of individuals in maintaining the functional community and ensuring its survival), shelter (house sites, construction materials), education (intergenerational knowledge required to ensure sustainable survival throughout time and maintain personal and community identity), commerce (barter items and stability of extended trade networks), hospitality, energy (fuel), transportation (land and water travel, waystops, navigational guides), recreation (scenic visitation areas), and economic support for specialized roles such as religious leaders and teachers.

7.5.7 Ecology

The Nez Perce people have lived in these lands for a very long time and thus have learned about the resources and their ecological interrelationships. They knew about environmental indicators that foretold seasons and conditions that guided them. When Cliff Swallows first appeared in the spring, their arrival is an indicator that the fish are coming up the river. Doves are the fish counters, telling how many fish are coming. Many natural phenomena foretell when the earth is coming alive again in the spring, even if things are dormant underground. The Nez Perce have traditional ecological knowledge of this environment and Tribal people have ceremonies that acknowledge the arrival of spring. The winds bring information about what will happen. It provides guidance about how to bring balance back to the land.

7.5.7.1 Biodiversity on the National Monument

The Monument encompasses a biologically diverse landscape containing an irreplaceable natural and historic legacy. Limited development over approximately 70 years has allowed for the Monument to become a haven for important and increasingly scarce plants and animals of scientific, historic, and cultural interest. It supports a broad array of newly discovered or increasingly uncommon native plants and animals. Migrating salmon, birds, and hundreds of other native plant and animal species, some found nowhere else in the world, rely on its natural ecosystems. The Monument also includes 46.5 miles of the last free-flowing, nontidal stretch of the Columbia River, known as the "Hanford Reach."

7.5.7.2 Salmon

Columbia River salmon runs, once the largest in the world, have declined over 90% during the last century. The 7.4–12.5 million average annual numbers of fish above Bonneville Dam have dropped to 600,000. Of these, approximately 350,000 are produced in hatcheries. Many salmon stocks have been removed from major portions of their historic range (Columbia Basin Fish and Wildlife Authority 2009).

Multiple salmon runs reach the Hanford Nuclear Reservation. These runs include Spring Chinook, Fall Chinook, Sockeye, Silver, and Steelhead. The runs tend to begin in April and end in November. Salmon runs have been decimated as a result of loss and change of habitat. The changes include nontribal commercial fisheries, agriculture interests, and especially construction of hydro-projects on the Columbia River. Protection and preservation of anadromous fisheries were not a priority when the 227 Columbia River dams were constructed. Some dams were constructed without fish ladders and ultimately eliminated approximately half of the spawning habit available in the Columbia System.

The Hanford Reach is approximately 51 miles long and is the only place on the upper main stem of the Columbia River where Chinook salmon still spawn naturally. This reach is the last free-flowing section of the Columbia River above Bonneville Dam. It produces about 80–90% of the fall Chinook salmon run on the Columbia River.

The Columbia River Tribes, out of a deep commitment to the fisheries and in spite of the odds, plan to restore stocks of Chinook, Coho, Sockeye, Steelhead, Chum, Sturgeon, and Pacific Lamprey. This effort was united in 1995 under a recovery plan called the Wy-Kan-Ush-Mi Wa-Kish-Wit (Spirit of the Salmon). Member Tribes are the Nez Perce, Umatilla, Warm Springs, and Yakama. The Columbia River Tribes see themselves as the keepers of ancient truths and laws of nature. Respect and reverence for the perfection of Creation are the foundation of our cultures. Salmon are part of our spiritual and cultural identity. Tribal values are transferred from generation to generation through fishing and associated activities tied to the salmon returns. Without salmon, Columbia River Tribes would loose the foundation of their spiritual and cultural identity.

Tribes affected by the Hanford site are comanagers of Columbia River fisheries and assist in tagging fry and counting reds along the Hanford Reach for the purposes of estimating fish returns. This information is essential in the negotiation of fish harvest between the USA and Canada as well as between Indian and non-Indian fishermen.

In many ways, the loss of salmon mirrors the plight of native people along the Columbia. Elders remind us that the fate of humans and salmon are linked. The circle of life has been broken with the loss of traditional fishing sites and salmon runs on the Columbia River.

7.5.8 *Socioeconomics*

7.5.8.1 Modern Tribal Economy

A subsistence economy is one in which currency is limited because many goods and services are produced and consumed within families or bands, and currency is based as much on obligation and respect as on tangible symbols of wealth and immediate barter. It is well-recognized in anthropology that indigenous cultures include networks of materials interlinked with networks of obligation. Together these networks determine how materials and information flow within the community and from the environment. Today there exists with Tribal people an integrated interdependence between formal (cash-based) and informal (barter and subsistence-based) economic sectors. This relationship must be considered when thinking of economics and employment of Tribal people (http://www.ratical.org/many_worlds/6Nations/OLatUNin92.html; http://www.youtube.com/watch?v=hDF7ia23hVg).

Indian people engage in a complex web of exchanges that often involves traditional plants, minerals, and other natural resources. These exchanges are a foundation of community and intertribal relationships. Indian people catch salmon that become gifts to others living near and far. Sharing self-gathered food or self-made items is a part of establishing and maintaining reciprocal relationships. People have similar reciprocal relationships with mother earth including physical places and elements of nature. This mutual respect applies to all. Present contamination at Hanford, extended timelines for cleanup and proposals to place more waste at Hanford, may displace or limit traditional and contemporary Tribal use of resources, and thus direct production that permeates Indian life.

Use of the Hanford site and surrounding areas by Tribes was primarily tied to the robust Columbia River fishery. Tribal families and bands lived along the Columbia either year round or seasonally for catching, drying, and smoking salmon. Past associated activities included gatherings for such events like marriages, trading, ceremonial feasts, harvesting, fishing, and mineral collection. The loss of salmon runs, loss of fishing sites now under water, and Hanford land use restrictions has limited the once natural surplus that supported this gifting and barter system of our Tribal culture.

It is likely that the future of salmon in the Columbia system will be determined within the lifetime of Hanford cleanup and the lifecycle of the GTCC waste proposed for placement at Hanford. With the tremendous efforts to recover salmon (and other fish species) by Tribes, government agencies, and conservation organizations,

Tribal expectations are that these species will be recovered to stronger, healthy populations. If salmon, and other anadromous fish species were to recover, the regional economy and Tribal barter economy would likely change in the region, including at the Hanford Reach. These fish returns and the associated social and economic potential needs to be considered within the lifecycle of a waste repository at its inherent risk to the environment.

7.5.8.2 Direct Production

Direct production by Tribes is part of the economy that needs to be represented, especially considering the Tribe's emphasis on salmon recovery. This type of individual commerce in modern economics is termed and calculated as "direct production." The increase in direct production would be relational to the region's salmon recovery, yet there is no economic measure (within the NEPA process) to account for this robust element of a traditional economy.

In a traditional sense, direct production is a term of self and community reliance on the environment for existence as opposed to employment of modern economies. Direct production is use of salmon and raw plant materials for foods, ceremonial, and medicinal needs and the associated trading or gifting of these foods and materials. Direct production needs to be understood and should include the role of plant foods, ceremonial plants, medicinal plants, bead work, hide work, tule mats, and dried salmon.

The season prior to the flooding of Celilo Falls, an estimated 1,500 native fisherman assembled at the site during peak fishing. Trading among 1,500 fisherman and their families would have been substantial. It would make for a tremendous scene today to see that number fishing and drying meat. What would be the direct production generated from 1,500 fishermen and their families trading and gifting salmon, dentalia shells, mountain sheep horns, bows, horses, baskets, tule mats, buffalo robes, leather, rawhide, and hand-made art like bead work?

7.5.9 Environmental Justice

President Clinton signed Executive Order 12898 to address Environmental Justice issues and to commit each federal department and agency to "make achieving Environmental Justice part of its mission." (Environmental Biosciences Program 2001). According to the Executive Order, no single community should host disproportionate health and social burdens of society's polluting facilities. Many American Indians are concerned about the interpretation of "Environmental Justice" by the U.S. Federal Government in relation to Tribes. By this definition, Tribes are included as a minority group. However, the definition as a minority group fails to recognize Tribes' sovereign nation-state status, the federal trust responsibility, or protection of treaty and statutory rights of American Indians. Because of a lack of the these details, Tribal governments and federal agencies have not been able to develop a clear definition of Environmental Justice in Indian Country, and thus it is difficult to determine appropriate actions.

If federal decision-making does not fully protect trust resources to the degree necessary to protect aboriginal uses, those decisions could be interpreted to be a violation of aboriginal rights. Decisions that cause continued degradation of trust resources could place undue burden to Tribal people and could also be considered an Environmental Justice issue. Many federal and state environmental laws and regulations designed to protect the environment are not interpreted by regulators to fully address the concerns of Native Americans. This topic deserves more review and discussion to better define what constitutes a violation of federal trust responsibilities. When does a loss of protected Tribal use by government action(s), like those occurring at Hanford, become a violation of aboriginal rights and trigger an environmental justice issue? A review of existing case law might summon such an argument or opinion.

7.5.10 Land Use

The Nez Perce Tribe recommends that DOE continue efforts to identify special places and landscapes with spiritual significance. Newly identified sites would be added to those already requiring American Indian ceremonial access and protection through long-term stewardship. Native people maintain that aboriginal and treaty rights allow for the protection, access to, and use of resources. These rights were established at the origin of the Native People and persist forever. There are sites or locations within the existing Hanford reservation boundary with Tribal significance that are presently restricted through DOE's institutional controls and should be considered for special protections or set aside for traditional and contemporary ceremonial uses. Sites like the White Bluffs, Gable Mountain, Rattlesnake Mountain, Gable Butte, and the islands on the river are known to have special meaning to Tribes and should be part of the discussion for special access and protection. These locations should be placed in comanagement with DOE, fish and wildlife service (FWS), and the Tribes for long-term management and protection.

7.5.10.1 Tribal Access

There are several federal regulations, policies, and executive orders that define Tribal access at Hanford when hazard risk levels are acceptable. Institutional controls associated with the CLUP or the comprehensive conservation plan (CCP) should not override Tribal rights to access areas that no longer have human health hazards. The following is a brief summary of those legal references:

According to the *American Indian Religious Freedom Act*, Tribal members have a protected right to conduct religious ceremonies at locations on public lands where they are known to have occurred before. There has been an incomplete effort to research the full extent of Tribal ceremonial use at Hanford. *Executive Order* 13007 supports the American Religions Freedom Act by stating that Tribal members have the right to access ceremonial sites. This includes a directive for agencies to maintain existing trails or roads that provide access to these sites.

DOE managers that are considering moving waste or placement of new waste at Hanford must evaluate potential impacts to ceremonial access as part of DOE trust responsibility to Tribes. There are locations that have specific protections due to culturally significance, burial sites, artifact clusters, etc. These types of areas are further described under the Cultural Resources Section of this writing. As decommissioning and reclamation occurs across the Hanford site, findings of culturally significant areas will continue to expand the list of sites with special protections, and these protections override existing land use designation of the CLUP or other DOE documents.

7.5.10.2 Comprehensive Land Use Plan (CLUP)

The present DOE land use document for Hanford, called the CLUP, has institutional controls that limit present and future use by Native Americans. DOE plans to remove some institutional controls over time as the contamination footprint is reduced as a result of instituting their 2015 vision along the river and the proposed cleanup of the 200 area. With removal of institutional controls, the affected Tribes assume they can resume access to usual and accustomed areas.

Future decisions about land transfer must consider the implications for Usual and Accustomed uses (aboriginal and treaty reserved rights) in the long-term management of resource areas. The 50-year management time horizon of the CLUP does *not* create permanent land use designations. On the contrary, land use designations or their boundaries can be changed in the interim at the discretion of DOE and/or through requests to DOE by Hanford stakeholders. The CLUP is often misused by assuming designations are permanent. Also, it is important to note that the interim land use designations in the CLUP cannot abrogate treaty rights. That requires an act of Congress.

7.5.10.3 Hanford National Monument

A Presidential Proclamation established the Hanford Reach National Monument (Monument) (Presidential Proclamation 7319) and directed the DOE and the U.S. FWS to jointly manage the monument. The Monument covers an area of 196,000 acres on the DOEs Hanford Reservation. DOE agreements and permits delegate authorities to FWS for 165,000 acres while DOE still directly manages approximately 29,000 acres, and the Washington Department of Fish and Wildlife manages the remaining 800 acres (approximately) through a separate DOE permit.

The comanagement of the Monument directs each agency to fulfill several missions. The FWS is responsible for the protection and management of Monument resources and people's access to lands under FWS control. The FWS also has the responsibility to protect and recover threatened and endangered species; administer the Migratory Bird Treaty Act; and protect fish, wildlife and Native American trust resources, and other trust resources within and beyond the boundaries of the Monument.

The FWS developed a CCP for management of the Monument as part of the National Wildlife Refuge System as required under the National Wildlife Refuge System Improvement Act. The CCP is a guide to managing the Monument lands. It should be understood that FWS management of the Monument is through permits or agreements with the DOE.

Tribes participated in the development of the CCP with regard to protection of natural and cultural resources and Tribal access. Based on the Presidential Proclamation that established the Hanford Reach National Monument, Affected Tribes assume that all of Hanford will be restored and protected (Proclamation 7319).

7.5.10.4 Operable Units (OUs)

Hanford has delineated contamination areas called OUs for both surface and subsurface contamination. It is essential for the soils and groundwater sections of the Affected Environment Chapter to graphically illustrate and describe the surface and subsurface OUs. Land under consideration for long-term waste disposal like the area next to 200 West should not only describe the Land Use designation (according to the CLUP) but also describe the extent of surface and subsurface contamination that primarily dictated that designation. For example, the proposed GTCC site at 200 West lies over the 200 ZP-1 groundwater OU. This OU has contamination from uranium, technetium, iodine 129, and other radioactive and chemical constituents. The extent and timeframe for its cleanup should be understood within the context of proposing the placement of a waste repository immediately over these vadose contaminants and groundwater plumes.

7.5.11 Transportation

7.5.11.1 Traditional Transportation

Indian people have been traveling their homeland to usual and accustomed areas for a very long time. Early modes of transportation began with foot travel. Domesticated dogs were utilized to carry burdens. Dugout canoes were manufactured and used to traverse the waterways when the waters were amiable. Otherwise, trails following the waterways were best means for travel. With the arrival of the horse, it changed how people traveled. Numerous historians note that horses arrived to the Columbia Plateau in the late 1700s. That is incorrect according to Tribal history. The arrival of the horse was actually a full century earlier in the late 1600s. Their acquisition quickened Tribal movement on an already extant and heavily used travel network. This travel network was utilized by many Tribal groups on the Columbia Plateau and was paved by thousands of years of foot travel. Early explorers and surveyors utilized and referenced this extensive trail network. Some of the trails have become major highways and the Columbia and Snake Rivers are still a crucial part of the modern transportation network.

The Middle Columbia Plateau of the Hanford area is the crossroads of the Columbia Plateau located half way between the Great Plains and the Pacific Northwest Coast. Major Columbia River tributaries including the Walla Walla, Snake, and Yakima Rivers flow into this section of the main stem Columbia River. These rivers form a critical part of a complex transportation network through the region that includes the Hanford reach. The slow water at the Wallula Gap was one of the few places where horses could traverse the river year round. This river crossing provided access to a vast web of trails that crossed the region. Portions of these trails are known to cross Hanford.

7.5.11.2 Present Transportation

There are two interstate highways that are near the site [Interstate 90 (I-90) and Interstate 84 (I-84)]. There are estimates of as many as 12,000 shipments of GTCC waste that would need to be delivered to Hanford by rail, barge, or highway. The Nez Perce Tribe believes that decision-making criteria need to be presented in the EIS to clarify how rail, barge, or highway routing will be determined. Treaty resources and environmental protections are important criteria in determining a preferred repository location. The public needs to be assured that the public health and high valued resources like salmon and watersheds are going to be protected.

Northwest river systems have received significant federal and state resources over recent decades in an attempt to recover salmon and rehabilitate damaged watersheds. DOE needs to describe how public safety, salmon, and watersheds "fit" into the criteria selection process for determining a GTCC waste site and multiple shipping options. The protection and enhancement of existing river systems are critical to sustaining Tribal cultures along the Columbia River.

The interstate highway system is a primary transportation corridor for shipping nuclear waste through the states of Oregon, Washington, and Idaho. Waste moving across these states will cross many major salmon bearing rivers that are important to the Tribes. Major rail lines also cross multiple treaty resource areas.

7.5.12 Cultural Resources

From a Tribal perspective, all things of the natural environment are recognized as a cultural resource. This is a different perspective from those who think of cultural resources as artifacts or historic structures. The natural environment provides resources for a subsistence lifestyle for Tribal people. This daily connection to the land is crucial to Nez Perce culture and has been throughout time. All elements of nature therefore are the connection to Tribal religious beliefs and the foundation of their aboriginal rights recognized in the 1855 treaty. Oral histories confirm this cultural and religious connection.

> "According to our religion, everything is based on nature. Anything that grows or lives, like plants and animals, is part of our religion..."
>
> *Horace Axtell (Nez Perce Tribal Elder)*

7.5.12.1 Landscape and Ethno-Habitat

For thousands of years American Indians have utilized the lands in and around the Hanford Site. Historically, groups such as the Yakama, the Walla Walla, the Wanapum, the Palouse, the Nez Perce, the Columbia, and others had ties to the Hanford area. "The Hanford Reach and the greater Hanford Site, a geographic center for regional American Indian religious activities, is central to the practice of the Indian religion of the region and many believe the Creator made the first people here" (DOI 1994). Indian religious leaders such as Smoholla, a prophet of Priest Rapids who brought the Washani religion to the Wanapum and others during the late nineteenth century, began their teachings here (Relander 1986). Prominent landforms such as Rattlesnake Mountain, Gable Mountain, and Gable Butte, as well as various sites along and including the Columbia River, remain sacred. American Indian traditional cultural places within the Hanford Site include, but are not limited to, a wide variety of places and landscapes: archaeological sites, cemeteries, trails and pathways, campsites and villages, fisheries, hunting grounds, plant gathering areas, holy lands, landmarks, important places in Indian history and culture, places of persistence and resistance, and landscapes of the heart (Bard 1997). Since affected Tribal members consider these places sacred, many traditional cultural sites remain unidentified. NEPA 18 4.6.1.2 (p. 4.120).

7.5.12.2 Viewshed

The Nez Perce Tribe utilizes vantage points to maintain a spiritual connection to the land. Viewsheds tend to be panoramic and are made special when they contain prominent uncontaminated topography. The viewshed panorama is further enhanced by abrupt changes in topography and or habitats.

Nighttime viewsheds are also significant to indigenous people who still use the Hanford Reach. Each Tribe has stories about the night sky and why stars lie in their respective places. The patterns convey spiritual lessons which are conveyed through oral traditions. Often, light pollution from neighboring developments diminishes the view of the constellations.

There are several culturally significant viewsheds located on the Hanford site. The continued Tribal use of these sites brings spiritual renewal. The potential to impact viewsheds should be considered when accessing new DOE proposals.

Special travel considerations should be given to Tribal elders and youth to accommodate their desire to reach traditional ceremonial sites that have viewshed values.

7.5.12.3 Salmon as a Cultural Resource

Nez Perce life is perceived as being intertwined with the life of the salmon. Salmon remain a core part of oral traditions of Columbia Plateau Tribes and still maintains a presence in native peoples' diet just as it has for generations. Salmon are recognized as the first food at Tribal ceremonies and feasts. One example is the *ke'uyit*, which translates to "first bite." It is a Nez Perce ceremonial feast that is held in spring to recognize the foods that return to take care of the people. It is a long-standing ceremony that attendees immerse themselves in prayer, songs, and dancing throughout its activities.

A core tenant of the plateau people is to extend gratitude to the foods for sustaining their life. A parallel exists between the dwindling numbers of salmon returning to the Columbia and the struggle of the Nez Perce people (Landeen and Pinkham 1999).

7.5.13 Waste Management

The Nez Perce Tribe will continue to work with DOE through its cooperative agreement to ensure that cleanup decisions protect human health, the environment, and Tribal rights. The Nez Perce Tribe believes that the ultimate goal of the Hanford cleanup should be to restore the land to uncontaminated pre-Hanford conditions for unrestricted use. Our end-state vision would allow Tribal members to utilize the area in compliance with the Usual and Accustomed treaty rights reserved and guaranteed in the 1855 treaty (Nez Perce Tribe 2005).

7.5.14 Cumulative Impacts

As part of any EIS process, a cumulative risk assessment needs to be developed for Hanford. This risk assessment needs to utilize the three existing Hanford Tribal risk scenarios (CTUIR, Yakama Indian Nation, and DOE Hanford), and include existing values as part of Hanford risk to determine cumulative impacts.

The cumulative loss of Tribal access through fencing as part of the institutional controls needs to be clearly graphically displayed. This public and Tribal access limitation needs to be described as part of the existing environment. Any change to size and time extent of access restriction, especially Tribal access, needs to be clearly understood. For example, the proposed placement of a waste repository with 10,000-year half-life waste products would greatly extend access limitations.

The Natural Resource Damage Assessment and Restoration Program (NRDA) directs Federal Agencies like DOE to restore natural resources injured as a result of oil spills or hazardous substance releases into the environment. Damage assessments provide the basis for determining the restoration needs that address the public's loss and use of natural resources. If restoration is not met then compensation and mitigation will complete redress of loss of use.

This existing loss of use of the 200 East and 200 West areas from deep vadose and groundwater contamination has not yet been quantified. Present land use designation of industrial use by the CLUP could compromise and add complexity to the NRDA process. Industrial use is making the 200 East and 200 West areas the target for industrial uses like this long-term waste storage proposal with no regard or understanding of how this effects cleanup strategies andthe consequences of such proposals blur the lines of what is considered a loss of use from waste contamination verses loss of use due to access restrictions for safety reasons associated with the proposed waste storage.

Land Use designation is largely due to contamination and is leading to potential long-term waste storage that could extend length of time to correct contamination. There is no discussion of how surface uses may hinder cleanup strategies, placement of groundwater pump and treat systems or monitoring wells, etc. Overall, there is a need to describe how surface uses like waste storage may infringe on future treatment and removal of waste under the 200 area.

Acknowledgments Contributors include:
Gabe Bohnee-Director of the Environmental Restoration and Waste Management (ERWM) Program, Nez Perce Tribe
John Stanfill, Hanford Coordinator, ERWM Program, Nez Perce Tribe
Tony Smith, Research and Design Specialist, ERWM Program, Nez Perce Tribe
Josiah Pinkham, Ethnographer, Cultural Resources Program, Nez Perce Tribe
Mike Lopez, Attorney, Legal Council to the Nez Perce Tribe

7.6 Appendix: Legal Framework

7.6.1 Treaty Rights and Obligations

The Nez Perce Tribe is a sovereign government whose territory comprises over 13 million acres of what are today's northeast Oregon, southeast Washington, and north-central Idaho. In 1855, the Nez Perce Tribe entered into a treaty with the United States, securing, among other guarantees a permanent homeland, as well as fishing, hunting, gathering, and pasturing rights (Treaty with the Nez Perces, June 11, 1855; 12 Stat. 957).

Since 1855, many federal and state actions have recognized and reaffirmed the Tribe's treaty reserved rights. Since these rights are of enormous importance to the Tribe's subsistence and cultural fabric, the ecosystems that support fish and wildlife must remain undamaged and productive. DOE recognizes the existence of reserved treaty rights and has shown a commitment to identifying and assessing impacts of all DOE activities to both on and off-reservation lands.

The Nez Perce Tribe has the responsibility to protect the health, welfare, and safety of its members, and the environment and cultural resources of the Tribe.

Therefore, activities related to the Hanford operations and cleanup should avoid endangering the Tribe's environment and culture, or impairing their ability to protect the health and welfare of Tribal members.

7.6.1.1 The Nez Perce Tribe Treaty of 1855

The Nez Perce Tribe Treaty of 1855 promulgated articles of agreement between the United States and the Tribe. The Treaty is superior to any conflicting state laws or state constitutional provisions under the Supremacy Clause of the U.S. Constitution (Art. VI. cl. 2).

Under the Treaty of 1855, the Tribe ceded certain areas of its aboriginal lands to the United States and reserved for its exclusive use and occupation certain lands, rights, and privileges; and the United States assumed fiduciary responsibilities to the Tribe.

> Rights reserved under the Treaty of 1855 include those found in Article 3 of the Treaty, "The exclusive right of taking fish in all the streams where running through or bordering said reservation is further secured to said Indians; as also the right of taking fish at all usual and accustomed places in common with citizens of the Territory; and of erecting temporary buildings for curing, together with the privilege of hunting, gathering roots and berries, and pasturing their horses and cattle upon open and unclaimed land."

The reserved rights to the aforementioned areas are a fundamental concern to the Nez Perce Tribe. The fish, roots, wild game, religious sites, and ancestral burial and

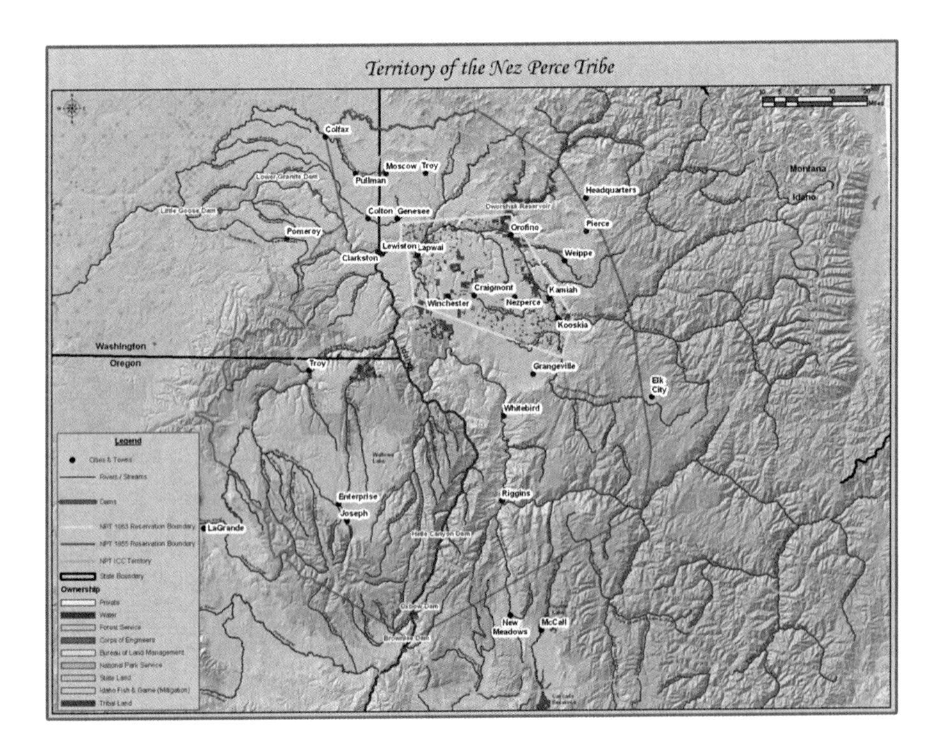

living sites remain integral to the Nez Perce culture. The Tribe expects, accordingly, to be the primary consulting party in all federal actions related to Hanford that stand to affect or implicate the Tribe's treaty reserved or cultural interests.

7.6.2 Treaty Reserved Resources

Treaty reserved resources situated on and off the Reservation (hereafter referred to as "Tribal Resources") includes but are not limited to:

Tribal water resources located within the Columbia, Snake, and Clearwater River Basins including those water resources associated with the Tribe's usual and accustomed fishing areas and Tribal springs and fountains described in Article 8 of the Nez Perce Tribe Treaty of 1863.

Fishery resources situated within the Reservation, as well as those resources associated with the Tribe's usual and accustomed fishing areas in the Columbia, Snake, and Clearwater River Basins.

Areas used for the gathering of roots and berries, hunting, and other cultural activities within open and unclaimed lands including lands along the Columbia, Clearwater, and Snake River Basins.

Open and unclaimed lands which are or may be suitable for grazing.

Forest resources situated on the Reservation and within the ceded areas of the Tribe.

Land holdings held in trust or otherwise located on and off the Nez Perce Reservation in the States of Idaho, Oregon, and Washington.

Culturally sensitive areas, including, but not limited to, areas of archaeological, religious, and historic significance, located both on and off the Reservation.

7.6.3 Federal Recognition of Tribal Sovereignty

A unique political relationship exists between the United States and Indian Tribes, as defined by treaties, the United States Constitution, statutes, federal policies, executive orders, court decisions, which recognize Tribes as separate sovereign governments.

As a fiduciary, the United States and all its agencies owe a trust duty to the Nez Perce Tribe and other federally recognized Tribes. *See United States v. Cherokee Nation of Oklahoma,* 480 U.S. 700, 707 (1987); *United States v. Mitchell,* 463 U.S. 206, 225 (1983); *Seminole Nation v. United States,* 316 U.S. 286, 296–97 (1942). This trust relationship has been described as "one of the primary cornerstones of Indian law," Felix Cohen, *Handbook of Federal Indian Law* 221 (1982), and has been compared to one existing under the common law of trusts, with the United States as trustee, the Tribes as beneficiaries, and the property and natural resources managed by the United States as the trust corpus. *See, e.g., Mitchell,* 463 U.S. at 225.

The United States' trust obligation includes a substantive duty to consult with a Tribe in decision-making to avoid adverse impacts on treaty resources and a duty to protect Tribal treaty reserved rights "and the resources on which those rights depend." *Klamath Tribes v. U.S.*, 24 Ind. Law Rep. 3017, 3020 (D.Or. 1996). The duty ensures that the United States conduct meaningful consultation "in advance with the decision maker or with intermediaries with clear authority to present Tribal views to the ... decision maker." *Lower Brule Sioux Tribe v. Deer*, 911 F. Supp 395, 401 (D. S.D. 1995).

Consistent with the United States' trust obligation to Tribes, Congress has enacted numerous laws to protect Tribal resources and cultural interests, including, but not limited to the National Historic Preservation Act (NHPA) of 1966; the Archaeological Resources Protection Act of 1979; the Native American Graves Protection and Repatriation Act (NAPRA) of 1990; and the American Indian Religious Freedom Act (AIRFA) of 1978.

7.6.4 Executive Orders

Executive order, 13007, May 24, 1996. Updated April 30, 2002.

Section 1. Accommodation of Sacred Sites. (a) In managing Federal lands, each executive branch agency with statutory or administrative responsibility for the management of Federal lands shall – to the extent practicable, permitted by law, and not clearly inconsistent with essential agency functions – (1) accommodate access to and ceremonial use of Indian sacred sites by Indian religious practitioners and (2) avoid adversely affecting the physical integrity of such sacred sites. Where appropriate, agencies shall maintain the confidentiality of sacred sites.

This Executive Order directs Federal land-managing agencies to accommodate Native Americans' use of sacred sites for religious purposes and to avoid adversely affecting the physical integrity of sacred sites. {267} Some sacred sites may be considered traditional cultural properties and, if older than 50 years, may be eligible for the National Register of Historic Places. Thus, compliance with the Executive Order may overlap with Section 106 and Section 110 of NHPA. Under the Executive Order, Federal agencies managing lands must implement procedures to carry out the directive's intent. Procedures must provide for reasonable notice where an agency's action may restrict ceremonial use of a sacred site or adversely affect its physical integrity. {268} Federal agencies with land-managing responsibilities must provide the President with a report on implementation of Executive Order No. 13007 1 year from its issuance.

Executive Order No. 13007 builds upon a 1994 Presidential Memorandum concerning government-to-government relations with Native American Tribal governments. The Memorandum outlined principles Federal agencies must follow in interacting with federally recognized Native American Tribes in deference to Native Americans' rights to self-governance. {269} Specifically, Federal agencies are directed to consult with Tribal governments prior to taking actions that affect federally recognized Tribes and to ensure that Native American concerns receive consideration during the development of Federal projects and programs. The 1994 Memorandum amplified provisions in the 1992 amendments to NHPA enhancing the rights of Native Americans with regard to historic properties.

7.6.4.1 Executive Order 11593

Section 1. Policy. The Federal Government shall provide leadership in preserving, restoring, and maintaining the historic and cultural environment of the Nation. Agencies of the executive branch of the Government (hereinafter referred to as "Federal agencies") shall: (1) administer the cultural properties under their control in a spirit of stewardship and trusteeship for future generations, (2) initiate measures necessary to direct their policies, plans, and programs in such a way that federally owned sites, structures, and objects of historical, architectural or archaeological significance are preserved, restored, and maintained for the inspiration and benefit of the people, and (3), in consultation with the Advisory Council on Historic Preservation (16 U.S.C. 4701), institute procedures to assure that Federal plans and programs contribute to the preservation and enhancement of nonfederally owned sites, structures, and objects of historical, architectural, or archaeological significance.

The Executive Order requires Federal agencies to administer cultural properties under their control and direct their policies, plans, and programs in such a way that federally owned sites, structures, and objects of historical, architectural, or archeological significance were preserved, restored, and maintained. {250} To achieve this goal, Federal agencies are required to locate, inventory, and nominate to the National Register of Historic Places all properties under their jurisdiction or control that appear to qualify for listing in the National Register. {251} The courts have held that Executive Order No. 11593 obligates agencies to conduct adequate surveys to locate "any" and "all" sites of historic value, {252} although this requirement applies only to federally owned or federally controlled properties. {253} Moreover, the Executive Order directs agencies to reconsider any plans to transfer, sell, demolish, or substantially alter any property determined to be eligible for the National Register and to afford the Council an opportunity to comment on any such proposal. {254} Again, the requirement applies only to properties within Federal control or ownership. {255} Finally, the Executive Order requires agencies to record any listed property that may be substantially altered or demolished as a result of Federal action or assistance and to take necessary measures to provide for maintenance of and future planning for historic properties {256}.

7.6.4.2 Executive Order 13175, November 6, 2000

Executive Order 13175 establishes regular and meaningful consultation and collaboration with Tribal officials in the development of Federal policies that have Tribal implications, to strengthen the United States government-to-government relationships with Indian Tribes, and to reduce the imposition of unfunded mandates upon Indian Tribes. The executive Order applies to all federal programs, projects, regulations, and policies that have Tribal Implications.

E.O. further provides that each "agency shall have an accountable process to ensure meaningful and timely input by Tribal officials in the development of regulatory policies that have Tribal implications." According to the President's April 29, 1994 memorandum regarding government-to-government relations with Native American Tribal Governments, federal agencies "shall assess the impacts of Federal Government plans, projects, programs, and activities on Tribal trust resources and assure that Tribal government rights and concerns are considered during the development of such plans, projects, programs, and activities." As a result, Federal agencies must proactively protect Tribal interest, including those associated with Tribal culture, religion, subsistence, and commerce. Meaningful consultation with the Nez Perce Tribe is a vital component of this process.

On November 5, 2009 President Obama issued a Presidential Memorandum for the Heads of Executive Departments and Agencies. That Memorandum affirms the United States' government-to-government relationship with Tribes, and directs each agency to submit to the Office of Management and Budget (OMB), within 90 days and following consultation with Tribal governments, "a detailed plan of actions the agency will take to implement the policies and directives of Executive Order 13175."

7.6.4.3 U.S. Department of Energy American Indian Policy

On November 29, 1991, DOE announced a seven-point American Indian Policy, which formalizes the government-to-government relationship between DOE and federally recognized Indian Tribes. A key policy element pledges prior consultation with Tribes where their interests or reserved treaty rights might be affected by DOE activities. The DOE American Indian Policy provides another basis for the Cooperative Agreement. The Cooperative Agreement will also serve as an Office of Environmental Management Implementation Plan for the DOE American Indian Policy regarding interactions with the Nez Perce Tribe.

7.6.5 The Roles of the Nez Perce Tribe at Hanford

The Tribe has a duty to protect its reserved treaty rights and privileges, environment, culture, and welfare as well as to educate its members and neighboring public to its activities. The Tribe assumes many different roles. It is a governmental entity with powers and authorities derived from its inherent sovereignty, from its status as the owner of land, and from legislative delegations from the Federal government. The Tribe exercises its powers and authority to serve its members and to regulate activities occurring within the reservation. The Tribe is also a cultural entity and is accordingly charged with the responsibility of protecting and transmitting that culture which is uniquely Nez Perce. The Tribe is also a beneficiary within the context of federal trust relationship with, and obligations to Indian Tribes. The Tribe is a trustee responsible for the protection and betterment of its members and the protection of its and their rights and privileges. The Tribe is also party to treaties between itself and the United States government.

7.6.5.1 Nez Perce and DOE Relationship

The relationship between the Tribe and DOE is defined by the trust relationship that exists between the Federal government and the Tribe, by treaty, federal statute, executive orders, administrative rules, caselaw, DOE's American Indian Policy, and by the mutual and generally convergent interests of the parties in the efficient and expeditious cleanup of the DOE weapons complex, and by the Cooperative Agreement. The structured relationship embodied by the Cooperative Agreement can best be described as a partnership grounded in the site-specific cleanup of Hanford, and extends to all trust-related activities of the Department.

The Tribe sees itself not only as an advisor to DOE, but also as a technical resource available to assist DOE. The Tribe sees its members and employees as a source of technically trained and certified labor for environmental restoration and decontamination and decommissioning work. The continuation of the Cooperative Agreement contemplates an approach that will integrate these and other roles into a comprehensive Nez Perce-DOE program.

The Tribe is asked to review and comment on documents and activities by DOE implicates our Treaty reserved rights and DOE's acknowledgement of other federal statutes, laws, regulations, executive orders, and memoranda governing the United States' relationship with Native Americans and the Nez Perce people. Several Tribal departments lend their respective technical expertise to DOE Hanford issues and present recommendations to the NPTEC, for consideration and guidance. The NPTEC also may requests formal consultation with the federal agency to discuss a proposal or issue further.

7.6.5.2 Consultation with Native Americans

DOE's consultation responsibilities to the Tribe are enumerated generally in the document entitled, Consultation with Native Americans. This policy defines consultation in relevant part:

> "Consultation includes, but is not limited to: prior to taking any action with potential impacts upon American Indian and Alaska Native nations, providing for mutually agreed protocols for timely communication, coordination, cooperation, and collaboration to determine the impact on traditional and cultural lifeways, natural resources, treaty and other federally reserved rights involving appropriate Tribal officials and representatives through the decision making process."

In regard to security clearance, none of the various provisions of the continuation of the Cooperative Agreement shall be construed as providing for the release of reports or other classified information designated as "classified" or "Unclassified Controlled

Nuclear Information" to the Nez Perce Tribe, or as waiving any other security requirements. Classified information includes National Security Information (10 CFR Part 1045) and Restricted Data (10 CFR Part 1016). Unclassified Controlled Nuclear Information is described in 10 CFR Ch. X, Part 1017.

In the event that reports or information requested under the provisions of the continuation of the Cooperative Agreement, while not "classified" or "Unclassified Controlled Nuclear Information," are determined by DOE-RL to be subject to the provisions of the Privacy Act, or the exemptions provided under the Freedom of Information Act, DOE-RL may, to the extent authorized by law, provide such reports or information to the Tribes upon receipt of the Tribe's written assurance that the Nez Perce Tribe will maintain the confidentiality of such data.

References

Bard JC (1997) Ethnographic/Contact Period (Lewis and Clark 1805 – Hanford Engineer Works 1943) of the Hanford Site, Washington. In National Register of Historic Places Multiple Property Documentation Form – Historic, Archaeological and Traditional Cultural Properties of the Hanford Site, Washington. DOE/RL-97-02, Rev. 0, pp. 3.1–3.155, Richland, Washington

Benton County (2009) Benton County Hazard Mitigation Plan – Section 4.3 Windstorms. http://www.bces.wa.gov/windstorms.pdf. Accessed June 18, 2009

Browman DL, Munsell DA (1969) Columbia Plateau prehistory: Cultural development and impinging influences. American Antiquity, 34:249–264

Columbia Basin Fish and Wildlife Authority (2009) 2008 Status of Fish and Wildlife Resources in the Columbia River Basin. Columbia Basin Fish and Wildlife Authority, Portland, Oregon

Campbell NP, Ring T, Repasky TR (1995) 1994 NEHRP Grant Earthquake Hazard Study in the Vicinity of Toppenish Basin, South Central Washington, United States Geologic Survey NEHRP Award Number 1434 94 G-2490

DOI (1994) Hanford Reach of the Columbia River: Comprehensive River Conservation Study and Environmental Impact Statement - Final. Volumes I and II. U.S. Department of Interior, Washington, D.C

Greider T (1993) Aircraft noise and the practice of Indian medicine: The symbolic transformation of the environment. Human Organization, 52(1):76–82

Harper BL, Harris SG (1999) A "Reference Indian" for use in Radiological and Chemical Risk Assessment, CTUIR

Harper BL, Harris SG (2000) Using Eco-Cultural Dependency Webs in Risk Assessment and Characterization. Environ Sci Pollut Res 7(Special 2): 91–100 Federal Register. 36(23): 1271–1329

Harris SG, Harper BL (1997) A Native American Exposure Scenario. Risk Anal 17(6):789–795

Hicks B (2004) Marmes Rockshelter: A Final Report on 11,000 Years of Cultural Use. Pullman: Washington State University

Landeen D, Pinkham A (1999) Salmon and His People: Fish and Fishing in Nez Perce Culture. Lewiston, Idaho: Confluence Press

Nez Perce Tribe (2005) End-State Vision, Resolution NP-05-411, Sept 27, 2005 (Revised 2009)

Repasky TR, Campbell NP, Busacca AJ (1998) "Earthquake Hazards Study in the Vicinity of Toppenish Basin, South-Central Washington" United States Geologic Survey NEHRP Award Number 1434-HQ-97-GR-03013

Relander C (1986) Drummers and Dreamers. Caldwell, ID

Roberts N (1998) The Holocene: An Environmental History. Blackwell, Oxford

Chapter 8
Amchitka Island: Melding Science and Stakeholders to Achieve Solutions for a Former Department of Energy Site

Joanna Burger, Michael Gochfeld, Charles W. Powers, and David S. Kosson

Contents

Abstract Traditional scientific research proceeds from development of a hypothesis, through data gathering to final conclusions, and without much input from stakeholders. This chapter proposes that the melding of scientists and stakeholders throughout the process can reduce conflicts and lead to acceptable solutions for problems that are inherently complex and have eluded resolution. We use the closure of the Department of Energy's Amchitka Island, where three underground nuclear tests were conducted from 1965 to 1971, as a case study to illustrate how stakeholders

---

J. Burger (✉)
Division of Life Sciences, Environmental and Occupational Health Sciences Institute (EOHSI), Consortium for Risk Evaluation with Stakeholder Participation (CRESP), and Rutgers University,
604 Allison Road, Piscataway, NJ 08854, USA
e-mail: burger@biology.rutgers.edu

J. Burger (ed.), *Stakeholders and Scientists: Achieving Implementable Solutions to Energy and Environmental Issues*, DOI 10.1007/978-1-4419-8813-3_8,

can be included as participants throughout the process, leading to acceptance and incorporation of the science, and a path forward. Success was dependent upon interactions to stimulate relevant science investigations, in a participatory process. Without such inclusion, well-intended policies and practices may be ineffective and may not lead to a solution, particularly to such difficult problems as closure of chemical and radioactive waste sites, and the handling of civilian and military nuclear wastes in the future, both of which influence the future of nuclear energy in the United States.

8.1 Introduction

Before moving forward with more extensive, commercial nuclear power, the public and public policy makers need assurances that the current nuclear legacy can be managed, particularly legacy wastes remaining from weapons productions. Following the end of the Cold War, the United States and other nations were faced with cleaning up nuclear and chemical wastes, a difficult task in terms of time, costs, and public attitudes.

In the United States, the Department of Energy (DOE), responsible for this cleanup, faced a distrustful and wary public suggesting a comprehensive and compelling rationale for the inclusion of stakeholders in the decision-making process (NRC 1994, 1995, 2000; DOE 1994; Leslie et al. 1996; PCCRARM 1997). Despite colossal expenditures, very few DOE sites have been cleaned up or closed down. DOE faced a crisis in which the annual cost of maintaining radioactive waste in place was so high that funds for actual remediation were limited. This has been likened to a long-term mortgage where most of the payment covers only interest. These costs need to be reduced and more money needs to be spent on cleanup and closure, rather than marking time. The Nation is facing a crisis in public policy, management, and research with nuclear wastes. Partly this crisis reflects failure to meld stakeholder needs with science, technology, and public policy.

The Nation faces two distinct nuclear waste problems, dealing with the cleanup of past nuclear wastes (the legacy) from both domestic nuclear power and military buildup, and the possible siting of new commercial nuclear energy facilities (with their attendant wastes), and both require public support and inclusion in the decision-making process. Public concern partly revolves around the potential for nuclear accidents, such as Chernobyl (Jeschki 1989), and the recent disaster in Fukushima, Japan, where the probability may be low, but the adverse consequences are high. Fukushima was particularly worrisome because it was caused by a natural disaster (rather than being human error), and the full extent of the radionuclide contamination is still unkown. The public is understandably risk averse.

To address the question of both past and future nuclear waste, governmental agencies must address the divergent concerns of managers, regulators, scientists and the public, as well as those of Native Americans and Alaskan Natives (Slovic 1987, 1993; Slovic et al. 1991; Flynn et al. 1994). Relationships between scientists and nonscientists must be forged that are based on trust, mutual respect, and a true willingness to modify research to address specific needs (Rhoads et al. 1999).

But more importantly, open relationships and communication channels must be forged between DOE and its stakeholders (including Native Americans) within a framework of both sound science and the public's needs. The difficulty is partly that the public does not have an active role in framing the questions, selecting the scientific information needed to address the questions to their satisfaction, and in shaping the research and analysis itself (PCCRARM 1997). Thus, stakeholders and Native Americans/Alaska Natives tend to be suspicious of proposed solutions. We suggest that stakeholder involvement throughout the entire process can lead to buy-in, therefore moving the country toward closure of some of the legacy waste sites, as well as dealing with civilian nuclear waste. Such participatory research leads to better decisions.

8.1.1 Objectives for This Chapter

This chapter reports on a multiyear, stakeholder-driven process that led to providing the science and mechanisms for closure of DOE's Amchitka Island, the site of three underground nuclear test from 1965 to 1971, including the largest U.S. underground test (Kohlhoff 2002; Burger et al. 2005, 2007a). We describe a consensus process that was iterative and interactive, that began in an environment of distrust, anger, frustration, and disagreement over what to do about Amchitka Island, and what "closure" even meant. The process included inviting divergent groups to express their viewpoints and concerns, developing and funding of the Amchitka Science Plan, conducting expeditions to Amchitka that collected biota for radionuclide analysis, and developing a biomonitoring plan that served as the basis for a long-term stewardship plan for Amchitka (Fig. 8.1). At every step, mechanisms had to be found to meld very different views into a widely agreed upon path forward, which in turn led to development of new mechanisms to address the next problem. This chapter illustrates that science can form the basis for iterative, interactive, and novel methods of solving large-scale environmental problems. While the specific issues and mechanisms may differ among DOE sites, and among different energy sources, the process can form a blueprint for breaking the log-jam when stakeholders do not trust or accept environmental policy decisions.

8.1.2 The Department of Energy

Following the ending of the Cold War in 1989, the DOE created Environmental Management (EM) to remediate nuclear and hazardous wastes, render sites safe for future use, and allow the Federal government to de-accession large areas of land and other facilities, creating assets out of liabilities. It has been over 20 years since its creation, yet very few sites have been cleaned up, restored, and put to productive uses, despite the expenditure of billions of dollars. Amchitka Island was one of over 100 sites in 34 states that comprise the DOE's "Complex" (Barke and Jenkins-Smith 1993; Crowley and Ahearne 2002). Some of DOEs sites required cleanup and closure,

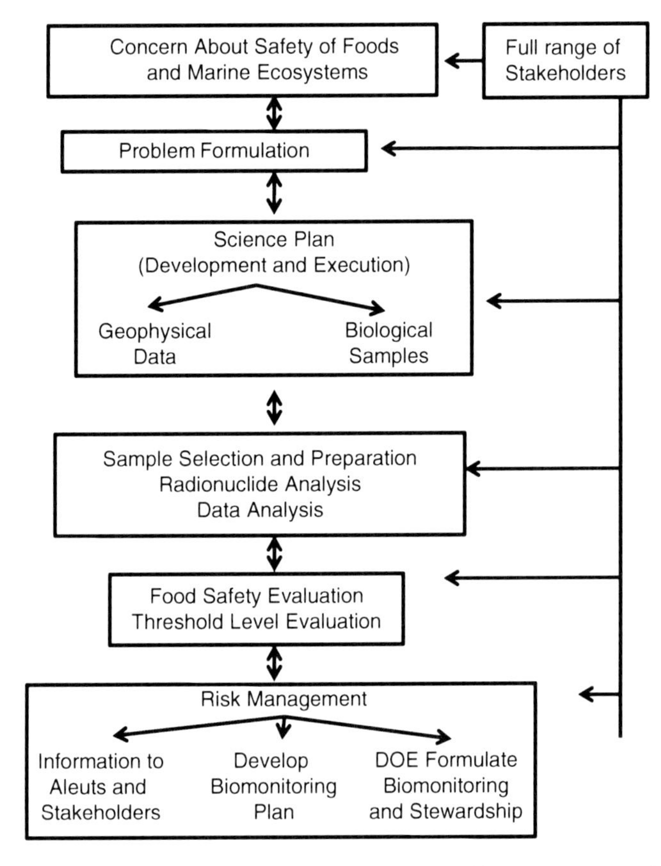

Fig. 8.1 Overall process of involving stakeholders in reaching management decisions for Amchitka Island

while others are slated for future missions. Several other chapters in this book describe stakeholder involvement in DOE cleanup and risk decisions. The DOE, and therefore the Nation, face a growing challenge to close some of these sites, averting the multibillion dollar maintenance and mortgage costs. Yet, time and time again solutions elude DOE, regulators and managers, who continue to forge either temporary agreements or interim actions. Amchitka Island was a case in point.

8.2 Background on Amchitka Island

Amchitka Island (51°N lat, 179°E long) was the site of three underground nuclear tests (1965–1971) by the Atomic Energy Commission (AEC) and the Energy Research and Development Administration (ERDA), predecessors of the DOE (Fig. 8.2). Responsibility for its cleanup rested with the National Nuclear Security Administration (NNSA). The U.S. Fish and Wildlife Service (USFWS) has landowner

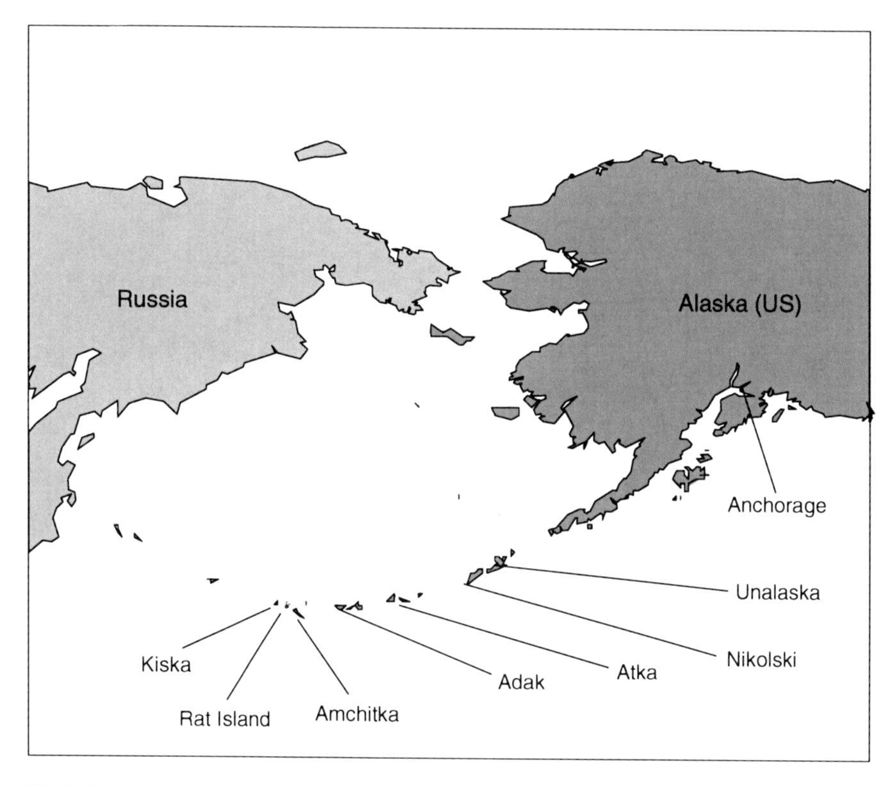

Fig. 8.2 Map showing the location of Amchitka Island, and the relative isolation of the islands where Aleuts live in the Aleutian Chain of Alaska

responsibility for Amchitka Island, which remains part of the Alaska Maritime National Wildlife Refuge.

There was considerable controversy about nuclear testing at Amchitka, including the potential health risks to humans, particularly the local Aleuts, the serious damage to the marine ecosystem, and the possible generation of tsunami activity (Greenpeace 1996; Kohlhoff 2002). The small releases of radiation to the surface from the tests were not considered to pose serious health risks at the time (Seymour and Nelson 1977; Faller and Farmer 1998), partly because most of the radioactive material was probably spontaneously vitrified when the intense heat of the blast melted the surrounding rock (DOE 2002a). The present controversy resulted from increasing concerns about the possibility of subsurface transport of radionuclides from the three cavities to the marine environment in light of the region being one of the most seismically active and dynamic subduction zones on earth (Eichelberger et al. 2002).

One of the primary concerns was whether the subsistence foods of the Aleuts, and the commercial fish and shellfish from the island vicinity, are safe to eat. Dutch Harbor in the Aleutians, the port for commercial fish in the Bering Sea, had the highest tonnage of fish landings in the world in the last several years, and provides 17% of

Alaska's $811 million fish landings (2.3 million metric tons of fish, NOAA 2005). Understanding baseline concentrations of contaminants is particularly important for the Bering Sea region, where there is intense commercial fishing. Over 90% of the world's fish catch comes from 10% of the world's oceans (including the Bering Sea, Waldichuk 1974).

8.2.1 The Controversy

The controversy surrounding Amchitka began with the initial decision to conduct underground nuclear tests at Amchitka (Kohlhoff 2002). In the early 1960s, many of the players described in this chapter had strongly and vocally objected to the siting of the tests at Amchitka in the first place. The State of Alaska and the Aleuts had been particularly vocal. In the interim, the AEC did little to allay fears about the site, to foster alliances, or to ascertain the information (or remediation) needs of different stakeholders. "Stakeholder" was not yet a word in the lexicon on public policy. Public outrage peaked in the months before the Cannikin test in 1971.

During an underground test, intense heat melts adjacent rock, creating a cavity of molten rock (Laczniak et al. 1996). Rapid cooling turns the molten rock into glass, and some of the radioactive material is trapped in the glass, while other radionuclides reside outside the glass, and are potentially mobile (Smith 1995). The resulting glass is subject to slow dissolution in groundwater and to mechanical breakdown, but it retards the rapid transport of chemicals (Kersting et al. 1999; Haschke et al. 2000). Transport of the material depends upon the physical state of the source, local geochemistry, the extent of fractures or fissures, and local hydrology.

In 2001, the DOE removed all structures and remediated the surface contamination, and the surface was closed as part of the Alaska Department of Environmental Conservation's (ADECs) contaminated sites program (DOE 2002a). Although Greenpeace (1996) concluded that surface radionuclide contamination existed, Dasher et al. (2002) did not confirm this. At the same time DOE announced that Amchitka remediation was complete, and years of frustration resurfaced as participants brought to the process their anger over the initial siting, their frustration over what they saw as DOEs lack of concern for subsistence people, commercial fisheries, or ecosystem well-being and health, and DOE's lack of understanding or disregard about the science base that stakeholders felt was necessary to assure peace of mind about seafood, human, and ecosystem health. Chaos reigned.

8.3 Amchitka, Stakeholders, and Solutions

Letters exchanged between the DOE, state officials, other federal agencies, APIA, environmental groups, and other members of the public were contentious, hostile, and demanding. DOE could not understand why everyone could not take its word that there was no risk. DOE offered its sophisticated computer models as a means of showing that there was no risk, and without risk, felt they could depart without future monitoring (DOE 2002a, b). Other stakeholders felt there was a clear need for scientific data to answer a range of biological and geophysical questions about Amchitka, now, and in the future. A range of stakeholders, including the State of Alaska, U.S. Fish & Wildlife Service (USFWS) and Aleut representatives, did not feel that there was enough data about levels of radionuclides in biota to know whether there was any risk. There was no common ground and no clear path forward. Distrust and disagreements about the assumptions and models reigned, and ultimately, the DOE models and human health risk assessment remained in draft form (DOE 2002a, b) and were never finalized or used in the final stewardship plan. No one agreed on a path forward.

8.3.1 Identification of Stakeholders and Their Role

Identification of relevant stakeholders is a key aspect of problem formulation. Although it is often a difficult step, it was easier for Amchitka because the major stakeholders had been involved for over 20 years in heated discussions about the DOE's handling of Amchitka. The major stakeholders were the State of Alaska (ADEC), USFWS, and the Aleut and Pribilof Islanders (APIA, Table 8.1). While these were the formal designated stakeholders, each entity represented specific groups of people or interests, which had to be considered separately as well as jointly. For example, although APIA represents the Aleuts, the Aleuts themselves needed a voice in the process; Aleuts on remote islands relied heavily on subsistence resources (Fig. 8.3). Similarly, although ADEC represented the State of Alaska, there were also human health concerns separate from environmental conservation. And the State of Alaska was responsible for the Aleuts as well as its other citizens.

8.3.2 Formulation of the Problem and Defining a Path Forward

The full range of stakeholders wanted answers to questions of seafood safety (and data to support those answers), and DOE wanted to close Amchitka. By closure, DOE meant that it needed no further action for remediation, and that it would not need to monitor the Amchitka environment in the future. DOE would turn full responsibility for Amchitka to the USFWS (although they still would retain responsibility for radionuclides). To support their contention, DOE funded a complex groundwater modeling effort and also a human health risk modeling (DOE 2002a, b). Both of these had significant limitations, and did not provide assurance to stakeholders who wanted to know whether the foods were safe and the ecosystems free from radionuclides. Basically the issue came down to the following dilemma: DOE

Table 8.1 Key stakeholders involved in formulating the problem, holding a workshop, developing a science plan, and moving forward to closure and long-term stewardship

Agency or group	Role and interests
Aleuts	Indigenous peoples who live on isolated islands in the Aleutian Chain off the coast of Alaska. Their interests are in having safe subsistence foods (including marine mammals, birds [and eggs], fish, shellfish, other invertebrates, algae), and in the health of the marine ecosystem
Other consumers of marine foods	This includes not only other residents of the Aleutian Chain and citizens of Alaska, but the people of the U.S. (a substantial portion of fish sold in the U.S. comes from the Bering Sea)
General public	People who are interested in the health and well-being of the Aleutian Chain and the Bering Sea, including its biota and ecosystem health
Greenpeace	This and other environmental groups have an interest in assuring that the Aleutian Chain are free from harm due to radionuclide exposure from the underground nuclear tests
Aleutian Pribilof Island Association (APIA)	This is a nonprofit Tribal organization of the Aleut people in Alaska providing services including cultural heritage, health, education, social, psychological, employment, vocational training, environment, natural resources, and public safety services. Their interests were in .providing information to the Aleuts about the safety of their subsistence foods, their commercial products, and the marine ecosystem
U.S. Department of Energy (DOE)	Agency responsible for the underground nuclear tests (1965–1971). Interested in closure of Amchitka and transferring the responsibility for radionuclides on Amchitka to their long-term stewardship program
National Nuclear Security Administration (NNSA)	This is the unit within DOE that is responsible for the nuclear test sites, including Amchitka Island. They were interested in finding a path forward to closure for Amchitka
U.S. Fish & Wildlife Service (F&WS)	Land owners of Amchitka, part of the National Wildlife Refuge system. They are interested in overall ecosystem health, radionuclide levels in biota, and ecosystem integrity on Amchitka and surrounding marine environments
National Oceanographic and Atmospheric Administration (NOAA)	Responsible for marine mammals in the Bering Sea and other marine waters. They are interested in radionuclide levels in marine mammals
Alaska Department of Environmental Conservation (ADEC)	The lead Alaskan agency responsible for the health of ecosystems, including the organisms within them. They are interested in the preservation of biota and ecosystems, both terrestrial and marine
Alaska Department of Health and Social Services	The Alaskan agency is responsible for human health. They are interested in possible human effects from exposure to radionuclides, and in seafood safety generally

(continued)

Table 8.1 (continued)

Agency or group	Role and interests
Environmental Groups	Greenpeace, Alaska Community Action on Toxics (ACAT) and others interested in the environmental legacy of DOE on the Aleutian Islands, and in toxics in general
Consortium for Risk Evaluation with Stakeholder Participation (CRESP)	A DOE funded, independent university-based research entity that conducts original research, reviews documents and reports, and involves stakeholders in research and management processes. They are interested in facilitating solutions to environmental problems, involving stakeholders in all processes, and collaborating with stakeholders whenever possible. Their interest in Amchitka was in providing credible science to move forward

Fig. 8.3 Aleuts rely on subsistence foods because they live on remote islands, such as Nikolski (*center*, and right photo)

believed their models showed that there was no risk to humans or the ecosystem, and other stakeholders believed you could not know this without the data to test biota or validate the models. DOE staffers were dismayed and disappointed that the public did not believe them when they said there was no future risk from the residual radiation under Amchitka. The State of Alaska, USFWS, APIA, environmental groups, and other stakeholders did not believe the models, questioned the assumptions underlying them, and wanted further assurance.

A critical mechanism for fostering understanding, if not agreement or consensus, among the different stakeholders was a 2-day workshop held in Fairbanks in February 2002, and organized by the University of Alaska, Fairbanks, and the Consortium for Risk Evaluation with Stakeholder Participation (CRESP).

CRESP is an independent multiuniversity consortium consisting of environmental, biological and social scientists, risk assessors, and public policy analysts who had been working together for nearly 10 years to address environmental and risk problems faced by the DOE. CRESPs involvement had been suggested by staff of the ADEC, who searched the literature and had found CRESP's earlier research examining stakeholder needs, consumption patterns, and contaminant levels in fish or other subsistence foods at the DOE's Savannah River Site (see Chap. 5). The State of Alaska was searching for an independent science team that could formulate the science necessary to address the divergent concerns of the state and other stakeholders.

Considerable groundwork went into both the invitation list and the formal speakers for the Workshop, and CRESP sought the advice of a wide range of state and federal regulators, land owners and managers (USFWS), Aleuts, and other stakeholders. At the Workshop, stakeholders were able to express their views about what ought to be done about Amchitka, what science was needed to answer questions about possible risk to humans and the environment around Amchitka, and what DOEs responsibilities were. While the discussions were sometimes heated and hostile, views were nonetheless aired in an open forum, and everyone expressed their views. Holding such a meeting in Alaska, rather than at NNSA offices in Nevada or in Washington DC made it possible for a range of stakeholders to attend and to be empowered on their own turf.

The major problem facing stakeholders and DOE, which emerged from the workshop, was how to move from contention and disagreements to: (1) reassure the Aleuts, agencies, and the public that the foods from the marine environment around Amchitka were safe to eat, (2) reassure the stakeholders that there was no evidence of harm to the marine ecosystem around Amchitka, and (3) find a solution so that DOE could close Amchitka and move toward long-term stewardship of the island. The Workshop led to discussions between NNSA, DOE officials in Las Vegas, and the State of Alaska (ADEC) about how to proceed (Fig. 8.1), which culminated in the signing of a Letter of Intent (LOI) by the relevant parties in June 2002 that stipulated: (1) closure would leave residual contamination in place, (2) inclusion of four primary stakeholders (ADEC, APIA, USFWS, DOE), (3) development of a Science Plan to provide the science basis for long-term stewardship, (4) designation of CRESP to develop the Science Plan independently, and conduct an expedition under the oversight committee (5) development of a groundwater model by DOE, and (6) continued involvement of stakeholders.

The LOI effectively puts in place a process for moving toward closure that involved a range of primary stakeholders, but it did not assure its success. Success required the development of a number of different mechanisms to address specific aspects of the stakeholder controversies; each subsequent mechanism developed was in direct response to previous concerns and mechanisms. The Science Plan would, when implemented, provide credible scientific data on the geophysical conditions on Amchitka, and on the radionuclide levels in biota, which would in turn allow for closure of Amchitka.

8.3.3 The Multiyear Process

8.3.3.1 The Science Plan

Following the Workshop, and the development of the LOI, the Science Plan was to be developed. The Amchitka Independent Science Plan was developed by CRESP to provide the science necessary to understand fully the biological and geophysical aspects of potential radionuclide seepage at Amchitka. Although the Science Plan was to be developed independently, and to represent the best science necessary to understand possible seepage, transport, receptors, and health risks to humans and the environment, the actual approval of the plan prior to its implementation had to be given to each of the four designated parties (the State of Alaska [ADEC], USFWS, DOE, and APIA).

Deciding on the best science needed to address the concerns of such a wide range of stakeholders was a monumental task. Concerns included food safety, risks to humans consuming foods from the region, the health of marine ecosystems, potential impacts on marine life (invertebrates, fish, birds, marine mammals), pathways and timing of potential seepage from the underground test shots, potential disruptions of radionuclides by earthquakes or volcanoes, and future risks to people and the environment. Even among the scientists there were lively discussions about approaches, types of new information required, and relative weighting of biological vs. geophysical information. And the designated stakeholders watching the process had somewhat divergent views, but all stressed the importance of biota testing. In the end, we in effect proposed that information about radionuclides in biota was to receive the primary emphasis (food safety, marine ecosystem health), with several types of tests to gather geophysical information in hope that it might reduce some uncertainties about pathways and possible transport times by examining both the marine floor and the island massif itself.

The plan was developed in stages, presented to the four primary stakeholders in multiday workshops, revised, expanded, contracted, and presented again (Fig. 8.4). One of the difficult tasks was initial selection of species for the marine assessment, and this phase required input from the full range of stakeholders (Burger et al. 2006). When complete, the plan was evaluated, discussed, and projects ranked in collaboration with parties designated by the LOI, (described in Burger et al. 2005, 2007a). But its success depended upon inclusion of these and other stakeholders throughout in an interactive and iterative process. This included visits to Aleut villages to present the proposed plan and modify it to account for local interests, concerns, and advice. The process thus required the development and approval of a Science Plan; the designation of four primary players in the LOI narrowed the field of discussions, although CRESP, at the request of APIA and out of our concern to really understand how Aleuts understood the effort, continued to meet with Aleuts, agency personnel, and others during Science Plan development (Burger et al. 2007a).

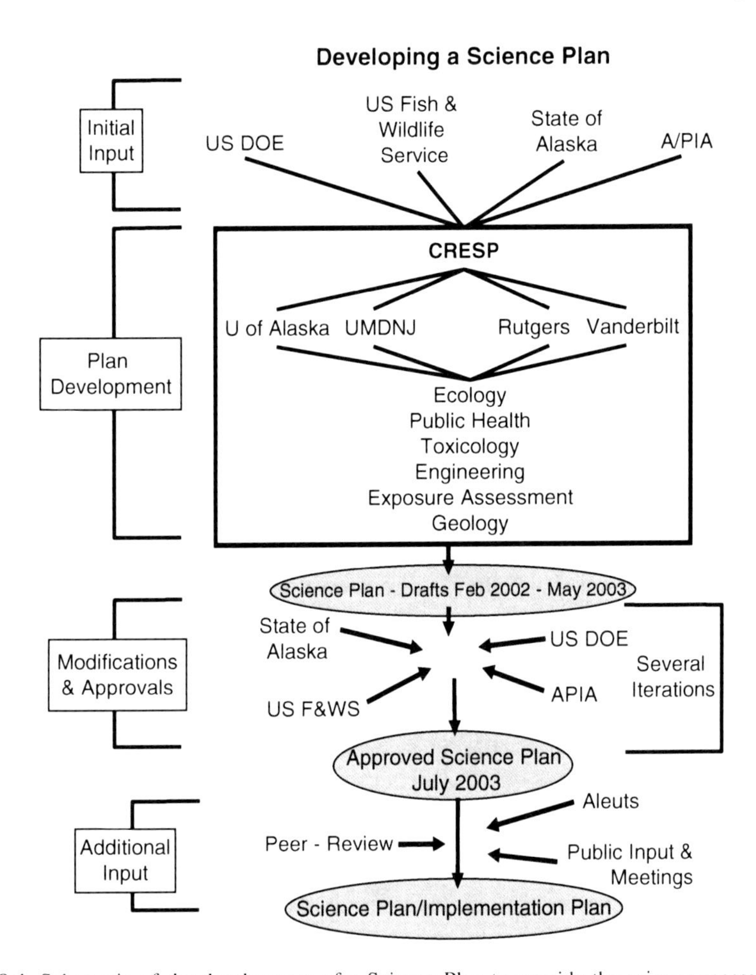

Fig. 8.4 Schematic of the development of a Science Plan to provide the science necessary to assure the Aleuts and other stakeholders (including the State of Alaska and the U.S. Fish and Wildlife Service) that the marine environment and subsistence foods are safe from anthropogenic radionuclides

Stakeholders participated in the process at every stage, and collaborated with the CRESP team on many of the phases (Table 8.2). Meetings with Aleuts in their home villages resulted in several additions to the expeditions, including adding new species and suggesting the addition of Aleuts to the expedition itself. Four Aleuts and representatives of APIA went on the expedition to collect foods in their traditional subsistence methods. Tables 8.3 and 8.4 show some of the additions made to the biological expedition as a function of Aleut and stakeholder collaboration.

Table 8.2 Decision and research phases for the CRESP Amchitka independent science assessment

Problem	Mechanism	Outcome	Stakeholders and participants[a]
Complete lack of coordination, cooperation, and open discussion	Free flowing discussions among interested parties	Recognition that a workshop would facilitate identification of issues, problems and stakeholders	Alaska state government, DOE, and other regulators, managers, scientists, APIA
Recognizing the issues, problems, and stakeholders	Open workshop in Fairbanks	Presentations, direct interactions, and exchange of viewpoints. Workshop proceedings	Scientists, regulators, managers, affected parties, and environmentalists
Finding a path forward	Letter of Intent signed by DOE and State of Alaska	Determined a path of obtaining the science needed to ascertain the risks from the nuclear tests. Stipulated that CRESP should be involved	DOE State of Alaska (ADEC) APIA U.S. Fish & Wildlife Service (USFWS)
Agreeing on the science necessary to move forward	CRESP to develop a science plan. Identified the four parties as necessary signatories	Developed a plan that was modified and approved by the four designated stakeholders after extensive revisions	DOE ADEC APIA USFWS CRESP
Refining the plan and organizing the expedition	CRESP to organize the expedition	Continued refinement of plan, including additional target species, reference site, and visits to Aleut villages; inclusion of stakeholders on expedition	DOE ADEC APIA Aleuts USFWS NOAA Commercial fisheries SeaGrant ACAT CRESP

(continued)

Table 8.2 (continued)

Problem	Mechanism	Outcome	Stakeholders and participants[a]
Conducting the expedition	CRESP to conduct three expeditions[b]	Collected geophysical and biological data, 4,500 pounds of biota	APIA and Aleuts NOAA CRESP
Radiological and data analysis	CRESP to conduct and oversee analyses	Analyzed biota and synthesized data to address questions of food and ecosystem safety	CRESP laboratories, Idaho National Laboratory
Conclusions and report writing	CRESP to write a report for DOE and stakeholders	Developed conclusions and wrote report	CRESP
Dialogue with stakeholders	CRESP to present results to stakeholders and the public	Presentations and discussions with all major stakeholders	DOE ADEC APIA Aleuts USFWS NOAA The general public
Developing a biomonitoring plan	DOE requested CRESP to develop the plan	CRESP wrote a plan, after interactions with primary stakeholders, and presented it to the primary stakeholders for further comments	DOE ADEC APIA USFWS CRESP
Developing long-term stewardship plans	DOE had responsibility for plan development	DOE wrote a plan, using CRESP biomonitoring plan as base	DOE in consultation with primary stakeholders

[a] Primary participants were DOE, ADEC, APIA, USFWS

[b] There were two biological expeditions and one geophysical. This framework largely relates to the biological aspects of the study

Table 8.3 Examples of collaboration in the Amchitka independent science assessment and biomonitoring plan for Amchitka

Step	Stakeholder	Input
Workshop	All parties listed	Expressed their concerns and ideas about what to include in the science plan. Served as the basis for the outline of the science plan research
Science plan	APIA	Suggested focus on subsistence foods
	USFWS	Suggested inclusion of specific ecological receptors
	ADEC	Suggested food chain approach and possible reference sites
	USDOE	Described data necessary for their models and established limits of funding
Expedition refinement (after funding approval, but before expedition)	Aleuts	Suggested that they help collect in their traditional manner (resulted in Aleut participation in the expedition). Also suggested additional subsistence foods to collect
	APIA	Provided mechanism for hiring Aleuts and villagers for the expedition
	USFWS	Helped select the reference site, helped refine species to be collected, provided permits
	ADEC	Helped select reference site, helped refine species list, suggested additional species for inclusion, provided permits
Expedition	Aleuts	Collected subsistence foods in traditional manner. Provided knowledge of subsistence foods and parts eaten. Provided local knowledge of where and how to collect marine species. Helped keep expedition members safe in rough collecting conditions (local knowledge of winds, rain, storms, species behavior and locations)
	APIA	Provided advice on the expedition about Amchitka, Aleuts, and subsistence methods. Representative on board the ocean explorer
	NOAA	Provided a trawling ship and help in collecting commercial fish for analysis. Aged otoliths of Pacific Cod
Data analysis and report writing	Aleuts	Provided information on which species were most important for their subsistence diets
	APIA	Provided corroborative information from other Aleut villages on subsistence diets
	USFWS and ADEC	Provided information on which species they were interested in from a food chain and ecosystem perspective
Biomonitoring plan	Aleuts and APIA	Provided information on which species they wanted to be included based on subsistence lifestyle
	USFWS and ADEC	Provided information on which species fit their needs for biomonitoring and management of Amchitka Islands' natural resources
	General public	Provided information on timing of biomonitoring and mechanism, including concern for volcanic activity
	DOE	Provided input on possible limits and needs of their long-term stewardship plans for closure of DOE sites, including usual monitoring intervals

This provides a sampling of contributions and collaborations that improved the science assessment, and is not meant to be exhaustive

Table 8.4 Collaborative role of stakeholders in the independent science assessment (after Burger et al. 2007a)

Collaboration	USDOE	USF&WS	ADEC	APIA	Aleuts	NOAA or Fishermen
Workshop	X	X	X	X	X	
Formulating the problem	X	X	X	X	X	
Signatory to LOI	X	X	X	X		
Science plan development	X	X	X	X	X	
Refining the science plan						
Selection of a reference site		X	X	X		
Adding NOAA vessel for commercial fish sampling						X
Removal of independent collection by Aleuts on their own vessel				X	X	
Adding species		X	X	X	X	X
Substituting species		X			X	
Deletion of species[a]		X				X
Adding personnel				X	X	X
Intertidal sampling		X			X	
Conducting the expeditions				X	X	X
Refining sampling while on the expedition			X	X	X	X
Refining data analysis	X	X	X	X	X	
Developing biomonitoring plan	X[b]	X	X	X	X	

CRESP is not included because they participated in all phases

[a]Both USF&WS and NOAA discouraged the collection of marine mammals and USF&WS suggested deletion of some birds whose populations were declining

[b]CRESP initially developed a biomonitoring plan based on the data from the expeditions. DOE then subsequently incorporated the major components of this monitoring plan into their long-term stewardship plan for Amchitka (DOE 2008)

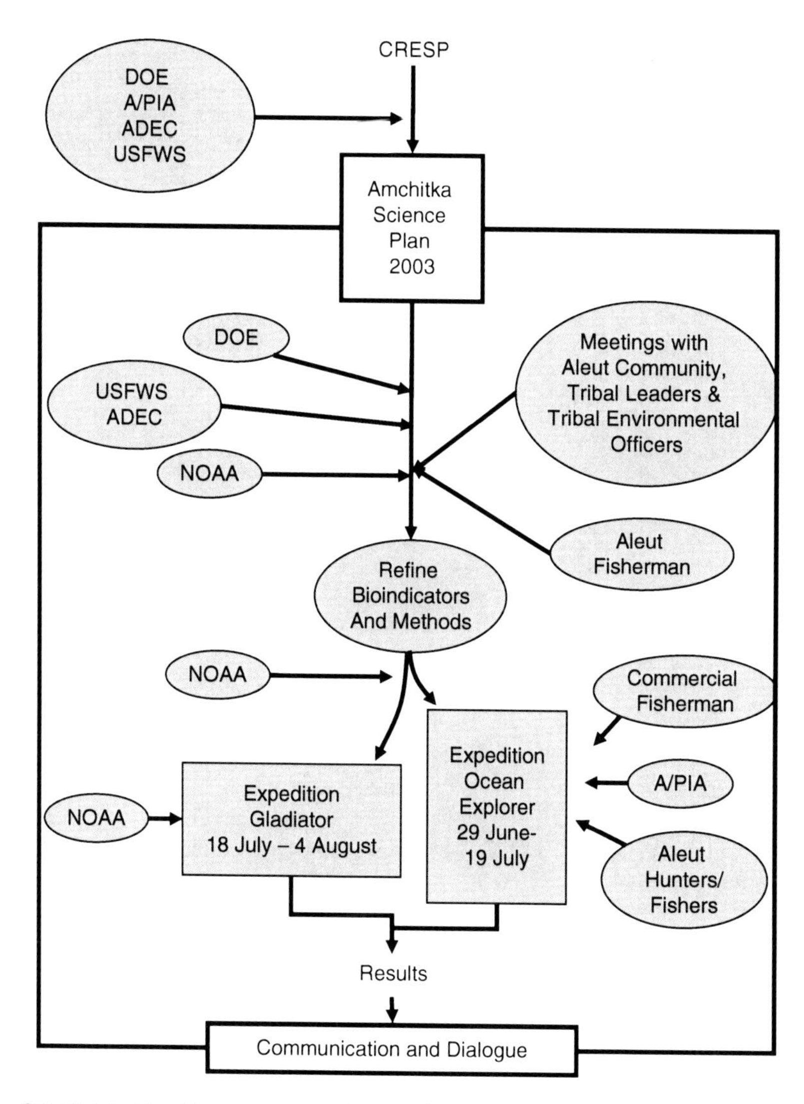

Fig. 8.5 Stakeholder-driven process to select species for collection around Amchitka Island and at a reference site (Kiska Island), conduct the expeditions to collect biota, and provide the results to the Aleuts and other stakeholders

8.3.3.2 Science Plan Execution

The execution of the Science Plan, data analysis and report writing were mainly the responsibility of the CRESP science team, interacting with, integrating ideas and concerns from, and collaborating with, a range of stakeholders during each phase (Table 8.1, Fig. 8.5). A team from the APIA accompanied the scientists on the expedition to Amchitka to provide information on subsistence foods and to collect in their traditional manner, and changes and modifications were made on the expedition itself (Burger et al. 2005).

Another aspect bearing comment is the inclusion of commercial fishermen in the process. In addition to consultation with commercial fishermen in Unalaska (and Dutch Harbor) before and after the expedition, CRESP had a researcher on a second expedition on the Gladiator, a NOAA-operated research vessel (Fig. 8.5). Additional advice and collaboration occurred between the personnel on this vessel and CRESP, and this resulted in collecting fish of commercial interest for radionuclide analysis.

8.3.3.3 Data Analysis and Safety Evaluation

Once the biological material was returned to Rutgers University laboratories for analysis, there were continued discussion about which biological samples to analyze, and a reevaluation of the radionuclides to be analyzed. This allowed continued examination of the overall biological sampling design. Following radionuclide analysis, CRESP analyzed the data and wrote a report, although there were preliminary meetings with the four primary stakeholder groups before the final report was written to report on progress and the material collected. This report was formally presented to the four LOI-named stakeholders in an open public meeting in Anchorage on 1 August 2005. This meeting, also attended by the median and public, resulted in follow-up data analyses and some additional radionuclide analysis in response to Aleut and stakeholder input.

Continuing the dialogue with stakeholders in meetings in Anchorage, Homer, and the Aleut villages of Atka, Nikolski and Unalaska was important in discussing the findings and listening to advice about future biomonitoring and continued concerns about Amchitka. As with CRESP's initial meetings in the Aleut villages, the follow-up meetings to discuss the findings were well attended. Usually nearly the whole village turned out to discuss the findings (Fig. 8.6), and because of the Aleut participation in the expedition itself, they were generally aware of the methods and the expedition itself. In addition to holding meetings with village elders, and the adult community, we met with the children and teenagers of the villages, and they were extremely interested in both the science methods and the findings. Mainly questions and comments related to the information provided about radionuclides and their levels in biota, possible risk from radionuclides, future monitoring to assure subsistence food safety, and an overwhelming concern about mercury and its effects. Many concerns were expressed about the subsequent development of a biomonitoring plan.

The DOE then formally asked CRESP to write a biomonitoring plan to serve as the centerpiece for the long-term stewardship plan stipulated in the LOI (Burger et al. 2007b). The full range of stakeholders endorsed the biomonitoring plan because it was based on sound scientific data and not on computer-generated risk models in the absence of radionuclide data from biota at Amchitka. This biomonitoring plan was used by DOE to develop its long-term stewardship plan (DOE 2008). As with the other steps, development of a biomonitoring plan involved

Fig. 8.6 There was nearly 100% attendance at meetings that occurred between CRESP and the Aleuts in their home villages, both before and after the expedition. Shown is Michael Gochfeld with the children and teenagers of Nikolski

extensive stakeholder involvement, including the following considerations (Burger et al. 2007b):

1. Occurrence at all three test shots and the reference site
2. Different receptor groups of interest as subsistence foods (Aleuts), commercial species (Aleut commercial fishery on Adak, other commercial fisheries out of Dutch Harbor), and food chain nodes (of interest to all stakeholders)
3. Different species groups (plants, invertebrates, fish, birds)
4. Different trophic levels (low to high)
5. An accumulator of one or several radionuclides of interest

8.4 Stakeholder Involvement During All Phases

The stakeholders involved in the steps in the process depended upon the nature of the step (Table 8.2). The process required recognizing and acknowledging the different involvements and responsibilities of the range of stakeholders. The DOE, USFWS, and ADEC have legal resource trustee responsibility for Amchitka Island; the State of Alaska has health responsibilities for potential human exposure at Amchitka; NOAA and ADEC have responsibility for the fisheries and marine mammals; and APIA represented the Aleut interests in subsistence, commercial, and ecological resources of Amchitka. In addition, however, a wide range of other stakeholders, including environmentalists and the general public were involved at key points, such as the Workshop to exchange ideas and concerns about Amchitka, review of the Science Plan itself, and the final report.

8.4.1 The Role of Stakeholders

We thus recognized three different aspects of stakeholder involvement: (1) the inclusion of stakeholders during every phase (Table 8.2, Figs. 8.1. and 8.2), (2) the relative role of different stakeholders during each phase (Table 8.2 and 8.3), and (3) the importance of collaboration rather than simply listening or informing stakeholders (Table 8.3). Stakeholders are often included in the problem formulation stage of risk assessments or environmental evaluations, but are seldom involved in other aspects of the research, analysis, or management options. At Amchitka, stakeholders were involved throughout the process. Inclusion and collaboration requires scientists to have a true willingness to modify research (Rhoads et al. 1999), and where possible, to include stakeholders (i.e., we included Aleuts on the expedition itself).

While Table 8.2 lists the steps in arriving at a path forward and solution for closure of DOE's responsibility for the marine environment at Amchitka, Tables 8.3 and 8.4 illustrate some of the ways that stakeholders collaborated. At all stages, CRESP worked to integrate the diverse needs and concerns of different stakeholders, acknowledging that each stakeholder groups concerns were integral to a viable Science Plan, Expedition, Data Analysis, and Report. Specific examples can be found in Burger et al. (2005, 2009). However, one example will suffice.

The selection of species to collect on the expedition for radionuclide analysis was one of the most crucial aspects of the Amchitka Independent Science Assessment (Burger et al. 2006). While all the major stakeholders had input into the initial species collection list in the approved Science Plan, after approval and funding, we went back to the major stakeholders to seek additional input to refine the collection list. While this step may at first seem to be unnecessary because stakeholders had input into the approved Science Plan list, we submit that once a plan becomes reality, stakeholders focus on the details more clearly, and need additional input to reflect this focus. The primary stakeholders had input into the target species for collection in the following ways: (1) USFWS suggested removal of some species because of population declines (i.e., Red-faced Cormorants, Tufted Puffins, Sea Lions) and added others because of population increases and ecological equivalency (i.e., Pigeon Guillemot, Horned Puffin), (2) Aleuts in villages suggested addition of some subsistence foods of importance to them (i.e., Octopus, Black Katy Chitons), and (3) ADEC suggested adding some marine invertebrates (i.e., Giant Chiton) and adjusted the number of different fish collected because of population or commercial concerns. Thus, although the species list to be collected was approved in the Science Plan, and CRESP scientists could have simply collected those species, CRESP went back to stakeholders to refine the list, and to include species that each stakeholder group felt were important (and necessary for their needs). This ensured that, in the end, everyone approved of the species selected, making bioindicator selection possible. Agencies and groups did not have to revisit the issue of species selection, but instead could let the radionuclide data help inform final bioindicator selection.

Similarly, a wide range of stakeholders made input into the species selected for the biomonitoring plan by CRESP, and this plan was largely used by DOE in its development of its long-term stewardship plan (DOE 2008). Although a wide range of stakeholders deemed the plan suitable and protective of human health and the environment, concerns were expressed that stronger oversight and continued research were essential to address remaining uncertainties (Benning et al. 2008).

8.4.2 The Role of Science

To CRESP, the disagreements among stakeholders revolved around obtaining the necessary science to assure the public that, with respect to radionuclides, the human foods obtained from the marine environment around Amchitka were safe to consume, the food chain was uncontaminated, and the marine environment was unimpaired, now and in the future. Stakeholders wanted to be sure that foods and the marine environment were not contaminated with radionuclides derived from the test site. The science necessary to address these assurances included understanding of the geophysical conditions under Amchitka and in the surrounding seafloor, and analyzing biota representative of the food chain in the marine environment around Amchitka, including subsistence and commercial foods. Assessing the food chain required examining radionuclide levels in organisms at all trophic levels, from plants and invertebrates, through shellfish and fish, to top predatory fish and birds. Models, in the absence of site-specific radionuclide data from Amchitka were never going to be acceptable. CRESP recognized, however, that the path forward required a melding of science and approaches of academics, agencies, and Aleuts. Further, the data gathered needed to be sufficient to provide the basis for future monitoring of Amchitka, an integral part of any long-term stewardship plan for Amchitka, and essential for DOE's departure from Amchitka.

8.4.3 Stakeholder: Science Collaborations and Conflict Resolution

The steps outlined (Table 8.2, Fig. 8.1) required 5 years to fully develop (from initial Workshop to exchange of ideas and concerns, to the development of a biomonitoring plan). While in the past, DOE and other agencies have tried to shortcut this long and seemingly cumbersome process by some exposure data gathering and model-building to assure peace of mind about the hazards to human and ecological receptors, we contend that such assurance can be achieved only by investing substantial time, effort, and money into providing the science base that stakeholders need to assure such peace of mind. Such an investment, however, can achieve solutions that lead to closure rather than more rounds of partial-solutions.

In the long run, the costs of conducting a complete science assessment that is agreeable to the parties concerned, although large, are less than continued expenditures of money to maintain the status quo without moving toward a solution. The cost of development and execution of the Amchitka Science Plan by CRESP was just over three million dollars. The DOE's models, lacking site-specific data from Amchitka, failed to provide the peace of mind about either food safety or ecosystem health necessary to move forward, and in some minds, enhanced suspicion. The models produced only a stalemate because the legally responsible parties did not agree to accept them as a basis for closure. In contrast, the process of assessing stakeholder concerns, and addressing them directly, as occurred in this example of Amchitka Island, resulted in a mutually agreeable solution. Site-specific data on radionuclide levels in the food chain provided the necessary peace of mind about food safety, and served as a basis for future biomonitoring to ensure early warning of any potential risks.

8.5 Conclusions: Melding Science and Stakeholders

The complexity of environmental problems, particularly at large-scale contaminated sites requiring restoration, remediation, and long-term stewardship increasingly requires consensus building, iterative science, and interactive dialogue with interested and affected parties (Burger et al. 2005, 2007a). The potential for true collaboration must be exploited; performance interactions should be the goal of stakeholder involvement, rather than mere capacity for communication (Fischoff 1995). The DOE simply must find ways to move forward the cleanup and closure of their EM sites. The Nation can no longer afford the huge mortgage costs involved in maintaining the status quo. Further the public's willingness to consider additional nuclear energy hinges, in part, on some assurance that the federal government knows how to maintain and contain the nuclear wastes.

In the case of Amchitka, DOE proposed relying on models to assure the public that there was no human or ecological risk from leakage from the underground nuclear tests (Fig. 8.7, top). DOE felt that conducting a groundwater model and screening risk assessment (DOE 2002a, b) should be sufficient, given that the natural resource

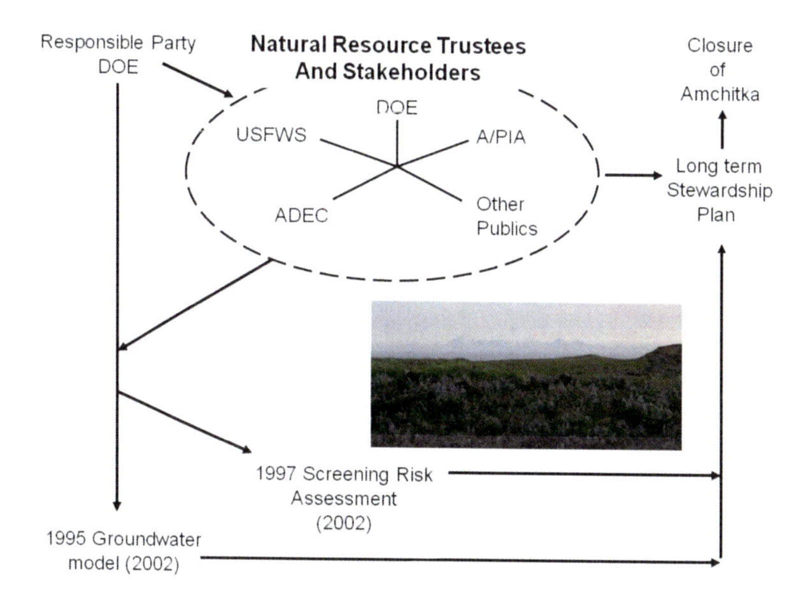

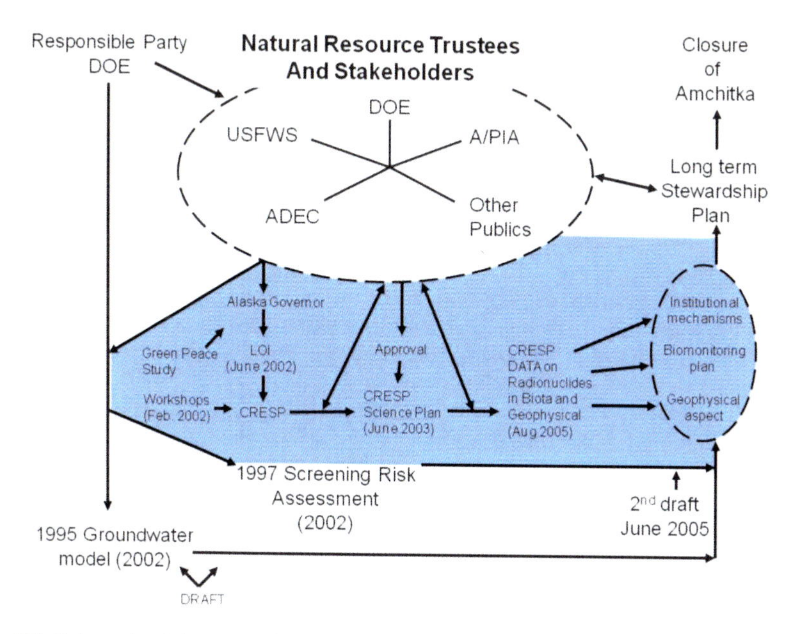

Fig. 8.7 Schematic of the approach to closure initially taken by DOE, and the final "stakeholder-collaborative" reported in this chapter. The blue area on the bottom figure denotes the processes put into place by CRESP to include stakeholders and obtain sufficient science to move forward.

trustees and stakeholders had an opportunity to respond to the two documents. Leaving aside whether answering questions and responses of stakeholders is the same as addressing their concerns, the public and other agencies remained unconvinced.

The public, APIA, and the State of Alaska questioned the assumptions of the models, required assurances that there was no risk, and believed that only data on radionuclide levels in biota could provide that assurance. The production of complex groundwater and human health risk models, although attractive and often believed to be cost-effective, did not lead to a solution. The cost of conducting the Amchitka screening risk assessment and ground water model was several million dollars, but was not sufficient for closure. The development and review of the models required several years from when they started working on the models or risk assessment to production of the documents, while the development of a stakeholder-approved Science Plan, and the gathering and analysis of data required only 4 years from the signing of the LOI to the development of a biomonitoring plan. Further, the models did not satisfy the stakeholders, and thus did not lead to a solution allowing DOE closure of Amchitka.

We suggest that working with stakeholders throughout the process led to agreement on a path forward, identification of the science needed to solve the problem, acceptance of the final science assessment report, and inclusion of the biomonitoring plan in a long-term stewardship plan that ultimately provided public confidence in the closure of Amchitka Island (Fig. 8.7, bottom). Amchitka was turned over to DOE's Office of Legacy Management in 2006, and that office developed a draft of long-term surveillance and maintenance plan for Amchitka (DOE 2008). EM, and public policy deriving from that research, is greatly improved by stakeholder involvement and collaboration throughout the process. We suggest that collaborative involvement affords all parties the opportunity to establish ownership, strengthen the research, and ensure that sound science is used in management and public policy decisions.

Acknowledgments We thank the many people who participated throughout this process, including scientists, managers, regulators, environmentalists, and the general public. Special thanks goes to L. Bliss, M. Greenberg, and B. Friedlander who were critical throughout the process, to the following from the primary stakeholders: Anne Morkill and Gregory Siekaniek (U.S. Fish & Wildlife Service), John Halverson (DEC, State of Alaska), Pete Sanders (U.S. DOE), Robert Patrick (APIA), and the people in the villages of Adak, Atka, Nikolski and Unalaska in the Aleutian Chain. This research was funded by the Consortium for Risk Evaluation with Stakeholder Participation (CRESP) through a grant from DOE (DE-FG 26-00NT 40938) to Vanderbilt University and Rutgers University. JB and MG were also partially supported by NIEHS ESO 5022, Wildlife Trust, and the Nuclear Regulatory Commission (NRC-38-07-502M02). The results, conclusions, and interpretations reported herein are the sole responsibility of the authors, and should not in any way be interpreted as representing the views of the funding agencies.

References

Barke RP, Jenkins-Smith HC (1993) Politics and scientific expertise: scientists, risk perception, and nuclear waste policy. Risk Anal 13:425–439

Benning JL, Barnes DL, Burger J, Kellye JJ (2008) Amchitka Island, Alaska: moving toward long-term stewardship. Polar Rec 45:1–14

Burger J, Gochfeld M, Kosson DS, Powers CW, Friedlander B, Eichelberger J, Barnes D, Duffy LK, Jewett SC, Volz CD (2005) Science, policy, and stakeholders: developing a consensus science plan for Amchitka Island, Aleutians, Alaska. Environ Monage 35:557–568

Burger J, Gochfeld M, Jewett S (2006) Selecting species for marine assessment of radionuclides around Amchitka: planning for diverse goals and interests. Environ Monit Assess 123:371–391
Burger J, Gochfeld M, et al (2007a) Scientific research, stakeholders, and policy: continuing dialogue during research on radionuclides on Amchitka Island, Alaska. Environ Manage 85: 232–244
Burger J, Gochfeld M, Kosson DS, Powers CW (2007b) A biomonitoring plan for assessing potential radionuclide exposure using Amchitka Island in the Aleutian Chain of Alaska as a case study. J Environ Radioact 98:315–328
Burger J, Gochfeld M, Pletnikoff K (2009) Collaboration versus communication: The Department of Energy's Amchitka Island and The Aleut Community. Environ Res 109:503–510
Crowley KD, Ahearne, JF (2002) Managing the environmental legacy of U.S. nuclear-weapons production. Am Scient 90:514–523
Dasher D, Hanson W, et al (2002) An assessment of the reported leakage of anthropogenic radionuclides from the underground nuclear test sites at Amchitka Island, Alaska, USA to the surface environment. J Environ Radioactiv 60:165–187
Department of Energy (DOE) (1994) How to design a public participation program. U.S. DOE, Office of Intergovernmental and Public Accountability (EM-22). Washington
Department of Energy (DOE):DRAFT (2002a) Modeling groundwater flow and transport of radionuclides at Amchitka Island's underground nuclear tests: Milrow, Long Shot, and Cannikan. DOE/NV-11508-51, Nevada Operations Office, Las Vegas
Department of Energy (DOE):DRAFT (2002b) Screening risk assessment for possible radionuclides in the Amchitka marine environment. DOE/NV-857, Nevada Operations Office. Las Vegas
Department of Energy (DOE) (2008) Long-term surveillance and maintenance plan for U.S. DOE Amchitka, Alaska site. DOE-LM/1121-2008
Eichelberger JC, Freymueller J, Hill G, Patrick M (2002) Nuclear Stewardship: Lessons from a not-so-remote Island. Geotimes 47:20–23
Faller SH, Farmer DE (1998) Long-term hydrological monitoring program: Amchitka, Alaska. EPA-402-R-98-002, U.S. Environmental Protection Agency. Washington
Fischoff B (1995) Risk perception and communication unplugged: twenty years of process. Risk Analysis 15:137-145. Risk Anal 15:137–145
Flynn J, Slovic P, Mertz C (1994) Decidedly Different: Expert and Public Views of Risks from a Radioactive Waste Repository. Risk Anal 6:643–648
Greenpeace (1996) Nuclear flashback: the return to Amchitka. Greenpeace, USA
Haschke JM, Allen TH, Morales LA (2000) Reaction of plutonium dioxide with water: formation and properties of PuO_{2+}. Science 287:285–287
Jeschki W (1989) The Chernobyl experience. Int J Radiat Applic Instrument 34:279–283
Kersting AB, Efurd DW, et al. (1999) Migration of plutonium in groundwater at the Nevada Test Site. Nature 397:56–59
Kohlhoff DW (2002) Amchitka and the bomb: Nuclear testing in Alaska. University of Washington Press Seattle, WA
Laczniak RJ, Cole JC, Sawyer DA, Trudeau DA (1996) Summary of hydrologic controls on ground-water flow at the Nevada Test Site, Nye County, Nevada, US Geological Survey Resources Investigations report 96-4109, pp 59
Leslie M, Meffe GK, Hardesty JL, Adams DL (1996) Conserving biodiversity on military lands: A handbook for natural resources managers. The Nature Conservancy, Arlington
National Research Council (1995) Improving the Environment: an evaluation of DOE's Environmental Management Program. National Academy Press, Washington
National Research Council (NRC) (1994) Building Consensus through Risk Assessment and Management of the Department of Energy's Environmental Remediation Program. National Academy Press, Washington
National Research Council (NRC) (2000) Long-term institutional management of U.S. Department of Energy legacy waste sites. National Academy Press, Washington

NOAA (National Oceanic and Atmospheric Administration). Dutch Harbor – Unalaska, in Alaska, Top U.S Port for landings in 2004. NOAA report 04-096 (2005) Available at: www.nmfs.noaa.gov/docs/04-096_top_ports.pdf. Accessed 26 May 2006

PCCRARM (1997) Risk Assessment and Management in Regulatory Decision-Making. Presidential/Congressional Commission on Risk Assessment and Risk Management. U.S. Government Printing Off, Washington

Rhoads BL, Wilson D, Urban M, Herricks EE (1999) Interaction between scientists and nonscientists in community-based watershed management: emergence of the concept of stream naturalization. Environ Manage 24:297–308

Seymour AH, Nelson VA (1977) Radionuclides in air, water, and biota. pp. 579–613 in The environment of Amchitka Island, Alaska (M. L. Merritt and R. G. Fuller, eds.). Report TID-26712, Technical Information Center, Energy Research and Development Administration, Washington

Slovic P (1987) Perception of risk. Science 236:280–287

Slovic P (1993) Perceived risk, trust, and democracy. Risk Anal 13:675–682

Slovic P, Layman M, Flynn J (1991) Lessons from Yucca Mountain. Environment 3:7–11, 28–30

Smith DK (1995) Characterization of nuclear explosion melt debris. Radiochim. Acta 69: 157–167

Waldichuk M (1974) Coastal marine pollution and fish. Ocean Manage 2:1–60

Chapter 9
Decommissioning of Nuclear Facilities and Stakeholder Concerns

James H. Clarke, Joanna Burger, Charles W. Powers, and David S. Kosson

Contents

Abstract The decommissioning of nuclear facilities provides good examples of stakeholder issues and concerns and approaches to resolution. Nuclear power stations, licensed by the Nuclear Regulatory Commission (NRC), typically strive for license termination conditions that would permit unrestricted use of the site so that, in some cases, reutilization of the site would be possible for a specific use. Former nuclear weapons production facilities are decommissioned to a variety of end states, consistent with the ongoing mission of the site. In all cases, stakeholder concerns must be

J.H. Clarke (✉)
Department of Civil and Environmental Engineering, Consortium for Risk Evaluation with Stakeholder Participation (CRESP), Vanderbilt University, 2301 Vanderbilt Place, Nashville, TN 37235, USA
e-mail: james.h.clarke@vanderbilt.edu

J. Burger (ed.), *Stakeholders and Scientists: Achieving Implementable Solutions to Energy and Environmental Issues*, DOI 10.1007/978-1-4419-8813-3_9,

factored into the decision and the overseeing agencies, the NRC, the Department of Energy (DOE), the Environmental Protection Agency (EPA), and State oversight agencies must be sensitive to their concerns.

In this chapter, we provide an overview of decommissioning activities, the types of facilities undergoing decommissioning and the different regulatory frameworks. Desired end states are discussed along with stakeholder concerns and issues. Examples of drivers and constraints for major decommissioning decision factors are examined. We end with a selected case study – the decommissioning of the Big Rock Point nuclear power station and lessons learned from this and other decommissioning activities.

9.1 Introduction

The legacy of the Cold War has left us with a number of facilities, including buildings that are no longer needed and must be decommissioned. The challenge of decommissioning such buildings safely is daunting and difficult, particularly when these buildings are contaminated with radionuclides or other hazardous substances. Decontamination and decommissioning[1] require a range of safety measures for both workers and the public, and must adhere to federal and state laws and regulations. Decommissioning of surplus facilities is particularly daunting for the Department of Energy (DOE) given the large number of facilities that were involved in nuclear weapons production. There are commercial nuclear facilities that face decommissioning and decontamination as well, and chemical facilities that face a similar challenge, including those that handle radionuclides e.g., uranium conversion facilities. Thus, consideration of the role of stakeholders in the decommissioning process is the key to finding viable, cost-effective, and consensual resolutions to the overall challenge of decommissioning.

Decommissioning and decontamination is of interest to site neighbors and other interested and affected parties. Whereas, such decontaminations and removals of buildings were once considered solely an agency or industry responsibility, it is now clear that involving stakeholders will improve the process, leading to better decisions.

9.1.1 Objectives for This Chapter

In this chapter, we look at the decommissioning of nuclear facilities and the importance of stakeholder involvement in these activities. We define "decommissioning" to include those activities, e.g., deactivation, decontamination, demolition, that are employed to achieve a particular end state for the facility or site e.g., unrestricted use, restricted use, or interim safe storage.

This chapter provides the following:

- Definitions for fundamental decommissioning activities and end states.
- Information concerning the regulatory frameworks under which decommissioning is conducted.

- Identification of major stakeholders and their concerns.
- Information on a case study in which all parties were engaged participants in the process and the outcome met the objectives of all parties including the stakeholders.

9.1.2 Stakeholders

We use the term "stakeholders" in the broadest sense to include any government agency, private entity, group, or individual interested in a decommissioning approach and potential outcomes (see Chap. 1). Specific concerns that are common to all stakeholders include the end state itself and acceptable land uses after decommissioning, the degree to which the decommissioning and resultant end state are protective of human health and the environment, and the cost-effectiveness of the effort. Failure to engage stakeholders in the discussions that lead to decommissioning decisions not only delays the process but can also generate adversity that could have been avoided if stakeholders were engaged early and throughout the decommissioning process.

9.2 Types of Nuclear Facilities

In this chapter, the term nuclear facilities is also used broadly to include nuclear power stations and other facilities e.g., complex materials sites, licensed by the Nuclear Regulatory Commission (NRC) and former nuclear weapons production facilities managed by the DOE.

9.2.1 Nuclear Power Stations and Complex Materials Sites Licensed by the NRC

Currently, there are approximately 104 commercial nuclear power reactors, at 64 plant sites, that are licensed by the NRC. Thirteen are permanently shut down and in some stage of decommissioning (NRC 2009). Through January 2008, ten additional nuclear power plants had completed decommissioning and had their licenses terminated (NRC Fact Sheet on Decommissioning Nuclear Power Plants 2010).

Also, as reported in the Status of the Decommissioning Program, Annual 2009 Report (NRC 2009), 18 complex materials sites are undergoing decommissioning as well. These include facilities, other than nuclear power stations, that operate under an NRC license e.g., the West Valley Demonstration Site in New York State that reprocessed spent fuel for a limited period, Kerr McKee in Cimarron OK, and Mallinckrodt Chemical in St. Louis Missouri.

Information about decommissioning and license termination activities for nuclear reactors, complex materials sites, and other types of facilities that are licensed by the NRC, such as fuel cycle facilities, can be found in the "Status of the Decommissioning Program" reports issued annually by the NRC.

Finally, it should be noted that there are approximately 37 "Agreement States" that oversee decommissioning activities in their state with the exception of nuclear power plant decommissioning. The decommissioning of nuclear power plants is overseen by the NRC.

The regulatory requirements for the process that the NRC follows to oversee the decommissioning of nuclear facilities that they license are given in the License Termination Rule (LTR). The LTR is described in 10CFR20 Subpart E Radiological Criteria for License Termination.

The following definitions are provided by the NRC in their on-line glossary of terms (http://www.nrc.gov).

Decommissioning The process of safely closing a nuclear power plant (or other facility where nuclear materials are handled) to retire it from service after its useful life has ended. This process primarily involves decontaminating the facility to reduce residual radioactivity and then releasing the property for unrestricted or (under certain conditions) restricted use. This often includes dismantling the facility or dedicating it to other purposes. Decommissioning begins after the nuclear fuel, coolant, and radioactive waste are removed. For additional information, see Decommissioning of Nuclear Facilities and Find Sites Undergoing Decommissioning (http://www.nrc.gov).

Decontamination A process used to reduce, remove, or neutralize radiological, chemical, or biological contamination to reduce the risk of exposure. Decontamination may be accomplished by cleaning or treating surfaces to reduce or remove the contamination; filtering contaminated air or water; subjecting contamination to evaporation and precipitation; or covering the contamination to shield or absorb the radiation. The process can also simply allow adequate time for natural radioactive decay to decrease the radioactivity.

Under the LTR, two types of decommissioning release options are available: unrestricted release and restricted release for which there are legal restrictions on future uses of the land with accompanying requirements for financial assurance and institutional controls. NRC retains oversight for the decommissioning of nuclear power stations, even in Agreement states. Agreement states, however, can oversee the decommissioning of complex materials sites, in addition to regularly overseeing noncomplex decommissioning.

Decommissioning guidance for nuclear power plants and complex materials sites can be found in the following references, (NRC, 2006a, b) and (NRC 2003a, b). Radiological criteria for License Termination can be found in the Code of Federal Regulations (10CFR20 Subpart E).

9.2.2 Former Nuclear Weapons Production Facilities That Will Be Decommissioned by the Department of Energy

The Office of Environmental Management of the DOE uses the term D&D (Deactivation and Decommissioning) where:

Deactivation includes the removal of radioactive materials and hazardous chemicals and the shutting down of process systems and equipment. Active security and utility systems are shutdown as well and personnel are relocated.

Decommissioning is used to mean the steps following deactivation that are taken to achieve the desired end state.

The number of DOE facilities, located at former nuclear weapons production sites, that are identified as surplus facilities and are yet to be decommissioned as of the completion of fiscal year (FY) 2009, is staggering – approximately 3,500 at an estimated cost of $20 to 30 billion (Collazo et al. 2010). This total includes over 1,000 nuclear and radiological buildings (see definitions given herein).

Approximately 1,800 facilities have been decommissioned through FY 2009 (see Department of Energy Office of Environmental Management Congressional Budget Request for FY 2011). The Office of Environmental Management (EM) places surplus facilities into one of three categories as follows: nuclear facilities, radioactive facilities, and industrial facilities. Definitions for these facility types are (see the DOE EM website):

Nuclear facilities are buildings that have contained nuclear materials that warranted robust designs both in structural mass and system redundancies. Examples are reactor buildings, separation process canyons, nuclear process testing buildings, and nuclear materials fabrication buildings. Disposition requires detailed characterization, rigorous planning, application of appropriate technologies, and thousands of man-hours of work in highly hazardous conditions.

Radiological facilities are buildings that handled radioactive materials whose types or quantities allowed less robust structures and systems. These are typically radiological support operations such as laboratories, test facilities, or waste handling and storage buildings.

Industrial facilities are buildings and structures that provide nonradiological support operations. These can have a wide variety of hazards including radioactive, chemical, and physical. These facilities include ancillary structures whose inventory, contamination, and other factors do not require categorization as nuclear or radiological hazards.

The Office of EM uses sixteen performance measures to track its environmental restoration and waste management progress. Three of these performance measures refer directly to facility decommissioning for the facility types discussed earlier. Progress with respect to decommissioning completions, is reported in the annual EM Congressional Budget Requests. Figures 9.1–9.3 are taken from the EM Congressional Budget Request for FY 2011 and show facility decommissioning progress, by number of facility completions, for each facility type.

Note that this approach to reporting progress does not distinguish among facilities of a particular category. Rather, the total number is used as the measure of progress raising the question of how priorities are set and the extent to which risk is a factor in priority setting. A university consortium led by Vanderbilt University, the Consortium for Risk Evaluation with Stakeholder Participation (CRESP), is currently engaged in a study for the DOE to develop a risk-informed approach to setting D&D priorities and has presented its preliminary findings (Clarke et al. 2010).

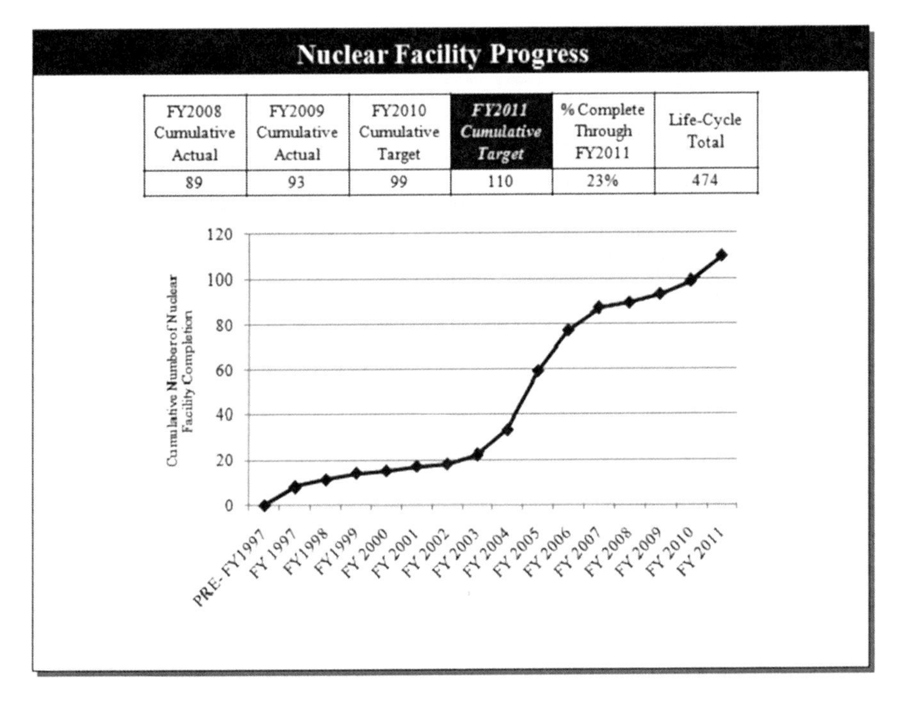

FY2008 Cumulative Actual	FY2009 Cumulative Actual	FY2010 Cumulative Target	*FY2011 Cumulative Target*	% Complete Through FY2011	Life-Cycle Total
89	93	99	110	23%	474

Fig. 9.1 Nuclear facility decommissioning progress

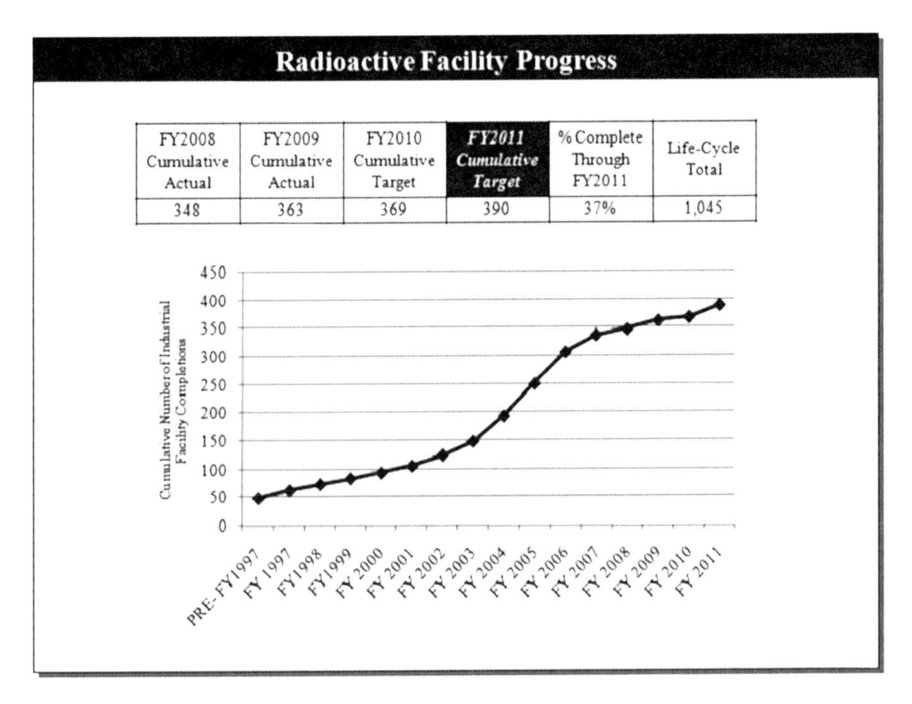

FY2008 Cumulative Actual	FY2009 Cumulative Actual	FY2010 Cumulative Target	*FY2011 Cumulative Target*	% Complete Through FY2011	Life-Cycle Total
348	363	369	390	37%	1,045

Fig. 9.2 Radioactive facility decommissioning progress

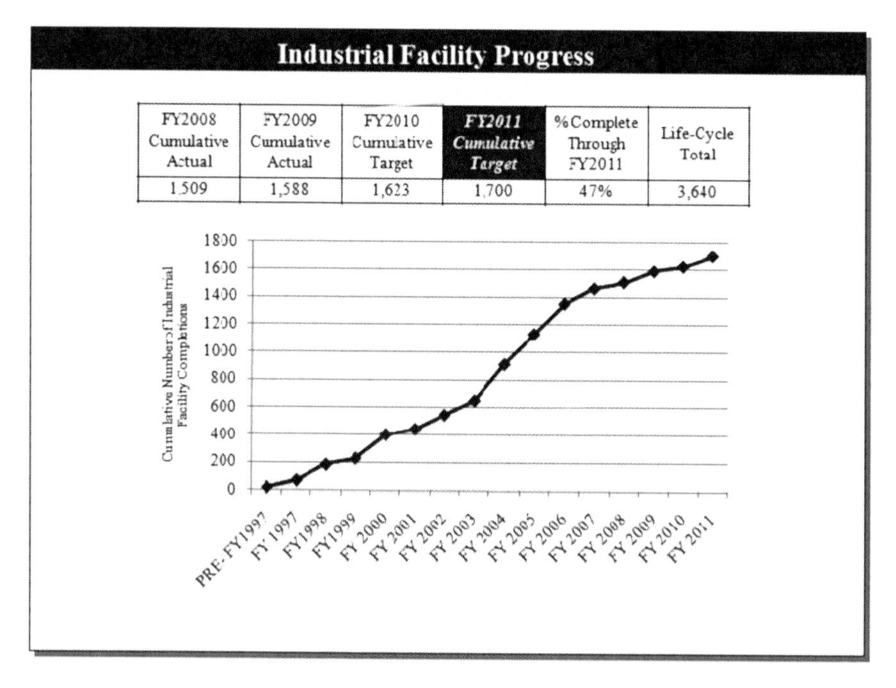

Fig. 9.3 Industrial facility decommissioning progress

Facilities managed by the DOE are often decommissioned under the Compre hensive Environmental Response Compensation and Liability Act (CERCLA) as a "nontime critical removal action."

9.3 Decommissioning End States

9.3.1 Facilities Licensed by the NRC

In their glossary, the NRC defines types of end states that could result, either permanently or temporarily, from the decommissioning of nuclear facilities. Additional information is provided in the NRC Fact Sheet on Decommissioning of Nuclear Power Plants.

9.3.1.1 NRC Decommissioning End States

DECON A method of decommissioning in which structures, systems, and components that contain radioactive contamination are removed from a site and safely disposed at a commercially operated low-level waste disposal facility, or decontaminated to a level that permits the site to be released for unrestricted use shortly after it ceases operation. Under DECON (immediate dismantlement), soon after the nuclear facility closes, equipment, structures, and portions of the facility containing radioactive contaminants are removed or decontaminated to a level that permits release of the property and termination of the NRC license.

SAFSTOR A method of decommissioning in which a nuclear facility is placed and maintained in a condition that allows the facility to be safely stored and subsequently decontaminated (deferred decontamination) to levels that permit release for unrestricted use. Under SAFSTOR, often considered "delayed DECON," a nuclear facility is maintained and monitored in a condition that allows the radioactivity to decay; afterwards, it is dismantled.

ENTOMB A method of decommissioning in which radioactive contaminants are encased in a structurally long-lived material, such as concrete. The entombed structure is maintained and surveillance is continued until the entombed radioactive waste decays to a level permitting termination of the license and unrestricted release of the property. During the entombment period, the licensee maintains the license previously issued by the NRC.

To date, the ENTOMB option has never been used. In some cases, a combination of the first two choices may be selected and ideally, stakeholders will have an opportunity to provide input that can affect the decisions concerning which end states are selected for a given situation. Some portions of the facility may be dismantled or decontaminated while other parts of the facility are left in SAFSTOR. This combination may be necessary when waste disposal options or spent fuel storage facilities are not available.

Nuclear power reactors and materials sites licensed by the NRC can meet the standard for unrestricted release if the decommissioning can achieve a dose limit of 25 mrem/year for all contaminants over all appropriate exposure pathways. It is NRC's policy to use realistic exposure scenarios that reflect expected land use and occupancy. Unrestricted release may be achieved through total or partial demolition, the latter being the end state of choice, if some of the site buildings are targeted for reuse. Figures 9.4 and 9.5 provide pre-decommissioning and the post-decommissioning photos for the Maine Yankee Nuclear Power Station. All facilities were demolished and all demolition debris was disposed of off-site. The nuclear power station at Rancho Seco, however, will leave some of the facilities intact for future industrial use (Watson et al. 2010).

The LTR provides the optional end state of restricted release, but typically the goal, for nuclear power stations and materials sites, is unrestricted release. For restricted release, the dose criterion is 25 mrem/year, but with restrictions on future use, and a dose criterion of 100/500 mrem/year, if institutional controls restricting land use are no longer in effect.

Also, licensing requirements require that facility owners enter into financial assurance mechanisms throughout the licensing period, so that when decommissioning is desired, financial resources, specified in 10CFR50.75(c), are available. Finally, the LTR contains provisions for stakeholder notification and engagement and 10CFR20.1403 requires licensees submitting decommissioning plans (DPs) for restricted release to document how stakeholders were engaged before the DP is submitted.

Fig. 9.4 The Maine Yankee nuclear power station before decommissioning (source: Watson et al. 2010)

Fig. 9.5 The site of the former Maine Yankee nuclear power station after decommissioning (source: Watson et al. 2010)

9.3.1.2 Stakeholder Opportunities

For nuclear power reactors, several opportunities are provided for public involvement during the decommissioning process. A public meeting is held in the vicinity of the facility after submittal to the NRC of a post-shutdown decommissioning activities report (PSDAR) for reactors or a DP for nonreactor facilities. Another public meeting is held when NRC receives the license termination plan (LTP). An opportunity for a public hearing is provided prior to issuance of a license amendment approving the LTP or any other license amendment request. In addition, when NRC holds a meeting with the licensee, members of the public are allowed to observe the meeting, except when the discussion involves proprietary, sensitive, safeguards, or classified information (taken from the NRC Fact Sheet on Decommissioning of Nuclear Power Plants).

For materials sites, upon receipt of a DP from a nonreactor licensee, the NRC is required to issue notifications to local and state governments in the vicinity of the site, any Tribal Nation or other indigenous people that have treaty or statutory rights that could be affected by the decommissioning and the Environmental Protection Agency when the decommissioning will not achieve unrestricted release (10CFR20.1405). 10CFR 20.1403 requires licensees submitting DPs for restricted release to document how stakeholders were engaged before the DP is submitted.

Also, while there is no requirement to notify and engage stakeholders before a DP is submitted for unrestricted release, it is the opinion of the NRC that all complex site decommissioning is "of public interest" and, as a practice, the NRC holds public hearings for all licensees. *In our opinion, there is always merit to early stakeholder engagement.*

9.3.2 *Facilities Managed by the DOE*

There are several DOE facilities decommissioning end states. The following end state definitions are taken from a document entitled *Facility Deactivation & Decommissioning Appendix B-D&D Project Basics* (DOE 2009).

9.3.2.1 Types of DOE Decommissioning End States

Long-Term Min-Safe Storage: This is a minimum safe, low-cost surveillance, and maintenance storage or "moth-balled condition." All significant quantities of nuclear material are removed and (the) facility is deactivated by shutting down the active safety and utility systems.... Not a final end state. Allows prioritization flexibility to delay final disposition costs.

Partial D&D to Cocoon: All significant quantities of nuclear material and waste are removed and facility support buildings and utility systems are demolished. The facilities central massive shield/containment structure is integrated with a new

Fig. 9.6 Initial reactor state (source: Hannah 2010)

containment "cocoon" structure for passive containment of highly radioactive portions over a minimum 50–100 year life with minimum monitoring. Not a final end state … allows hazard reduction through natural radiological decay.

Partial D&D to In Situ Disposition (ISD): Similar to cocooning but is considered a final end state. ISD (entombment) is designed to take advantage of robust contaminated structures and grouted cavities to ensure "encapsulation" of chemicals/radioactive materials as required by performance assessments…. Minimum 1,000-year life assumed in design.

Complete D&D All Structures: All structures and waste are removed. This is the most costly, but least controversial, end state. Two end states meet the complete D&D all structures definition:

Brownfield-end state: Property is retained for limited, controlled (limited access or industrial) use.

Greenfield-end state: property is cleaned to condition supporting unlimited reuse and can be released for private use.

In most cases, buildings are demolished, but for reactors and some other nuclear facilities such as spent fuel reprocessing canyons, an alternative end state may be selected. In these cases, components may be removed and disposed of, but a portion of the building is isolated for some period of time to take advantage of radionuclide decay time. Good examples of these approaches include the D and H-reactors along the 100 area Columbia River corridor of the Hanford site (Partial D&D to Cocoon option) and the planed ISD for the P-and R-reactors at the Savannah River site. Figures 9.6 and 9.7 depict the reactors at the Savannah River Site before and after ISD is completed.

Fig. 9.7 Reactor end state for the ISD decommissioning of the P-and R-reactors at the Savannah river site (source: Hannah 2010)

9.3.2.2 Stakeholder Opportunities

While the regulatory framework for the decommissioning of facilities managed by the DOE differs from that of the NRC, stakeholders interested in the decommissioning of facilities managed by the DOE have opportunities for engagement as well, e.g., through DOE Site Advisory Boards and public meetings. As expected, they share many of the same concerns that stakeholders for NRC decommissioning activities have.

9.4 What Are the Major Stakeholder Concerns for Decommissioning?

As discussed earlier, there are differences between the decommissioning of NRC-licensed nuclear power stations and the decommissioning of former nuclear weapons production facilities managed by the DOE. These differences include sources and availability of funding, e.g., costs borne by the licensee as opposed to funding from the Congressional budget, and unrestricted release from the license compared to regulatory requirements in many DOE decommissioning cases for ongoing monitoring, maintenance, and institutional controls depending on the decommissioning end state. NRC licenses terminated under restricted release conditions have requirements for ongoing monitoring, maintenance, and institutional controls as well.

In many cases, the stakeholder agencies and groups and stakeholder interests and concerns are the same or similar. Tables 9.1 and 9.2 provide information concerning

Table 9.1 Major stakeholders in decommissioning decisions and their interests and concerns decommissioning of nuclear power reactors and other NRC-licensed facilities

Agency or group	Interests and concerns
Nuclear Regulatory Commission	Decommissioning meets the requirements of the license termination rule Funds are available (licensee has financial assurance requirements) Decommissioning is conducted safely and desired end state is achieved Decommissioning is risk-informed Final end state is protective of human health and the environment
State Regulatory Agencies	Final end state is protective of human health and the environment Property is reutilized in a productive way
U.S. Environmental Protection Agency (if not unrestricted release)	Final end state is protective of human health and the environment
Potentially Affected Community	Final end state is protective of human health and the environment Property is reutilized in a productive way

Table 9.2 Major stakeholders in decommissioning decisions and their interests and concerns: decommissioning of DOE managed former nuclear weapons production facilities

Agency or group	Interests and concerns
Department of Energy	Decommissioning is conducted safely and desired end state is achieved Decommissioning is risk-informed and cost-effective Foot print reduction is achieved Final end state is protective of human health and the environment. Funds are available
U.S. Environmental Protection Agency (if decommissioning is conducted under CERCLA)	Decommissioning complies with regulations Final end state is protective of human health and the environment
Affected Communities	Minimum decommissioning waste remains on site Final end state is protective of human health and the environment
Tribal Nations (for sites on Native American land)	Final end state is protective of human health and the environment Risk assessments incorporate exposure scenarios specific to Native Americans e.g., differences in diet from non-Native Americans
Taxpayers in General	Decommissioning is risk-informed, cost-effective, and protective of the environment

stakeholders and their interests and concerns for the decommissioning of facilities licensed by the NRC and former nuclear weapons production facilities managed by the DOE, respectively.

9.5 What Are the Drivers and What Are the Constraints in Decommissioning?

Selections of remediation approaches for contaminated sites, typically are based on a determination of "what" should be done e.g., placing an engineered cover over a landfill; excavating waste and contaminated soil for either on-site or off-site management in an engineered disposal cell; or pumping and treating contaminated ground water.

However, decommissioning decisions depend not only on what e.g., demolition vs. interim safe storage, but also on "when" – now, later or possibly never. Also, while, ideally, the decision should be based on risk and cost, "exogeneous" factors, external to the decommissioning process, often impact the decision both positively and negatively (Clarke et al. 2010). Examples include the availability of funds and stakeholder concerns about the desired end state. Figure 9.8 provides examples of drivers and constraints for all of the major decision factors – risk, cost, and exogeneous.

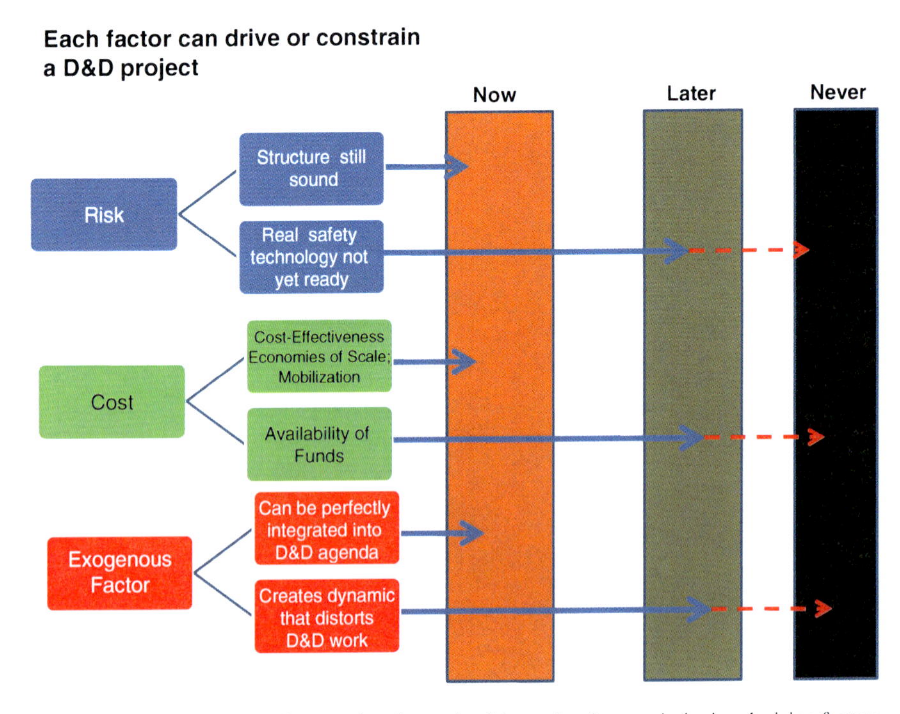

Fig. 9.8 Example drivers and constraints for each of the major decommissioning decision factors (source: Clarke et al. 2010)

In many respects, the decommissioning of an NRC-licensed facility is relatively straightforward, especially for nuclear power reactors, compared to decommissioning activities at DOE sites. NRC license requirements include a financial assurance provision intended to ensure that, when the time for decommissioning came, funds would be available. The mandated decommissioning funds for reactors do not include the funds needed for management of spent fuel, however.

There are exceptions for some NRC-licensed sites, notably complex sites such as the West Valley site in New York State where a commercial facility reprocessed spent nuclear fuel for a limited time and DOE has been given authority for the cleanup. Another example is the Shieldalloy site in New Jersey where large quantities of the slag from an alloy production process accumulated on-site. There is a desire on the part of many of the stakeholders that this material be removed and managed off-site and the licensee does not have sufficient funds for removal. In both cases, strong opposition by the States to restricted release, together with the very high costs of removal of materials continue to be challenges for completing the decommissioning process.

The DOE, on the other hand, relies on Congressional budget approvals to establish funds that can be used for decommissioning. A review of past Congressional budget requests reveals that decommissioning of surplus DOE facilities always has had the lowest priority and, as noted previously, is not risk-informed. Consequently, exogeneous factors can play a deciding role. For example the American Recovery and Reinvestment Act enabled many D&D projects at DOE sites to go forward. However, many D&D projects have been delayed resulting in concerns about the safety of aging buildings in poor repair and the cost of their upkeep.

9.6 Decommissioning of the Big Rock Point Nuclear Power Station: A Case Study

The Big Rock Nuclear Power Plant was the first nuclear power station to be licensed in the State of Michigan and the fifth to be licensed in the U.S. The plant was located near Charlevoix, Michigan and operated by Consumers Energy. The plant featured a 67-MW boiling water reactor and operated from 1962 until 1997. The plant was shut down on August 29, 1997 and on September 19, 1997, the NRC was notified that the plant would be decommissioned using the DECON (demolition) option. As, is the case with all nuclear reactor decommissioning, spent fuel would be removed and stored in an interim spent fuel storage installation (ISFSI) until the DOE could accept the fuel. Figures 9.9 and 9.10 provide photos of the Big Rock Power Station site before and after decommissioning

Major stakeholders included the NRC, the Utility – Consumers Energy, State and local governments, and the citizens of Charlevoix County.

A major issue in the decommissioning process is always the disposition of the demolition debris. If all of it must be managed as low-level waste, only a limited number of disposal facilities are available and, depending on the location of the

Fig. 9.9 The big rock point nuclear power station before decommissioning (source: Watson et al. 2010)

Fig. 9.10 The site of the big rock power station after decommissioning (source Watson et al. 2010)

decommissioning site, transportation costs can be a considerable addition to the disposal costs. Originally, given the cost of transporting all of the demolition waste to a low-level disposal site, the Big Rock Point utility wanted on-site disposal of the decommissioning waste. However, on-site disposal of demolition debris was not acceptable to the State of Michigan and other stakeholders, who wanted the site returned to a "Greenfields" condition.

As it turned out, the radiological activity of a significant portion of the demolition waste was low and this material was accepted for disposal in a solid waste landfill, located in the State of Michigan, at greatly reduced cost. The remainder went to a commercial low-level radioactive waste disposal facility.

The end result was a win-win for everyone. The savings that resulted from disposal of a portion of the waste in a Michigan state solid waste disposal facility made it possible to remove all the decommissioning waste to off-site waste disposal locations.

Throughout the decommissioning planning and execution process, the utility, Consumers Energy, actively engaged the stakeholders and brought them into the process. This approach is commendable and, in our opinion, is critical to a successful outcome in which all parties are satisfied with the decision.

9.7 Lessons Learned

Experience with decommissioning projects has provided many lessons learned that apply to the design and operation of new facilities. Of particular importance is the minimization of potential releases to soil and groundwater through improved waste management practices and emphasis on early detection when releases do occur. The NRC compiles information about lessons learned that is available through its website http://www.nrc.gov/about-nrc/regulatory/decommissioning/lessons-learned.html.

This website also provides information on lessons learned from the DOE.

With respect to the decommissioning process itself, early and continued engagement of stakeholders in the decision-making process is critical to success, whether or not there are regulatory requirements to do so. Stakeholders have legitimate concerns about the ultimate decommissioning end state and the degree to which it is safe and protective of human health and the environment. Cost-effectiveness is a factor as well, especially for DOE activities that are funded by Congressional appropriations. Tribal nations have additional concerns, as well, that reflect dietary practices and cultural values.

In summary, decommissioning experience has shown that, as in any major decision-making process that can affect stakeholders both positively and negatively, bringing in stakeholders early in the formative stages and keeping them actively engaged throughout the decision-making process, is the best way to ensure that the ultimate decisions are good decisions for all affected parties.

Acknowledgments We acknowledge support from the Department of Energy through Grant Number DE-FRC01-06EW07053 to the Consortium for Risk Evaluation with Stakeholder Participation

We acknowledge helpful discussions with James Shepherd and Robert Johnson from the Nuclear Regulatory Commission; Wade Whitaker and Ray Hannah (DOE Savannah River Site); Matt McCormick (DOE Hanford Site); Moses Jaraysi and Kurt Kehler (CH2M CPR Group at Hanford); and Paula Kirk and Andy Szilagyi (DOE EM Headquarters).

References

Clarke JH, Powers CW, and Kosson DS (2010) Development of a Risk-Informed Approach to Setting D&D Priorities. American Nuclear Society Topical Meeting and Technology Expo on Decommissioning, Decontamination & Reutilization, Idaho Falls, ID August 2010

Code of Federal Regulations, Title 10 Part 20 SubpartE Radiological Criteria for License Termination.

Collazo YT, Szilagyi AP, Frush SA et al (2010) Office of Deactivation and Decommissioning and Facility Engineering 2010 Prioritized Technology Initiatives to Improve D&D Operations., Waste Management 2010

Department of Energy, Office of Environmental Management, Congressional Budget Request for FY 2011. http://www.cfo.doe.gov/budget/11budget/Content/FY2011 Highlights. pdf Accessed 24 August 2010

DOE (2009) Facility Deactivation & Decommissioning Appendix B-D&D Project Basics, (http://www.em.doe.gov/EM20Pages/Presentations.aspx) Accessed 24 August 2010

Hannah R (2010) P Area Operable Unit (PAOU) and R Area Operable Unit (RAOU) Update, presented by Ray Hannah to the Savannah River Site Citizens Advisory Board, October 5, 2010

NRC (2009) Status of the Decommissioning Program, Annual Report, http://www.nrc.gov/reading-rm/doc-collections/commission/secys/2009/secy2009-0167/2009-0167scy.pdf Accessed 24 August 2010

NRC (2006a) NUREG 1757, Consolidated NMSS Decommissioning Guidance v1 rev.2 Decommissioning Process for Materials Licensees, U.S. Nuclear Regulatory Commission, Washington, DC 20555–0001

NRC (2006b) NUREG 1757, Consolidated NMSS Decommissioning Guidance v2 rev. 1 Characterization, Survey, and Determination of Radiological Criteria, U.S. Nuclear Regulatory Commission, Washington, DC 20555–0001

NRC (2003a) NUREG 1757, Consolidated NMSS Decommissioning Guidance v3 Financial Assurance, Recordkeeping, and Timeliness, U.S. Nuclear Regulatory Commission, Washington, DC 20555–0001

NRC (2003b) NUREG 1700, rev. 1 Standard Review Plan for Evaluating Nuclear Power Reactor License Termination Plans, U.S. Nuclear Regulatory Commission, Washington, DC 20555–0001

NRC Fact Sheet on Decommissioning Nuclear Power Plants, (2010) http://www.nrc.gov/reading-rm/doc-collections/fact-sheets/decommissioning.html Accessed 1 September 2010

Watson B et al (2010) Power Reactor Decommissioning – Regulatory Experience From Trojan to Rancho Seco and Plants In-Between. American Nuclear Society Topical Meeting and Technology Expo on Decommissioning, Decontamination & Reutilization, Idaho Falls, ID August 2010

Chapter 10
PSEG's Estuary Enhancement Program: An Innovative Solution to an Industry Problem

John H. Balletto and John M. Teal

Contents

J.H. Balletto (✉)
ARCADIS U.S., Inc., 8 South River Road, Cranbury, NJ 08512-3698, USA
e-mail: John.Balletto@arcadis-us.com

J. M. Teal
Professor Emeritus Woods Hole Oceanographic Institution, Teal Partners, 567 New Bedford Road, Rochester, Mass 02770-4116, USA
email: teal.john@comcast.net

J. Burger (ed.), *Stakeholders and Scientists: Achieving Implementable Solutions to Energy and Environmental Issues*, DOI 10.1007/978-1-4419-8813-3_10,

Abstract The Salem Generating Station is a nuclear power plant along the Delaware Bay in New Jersey that uses once through cooling. At the 1990 NJ NPDES permit renewal (required for operating), New Jersey DEP decided to require PSEG, the facility owner, to build cooling towers to address egg and larval fish loss in the cooling system that had been permitted a decade earlier. Rather than building cooling towers, PSEG proposed mitigation, part of which included salt marsh restoration to increase fish nursery habitat in Delaware Bay. In the years since the plant was first proposed in 1966, environmental education and awareness had expanded such that PSEG realized more public involvement would be needed than in the past. Hence the company implemented an outreach and education program that informed stakeholders.

10.1 Introduction

10.1.1 Power Generation

Electricity is a vital component of modern society, which is severely handicapped when its supply is interrupted (Makansi 2007). Most electricity is generated with steam power plants based upon the Rankine Cycle, a thermodynamic process that converts heat to work. A steam electric generation power plant (Fig. 10.1) uses a heat source to create steam to drive a turbine that turns an electric generator. After the steam passes through the turbine, it is condensed to create a vacuum increasing the pressure differential across the turbine and the efficiency of the electric generation. This condensation is accomplished by the plant's cooling system (Weston 1992).

10.1.2 Power Plant Cooling Systems

The cooling system consists of the condenser and associated pumps and piping. Water used to condense the steam while it is in the condenser generally is withdrawn from surface water sources such as lakes, rivers, and oceans. Power plants can require large volumes of water, many exceeding 100 million gallons per day (MGD) (378,500 m^3/day) (FR 67(68)17135). A screening process is usually installed upstream of the pump intake to protect the pump from potentially large debris in the waterbody. In addition to debris, various life stages of aquatic organisms are involved. An aquatic

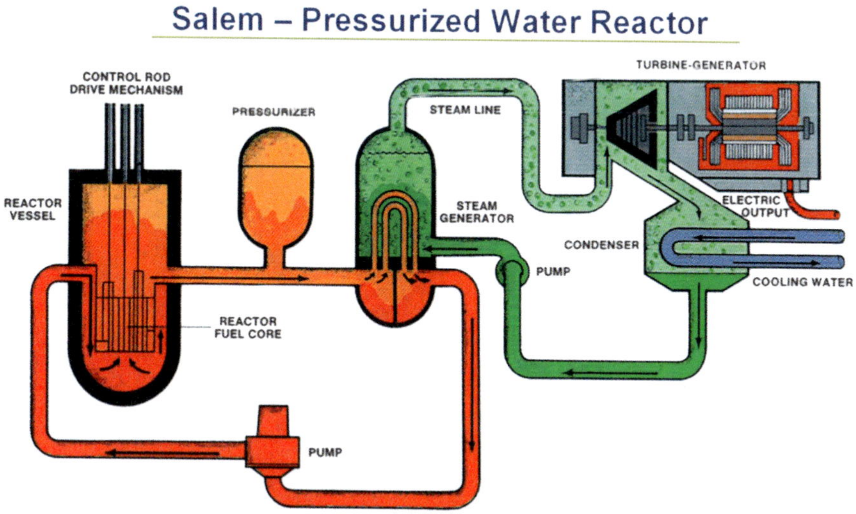

Fig. 10.1 Schematic of water systems at a pressurized water reactor nuclear electric generating station

organism drawn into the intake structure can either be retained by the screens (referred to as impingement) or, if small enough to pass through the screens, pass through the cooling system (referred to as entrainment). An aquatic organism that is impinged may be killed by either being pressed against the screen and suffocating or being washed off the screen into a collection basket and placed into a land fill with collected debris. Organisms that are entrained can be killed from thermal, pressure, or chemical (biofouling control agents) exposure. The effects of power plant cooling systems on aquatic organisms have been a hotly contested issue resulting in numerous regulations and court cases, recently reaching the US Supreme Court (Entergy Corp. v. Riverkeeper, Inc., et al. 556 U.S. 07–588).

10.2 Salem Generating Station

Salem Generating Station (Salem) is located in Lower Alloways Creek Township, Salem County, NJ, USA (Fig. 10.2) on the eastern shore of the Delaware River Estuary (Delaware Estuary) at River Mile (RM) 50. Salem is located in the mesohaline portion of the estuary that is used by the typical suite of mid-Atlantic estuarine species (Able et al. 2007). The edges the Delaware Estuary from its mouth north to Wilmington, DE (RM 64) are tidal salt marshes (Price and Beck 1988). Salem has two nuclear powered electric generation units (Fig. 10.3) with a net generating capacity of approximately 1,195 megawatts electric (MWe) for each unit (PSEG 2009). The Salem cooling system has two parts, the Circulating Water System (CWS) and the Service Water System (SWS). Salem is permitted to withdraw 3024 MGD (11,447,000 m^3) (New Jersey Pollutant Discharge Elimination System[1] ("NJPDES") Permit NJ0005622) with the vast majority of the flow directed through the CWS.

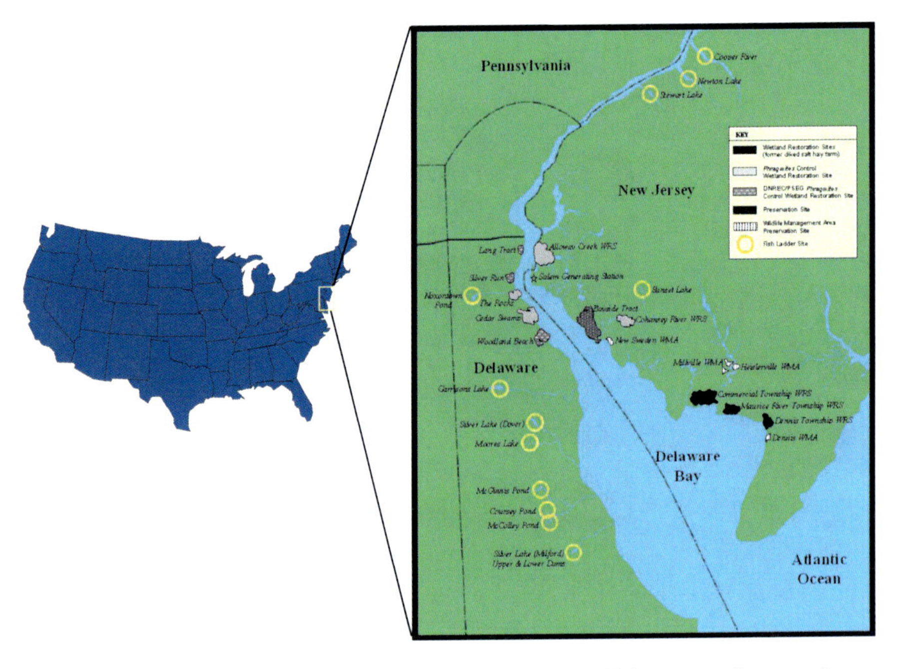

Fig. 10.2 Location map of Salem Generating Station and Estuary Enhancement Program sites

Fig. 10.3 Aerial photograph of Salem Generating Station (foreground) and Hope Creek Generating Station (background) with cooling tower

10.3 Brief History of the Issue

10.3.1 Regulatory Concern During Construction and Early Operation

Studies of the impact of the operation of Salem on the aquatic resources of the Delaware Estuary started in 1966 when plans for the construction of Salem were announced. The primary regulatory agencies were the US Nuclear Regulatory Agency (US Atomic Energy Commission at that time) and the US Army Corps of Engineers whose authority came from the Rivers and Harbors Act of 1899 (33 U.S.C. § 401). After the USEPA was created in 1970 and Federal Water Pollution Control Act of 1972 (33 U.S.C. §1251 *et seq.*) was passed, USEPA took the lead in regulating Salem and addressing its potential environmental impacts. A Technical Advisory Group consisting of federal and state scientists with knowledge of the Delaware Estuary annually reviewed monitoring plans and study results. USEPA delegated permitting authority to NJDEP and subsequently discontinued the Technical Advisory Group. Public concerns were addressed in the formal public hearings carried out by the regulatory agencies.

When each generating unit (Unit 1 in 1977 and Unit 2 in 1980) became operational, studies began to quantify losses of aquatic resources due to impingement and entrainment. Studies in the Delaware Estuary by PSEG contractors and state agencies were also performed to identify any impacts to fish populations. Over the years, the study area varied but predominantly was the area from the mouth of the Delaware Estuary to above the Delaware Memorial Bridge (RM 69). The potential impact issues related to the operation of the Salem cooling system are reviewed and considered each time PSEG files a NJPDES permit renewal application, approximately every 5 years.

10.3.2 The 1990 NJPDES Permit

In 1990, NJDEP issued a draft NPDES Permit that would have required Salem to shut down and not operate until closed cycle cooling (i.e., cooling towers) was installed (NJDEP 1990). NJDEP based their decision on §316(a) (the thermal discharge portion) of the Clean Water Act, judging that the impingement and entrainment losses were included in this section of the act.

USEPA commented on the draft permit and indicated that, while mitigation measures should not be used in lieu of control measures on the thermal component of the discharge to achieve balanced indigenous population, they were an appropriate consideration under §316(b) (relating to impingement and entrainment losses in the cooling water intake structure) and referred NJDEP to a number of NPDES permit proceedings where USEPA had also incorporated mitigation measures (Letter from Cynthia Doughterty, USEPA, to John Fields, NJDEP [January 14, 1991]).

PSEG responded to NJDEP during the public comment period that their estimates on the cost of closed cycle cooling and fish loss were in error. The estimated costs of retrofitting closed cycle cooling used by NJDEP in their decision making was an underestimate ($125 million vs. PSEG's estimated $1–2 billion) depending on outage scenario (no operation until cooling towers installed or a more reasonable expedited outage schedule, technology employed and the replacement power costs during construction). PSEG provided the results of a cumulative fish/larvae assessment that considered all of the data collected at Salem's intakes and included a decade of the river-wide data collected by PSEG, New Jersey Department of Environmental Protection (NJDEP) and the Delaware Department of Natural Resources and Environmental Control (DNREC) (PSEG 1993) that showed healthy fish and larval populations in the Estuary.

PSEG determined that it had three potential courses of action in response to that draft permit. First, litigation, a strategy used on the Hudson River (Barnthouse et al. 1988) that dragged on for more than 10 years. Second, build expensive cooling towers. Third, think outside the box and develop an innovative solution. Using USEPA's guidelines for potential mitigation, PSEG had a series of meetings with NJDEP which culminated in PSEG proposing the Estuary Enhancement Program (EEP).

EEP used a two-pronged approach to provide an environmentally workable solution: first reduce losses from entrainment and impingement as much as practical and second, offset the remaining losses. EEP consisted of the following:

- Upgraded fish protection technology at Salem's cooling water intake.
- Studies of the potential for deterring fish from Salem's intake area with underwater sound generation.
- Limitation on the amount of cooling water withdrawn from the Estuary.
- Restoration, enhancement, and/or preservation at least 4,000 ha of degraded salt marsh and upland buffers along the Estuary in both New Jersey and Delaware.
- Installation of fish ladders to increase spawning habitat for river herring.
- Biological monitoring of the Delaware Estuary.

To ensure implementation, NJDEP incorporated this into the Salem's 1994 NJPDES Permit. The Permit also required provisions that would ensure that the lands to be restored would be protected in perpetuity, establish financial guarantees that would enable the NJDEP to implement the program if PSEG failed to meet its commitments, and required the establishment of advisory committees to provide expert advice to PSEG on the design and implementation of the restoration measures and the biological monitoring program (Balletto et al. 2005).

10.4 Stakeholders

10.4.1 Open Communication with Stakeholders

There was regulatory and public opposition to Salem since plans to construct were first announced in the mid 1960s. Many of the arguments against Salem were antinuclear rather than environmental and a residue of antinuclear public sentiment influenced opposition to the EEP 15+ years after Salem started operating. To successfully implement EEP, PSEG recognized very early that public involvement in planning and implementation was a pivotal factor. The EEP project team created a stakeholder matrix to identify potential stakeholders, categorize them, and predict their position on EEP (Table 10.1). Stakeholder categories included regulators, legislators (state, county, and municipal), environmental organizations, local property owners, fishermen/watermen, and the general public. The groups were considered to be either decision makers (e.g., state and federal regulators and county and local officials) or interested parties. Since PSEG was proposing a very large, long-term, and complex project, the EEP project team held stakeholders meetings very early in the process. The EEP team not only wanted to deliver key messages early but also to solicit input regarding concerns during the design phase of the project. Thus, these early meetings were valuable because they encouraged discussions among the stakeholders and the project team and allowed the project team to incorporate ideas and suggestions into the planning process.

The EEP team recognized that decision makers could be adversely impacted by vocal opponents to any plan and so met individually with decision makers to describe EEP. The intent was to provide factual background information, so that the regulators and legislators could be knowledgeable about the scope and intent of the project before being contacted by their constituencies.

The EEP team also met with environmental organizations (such as The Nature Conservancy, NJ Audubon and NJ Conservation Foundation) to provide information regarding the EEP as well as to gauge their positions, address their questions or concerns, and encourage their support.

EEP covers three counties in NJ, two counties in DE, and numerous municipalities (Table 10.1). To communicate with local land owners and the general public, we held public participation meetings in each county that were organized like a trade show, with each "booth" providing specific information on various aspects of the EEP. One described the operation of Salem and its cooling system, another told of the importance of wetlands in general or fish production from wetlands, with another describing potential restoration techniques. Many of the landowners had their land passed down through generations of their families and they are extremely knowledgeable about local conditions including soil, groundwater, storm impacts, and wildlife. These meetings were an excellent vehicle for them to offer information and identify potential problems very early in the design phase.

The EEP team also met with potential or known opponents with the same intent of a dialogue. Many of these groups were opposed to nuclear power and were not going to be swayed from their position; others were opposed to mitigation believing that the Clean Water Act required technological remedies, i.e., cooling towers.

Some examples of the concerns raised by stakeholders were specific to the restoration sites: potential devaluation of adjacent land, flooding, and groundwater issues. Others were concerned with wetland restoration science and questioned whether the restoration would actually work. Others had very specific concerns about restoration management techniques, specifically the use of glyphosate® on *Phragmites australis* dominated marshes. Many interested parties and decision makers had general environmental concerns: whether or not appropriate fish species would benefit

Table 10.1 EEP stakeholder list

Regulatory Groups	
Appropriate Senior Management	NJ Department of Environmental Protection
Secretary	NJ Department of Agriculture
Senior Wetlands Specialist	NJ Meadowlands Commission
Chair	NJ Pinelands Commission
Executive Director, Deputy Executive Director, Manager – Project Review Branch	Delaware River Basin Commission (DRBC)
Secretary of Natural Resources	Delaware Department of Natural Resources and Environmental Control
Regional Director, Northeast Region; Manager NJ Field Office	US Fish and Wildlife Service
Project Director – Coastal Heritage Trail	National Park Service
Regional Administrator, Northeast Region; Chief, Ecosystem Processes Division	National Marine Fisheries Service
Appropriate Senior Management – Washington DC; Regional Administrator, Regions II and III	US Environmental Protection Agency
Legislators	
NJ Congressmen Districts 1 and 2	Governor's Office
DE Congressman	County Freeholders – Cumberland, Cape May, Gloucester and Salem
	Municipal Leaders Cohansey River, Commercial, Dennis, Elsinboro, Greenwich, Hopewell, Lower Alloways Creek, Maurice River, Salem
Environmental Groups	
Executive Director	Alliance for NJ
Executive Director	American Littoral Society
Executive Director	Association of NJ Environmental Commissions
Executive Director	Bayshores Discovery Project
President	Citizens United to Protect the Maurice River
Executive Director	Clean Ocean Action
President	Cohansey Area River Preservation
President	Delaware Audubon Society
Executive Director	Delaware Nature Society
Executive Director	Delaware Riverkeeper
Director	D&R Greenway Land Trust
Manager Conservation Programs	Ducks Unlimited
President	Delaware Wild Lands, Inc.
President	Episcopal Environmental Coalition
Director	Friends of the Nongame and Endangered Species Project
Executive Director	Great Swamp Watershed Association
Executive Director	Green Delaware
Executive Director	Hackensack Riverkeeper
Members	League of Woman Voters
Executive Director	NJ Conservation Foundation
Executive Director	NJ Environmental Lobby

(continued)

Table 10.1 (continued)

Executive Director	NJ Pinelands Coalition
Director	NJ Natural Lands Trust
Executive Director	Pinelands Preservation Alliance
Executive Director	Preservation NJ, Inc.
Associate Director	Natural Resources Conservation
President	National Wild Turkey Federation – West Jersey Chapter
President	NJ Audubon Society
Executive Director	NJ Commonwealth
Executive Director	NJ Environmental Federation
President	NJ Environmental Lobby
Executive Director	NJ Future
President	NJ Marine Sciences Consortium
Director	NJ Public Interest Research Group
Executive Director	Partners for Environmental Quality (Greenfaith)
Executive Director	Partnership for the Delaware Estuary Inc.
Executive Director	Passaic River Coalition
Director	Rutgers Environmental Law Clinic
President	Salem County Historical Society
President	Save the Delaware Coalition
Director	Sierra Club
Regional Manager	South Jersey Land Trust
Executive Director	Stony Brook-Millstone Watershed Association
Project Manager	Trust for Public Land
State Director	The Nature Conservancy (DE and NJ Chapters)
President	Water Resources Association of the Delaware River Basin
Executive Director	Watershed Association of the Delaware River
Executive Director	Wetlands Institute
Fishing Organizations	
	Cape May County Party and Charter Boat Association
	Jersey Coast Anglers Association
President	New Jersey State Federation of Sportsmen's Clubs, Inc.
Business Organizations	
Senior Vice President for Governmental Affairs	Delaware State Chamber of Commerce
President	Chamber of Commerce Southern NJ
Executive Director	New Jersey Business and Industry Association
Executive Director	NJ Chamber of Commerce
Executive Director	Southern NJ Development Council
Labor Groups	
Business Manager	International Brotherhood of Electrical Workers
Secretary Treasurer	State AFL-CIO

from the marsh restoration, the potential release of soil contaminants, possible increased mosquito production, and impacts to adjacent shellfish beds.

During the planning process, stakeholder concerns were addressed by reviewing and synthesizing existing scientific literature and by having recognized scientific experts provide information, comments, and critiques. As EEP progressed, the data from numerous monitoring studies and a formal scientific advisory committee were used to address the concerns. Several times EEP conducted special studies to address major concerns. For example, a field study of alternative *Phragmites* control methodologies was designed and implemented, including herbicide application, mowing, rhizome cutting, burning, and manipulating sulfide levels (Teal and Peterson 2005; Howes et al. 2005). Results of this study were discussed with stakeholders and led to general acceptance that, for the marshes where *Phragmites* had become a virtual monoculture, control was effective only with the use of herbicide. Other controls, some in combination with herbicide applications, were effective on small isolated stands of *Phragmites*. Additional details of this effort are discussed later.

PSEG also published a periodic newsletter during the design, construction, and post construction monitoring which described the processes being implemented and how they were linked to results. Great effort was taken to present the information in layman's terms, so a diverse stakeholder group could be reached with this publication. The newsletter included details about marsh restoration designs, construction progress, post construction restoration progress, some of the results of the monitoring programs, and issue discussion.

A hotline (1-888-MARSHES) was also established as a means for the public to contact EEP team members. All calls were logged and responses were sent to the specific member of the public. An EEP specific website was created on the primary corporate website (PSEG.com). This contained information on the project and other information of interest to the public such as directions to the sites, public access features recreational activities, and educational opportunities available. These actions led to an increase in public understanding of the EEP project and provided opportunities for PSEG to promote the positive benefits of marsh restoration to the communities.

10.5 PSEG's Strategy for Resolution

10.5.1 An Environmentally Sound Approach Would Still Be Controversial

PSEG recognized that a number of groups would oppose anything other than cooling towers (http://www.unplugsalem.org/marshproject.htm). The goal of antinuclear groups was a complete shutdown of Salem (http://www.unplugsalem.org/). Such groups were and are very active and vocal in their attempts to influence decision makers. PSEG concluded that there was a strong likelihood that one or more of these groups would intervene in the permitting process. To increase the likelihood of winning in an administrative proceeding such as an NJPDES adjudicatory hearing, an extremely defensible position had to be created.

PSEG was not in the business of restoring marshes and did not have the advanced technical knowledge required for salt marsh restoration, particularly at the scale proposed. To address this knowledge gap, PSEG assessed the types of scientists that would be required (e.g., wetland ecologists, sedimentologists, hydrologists, fisheries scientists) and performed a literature search to identify those scientists with the requisite skills who also had national or international reputations. The goal of this effort was to provide a high level of comfort with the nonscientific stakeholders and build confidence in PSEG's commitment to success and to assure that success.

Once the team of scientific experts was engaged, they were immediately immersed in the planning for the implementation of EEP. Internal meetings were held with the experts, or teams of experts as appropriate, to develop details for the various plans and the design principles required to implement such a complex project as EEP. As detailed designs were developed, these experts were actively involved in reviewing and commenting on them as part of a rigorous approval process. Throughout the process, meetings with various decision makers were held with the experts to assure the decision makers that the project was based on sound science and was following the appropriate path for successful implementation.

10.6 Advisory Committees and Community Involvement Committees

10.6.1 Original Committees

When NJDEP issued the final 1994 Permit, it required PSEG to establish a Management Plan Advisory Committee ("MPAC") to provide technical advice to PSEG concerning the development and implementation of the wetlands restoration. In addition, NJDEP required PSEG to establish a Monitoring Advisory Committee ("MAC") to provide technical advice to PSEG concerning the design and implementation of the biological monitoring programs. Each committee, consisting of scientists from the federal and state regulatory agencies, Delaware River Basin Commission and independent scientists with the requisite technical and/or Delaware Estuary specific knowledge, was chaired by PSEG (Table 10.2).

10.6.2 Estuary Enhancement Program Advisory Committee

When NJDEP issued the 2001 Permit, they modified the advisory committee requirement, consolidating the two committees into the Estuary Enhancement Program Advisory Committee (EEPAC). Since the principal marsh restoration activities were

Table 10.2 Management Plan Advisory Committee and Monitoring Advisory Committee membership

Management Plan Advisory Committee	
NJ Department of Environmental Protection	US Environmental Protection Agency (Region II)
National Marine Fisheries Service	US Fish and Wildlife Service
DE Department of Natural Resources and Environmental Control	US Army Corps of Engineers
Delaware Estuary Program	Salem, Cumberland and Cape May Counties
Michael S. Bruno, Ph.D., professor Stevens Institute of Technology	William S. Mitsch, Ph.D. professor Ohio State University
R. Eugene Turner, Ph.D., professor Louisiana State University Coastal Ecology Institute	Joseph Shisler, Ph.D., Shisler Environmental Consultants, Inc. (selected under a settlement with the Delaware Riverkeeper Network)
DRBC and US Geological Survey initially participated but their representatives resigned citing lack of time to participate	PSEG (Chair)
Monitoring Advisory Committee	
NJ Department of Environmental Protection	DRBC
National Marine Fisheries Service	Edward D. Houde, Ph.D., University of MD, Chesapeake Biological Laboratory Center for Environmental and Estuarine Studies
DE Department of Natural Resources and Environmental Control	Ronald T. Kneib, Ph.D., University of GA, Marine Institute
Delaware Estuary Program	Nancy Rabalais, Ph.D., Louisiana University Marine Consortium
Rick De Reisio, Ph.D., professor Scripps Institution of Oceanography (selected under a settlement with the Delaware Riverkeeper Network)	Joe Miller, retired USF&WS (selected under a settlement with the Delaware Riverkeeper Network)
PSEG (Chair)	

concluded by then and only monitoring and adaptive management of the marsh restoration was required, NJDEP reduced the number of independent scientists required and the frequency of EEPAC meetings to twice per year. Many of the same members on the original committees retained seats on the consolidated committee (Table 10.3).

The advisory committees not only provided technical advice to PSEG but also provided an opportunity for technical dialogue and scientific debate that was open to the public and allowed NJDEP and other regulators full and fair access to all of the available expertise including PSEG's expert scientists.

10.6.3 Community Involvement Committees

Community Involvement Committees (CICs) were established in most communities where restoration projects were implemented. The CICs met periodically and acted as conduits for information flow between PSEG and the local community

Table 10.3 Estuary Enhancement Program Advisory Committee membership

Estuary Enhancement Program Advisory Committee	
DNREC Division of Fish and Wildlife	Edward D. Houde, Ph.D. University of Maryland Center for Environmental and Estuaries Studies
US Dept. of the Interior, Fish and Wildlife Service	NJDEP Division of Fish and Wildlife
USEPA Region II Marine and Wetlands Protection Branch	NJDEP Mosquito Control Commission
Richard B. Deriso, Ph.D., Scripps Institute of Oceanography	DRBC
NJDEP Endangered and Nongame Species Advisory Committee	Ronald T. Kneib, Ph.D., University of Georgia, Marine Institute
Cape May County Mosquito Commission (Retired)	Cumberland County Department of Planning and Economic Development
Department of the Army Corps of Engineers	Salem County Freeholder
William J. Mitsch, Ph.D., Mitsch and Associates	R. Eugene Turner, Ph.D., Dept. of Oceanography and Coastal Sciences, Coastal Ecology Institute, Louisiana State University
NOAA, NMFS, NEFSC	NJDEP Division of Fish and Wildlife
Chairman, PSEG Manager – Biological Programs	Joseph K. Shisler, Ph.D., Shisler Environmental Consultants, Inc.

regarding the wetland restoration program. They provided advice on the development and implementation of the restoration sites, with an emphasis on both the salt marsh restorations and public use aspects of the sites.

10.7 Outcomes from the Advisory Committees

10.7.1 Introduction

The advisory committees proposed major improvements that were adopted by PSEG:

- Implementation of ecological engineering principles in the design and implementation of EEP
- Development of success criteria
- Use of adaptive management during the construction and postconstruction monitoring
- Development and performance of testing of alternative methodologies for controlling *Phragmites*.

10.7.2 Ecological Engineering

During the design and construction of the wetland restoration sites, EEP used ecological engineering as the integrating approach (Teal and Weishar 2005). This concept, described by Mitsch and Jørgensen (1989), showed ecological engineering (self-design) as a strategic tool to ensure sustainable interactions between humans and the environment. For complex environmental management actions such as PSEG's wetland restoration program, ecological engineering was the most effective and appropriate approach because it recognized the importance of using human engineering to initiate and encourage natural processes, which were then allowed to develop with little human intervention to complete the restoration.

Both hydroperiod and vegetation management are good examples of the application of ecological engineering to the wetland restoration part of EEP. In the diked salt hay farm restoration, rather than physically altering the entire surface of the marsh with machinery, hydroperiod was established by restoring the primary and secondary tributaries and then waiting to allow the tidal flows to develop smaller creeks and rivulets. Rather than planting vegetation plugs at regular intervals across the marsh plain, since the entire restoration was located in a landscape of healthy salt marshes, the plan relied on the abundant natural seed sources (both in situ and from adjacent marshes) to vegetate the mud flats.

10.7.3 Success Criteria

Performance criteria (Kentula et al. 1993) were developed by PSEG and the scientists engaged in the planning process and reviewed by MPAC and referred to as Success Criteria. These criteria were used to judge the restoration success and were also incorporated into the adaptive management program. The success criteria are described in Table 10.4.

10.7.4 Adaptive Management

Adaptive management is a process for identifying and meeting environmental management goals by an iterative process of monitoring and appropriate engineering response (Holling 1978). The ultimate objective of adaptive management is sustainable management of ecosystems in the context of human development (Thom 1996). Since ecosystems are highly complex and ecological processes are very site specific, it is only through the use of site specific data that effective management can be realized (Haney and Power 1996; Walters and Holling 1990). Adaptive management is the appropriate framework for use in a successful large-scale environmental restoration (Thom 1996). PSEG employed a team of wetland and estuarine experts to guide the adaptive management process and monitor all restoration sites on a regular basis.

Table 10.4 Estuary Enhancement Program Success Criteria

Diked Salt Hay Farms	*Phragmites* Dominated Sites
Interim Success Criteria	
>36% cover with desirable plant species in 7 years after construction	>36% cover with desirable plant species in 6 years after construction
Natural hydroperiod (daily tidal inundation without prolonged ponding)	
Final Success Criteria (applies 12 years after construction)	
>80% cover of total restoration area with desirable plant species	*Phragmites* coverage is <4% of total restoration area
<20% of total restoration area is open water	
Phragmites coverage is <4% of total restoration area	

10.8 Alternative Control Methodology for *Phragmites* and Herbicide Minimization Programs

10.8.1 Alternative Control Methodology for Phragmites Program

As noted earlier, some stakeholders objected to the use of any chemicals on the marsh, expressing concern that there was potential for site contamination and unpredictable site alteration. Glyphosate® is the active ingredient in an herbicide used for *Phragmites* control in many locations throughout the country (Weinstein et al. 2003), and is particularly useful in salt marshes because the desired vegetation such as *Spartina alterniflora* is less sensitive to it than *Phragmites* at the time of application.

Despite attempts to allay the fears of the stakeholders, their opposition to herbicide remained strong and vocal. Obviously, the dense stands of *Phragmites* inhibited the growth of desirable salt marsh vegetation and needed to be eradicated. MPAC proposed that PSEG study the efficacy of alternative treatment techniques. The EEP team developed a research protocol that tested various eradication techniques on replicate plots of varying acreage. The techniques included mowing (annual mowing, annual mowing combined with herbicide application, and mowing several times during the growing season), rhizome ripping (slicing the rhizomes below grade with a modified harrow with and without herbicide), surface scarification (to trap seed of more desirable marsh plants), burning, manipulating sulfide levels, and goats (don't ask; don't tell) (Teal and Peterson 2005). After 3 years, the conclusion was that the concentrated spraying was the best approach to *Phragmites* eradication.

10.8.2 Herbicide Minimization Program

EEPAC recognized that herbicide application was the most successful method for *Phragmites* control but also recognized the regulatory and public concern for the continued use of the herbicide. Thus PSEG agreed to a minimization program. It is important to recognize that minimization has delayed successful restoration of the *Phragmites* dominated marshes in some sites and prevented it in others. With a limit to the amount of herbicide that can be used each year, the choice is to (a) spray all the *Phragmites* areas with a low dose with the result that most of the *Phragmites* is minimally affected/deterred or (b) spray with enough herbicide to kill the *Phragmites* in selected areas with the result that the unsprayed areas are safe havens for *Phragmites* to recover and flourish.

10.9 Improved Science Resulting From Stakeholder Involvement

10.9.1 Additional Studies

PSEG decided to perform additional studies beyond those required by the NJPDES Permit to increase the scientific defensibility of EEP. The EEP team reviewed and identified the limitations of current knowledge and weaknesses on some issues and designed studies to further the science and gather data to address weaknesses.

10.9.2 Fisheries Impact Studies

A key issue regarding the operation of the Salem cooling system is the potential impact from the loss to the fish populations of the impinged and entrained aquatic organisms. The PSEG team developed a program which increased the field monitoring in the Delaware Estuary supplementing existing state fisheries monitoring programs. These studies addressed the first year of life of fish and blue crabs (*Callinectes sapidus*). The emphasis on the first year of life was because most of these species are coastal migrants and leave the Delaware Estuary where they are exposed to other potential anthropogenic impacts. By focusing on the first year of life, the suite of potential impacts was limited to those occurring in the Delaware Estuary.

PSEG collaborated with the state agencies in a mutual data sharing program. In addition, the state agencies provided historic data which had not been previously integrated and PSEG performed a time series analysis to determine any statistically significant anthropogenic fisheries impact.

10.9.3 Increased Studies Prior to Restoration

To reduce the risk of restoration failure, PSEG initiated other studies that supplemented those they anticipated might be required for permitting the restoration activities. These studies consisted of:

- Detailed topographic studies to develop digital terrain models to determine the optimum wetting and drying of the marsh surface likely under various design options.
- Groundwater monitoring to develop baseline data to address concerns regarding changes to the groundwater salinity and elevations.
- Threatened and endangered species surveys black rail (*Laterallus jamaicensis*), northern harrier (*Circus cyaneus*), and barred owl (*Strix varia*).
- Minimum sized area for threatened and endangered species habitat usage. This was done by constructing high marsh areas in triplicate of varying sizes (approximately 0.5 acres [0.2 ha], 1 acre [0.4 ha], 2 acres [0.8 ha], and 5 acres [2.0 ha]).
- Tidal hydrology studies within the marsh to determine potential throttling of tidal flow.
- Marsh vegetation test planting plots to determine optimum planting scheme.
- Sediment accretion studies.

10.9.4 Marsh Fish Production

A fundamental principle of the marsh restoration component of EEP was that increasing salt marsh area in Delaware Bay would result in an increase in fish production. A number of challenges were made to this concept. In addition to the concern that salt marsh restoration would be unsuccessful, opponents stated that if these marshes were restored, they would not benefit the correct fish species and not result in the equivalent quantity of fish production. As soon as restoration construction was completed, PSEG initiated an extensive fish study within and adjacent to the marsh restoration and control sites. The control sites were chosen among nearby marshes that appeared to be relatively unaltered or "naturally" restored. These studies monitored the occurrence and density of fish in the marshes and studied the marsh function for the fish. This was accomplished by studying growth, reproduction, feeding, distribution, and production of species (*Fundulus heteroclites* and *Cyprinodon variegatus*) which spent their entire lives in the marsh and comparing the results to those same species within the reference marshes (Chitty and Able 2004; Currin et al. 2003; Teo and Able 2003; Able et al. 2006).

10.9.5 Project Publications

To validate the science being produced by the EEP team, articles published in the peer-reviewed scientific literature are shown in Table 10.5.

Table 10.5 Estuary Enhancement Program publication listing

Able KW, Nemerson DM, Light PR, Bush RO (2000) Initial response of fishes to marsh restoration at a former salt hay farm bordering Delaware Bay. In: Weinstein MP, Kreeger DA (eds.) Concepts and Controversies in Tidal Marsh Ecology. Kluwer Academic Publishers, The Netherlands
Able KW, Nemerson DM, Bush RO, Light PR (2001) Spatial Variation in Delaware Bay (USA) Marsh Creek Fish Assemblages. Estuaries 24:441-452
Able KW, Nemerson DM, Grothues TM (2004) Evaluating Salt Marsh Restoration in Delaware Bay: Analysis of Fish Response at Former Salt Hay Farms. Estuaries 27:58-59
Able KW, Nemerson D, Light P, Bush R (1998) Spatial variation in Delaware Bay (USA) marsh creek fish assemblages (CM 1997/S:01 Spatial Gradients in Estuarine Systems). In. 1997 Annual Science Conference, Baltimore, MD. Tuckerton, New Jersey: Rutgers University Marine Field Station, Institute of Coastal and Marine Science
Able KW, Hagan SM (2000) Effects of common reed (Phragmites australis) invasion on marsh surface macrofauna: response of fishes and decapod crustaceans. Estuaries 23:633-646
Able KW, Hagan SM (2003) The impact of common reed, Phragmites australis, on Essential Fish Habitat: Influence on reproduction, embryological development and larval abundance of mummichog (*Fundulus heteroclitus*). Estuaries 26:40-50
Able KW, Hagan SM, McLellan J, Witting DA (2003) Characterization and comparison of benthic gears for sampling estuarine fishes and crustaceans Rutgers University, Institute of Marine and Coastal Sciences, Jacques Cousteau Technical Report #100-18.
Able KW, Hagan SM, Brown SA (2003) Mechanisms of marsh habitat alteration due to *Phragmites*: Response of young-of-the-year mummichog (*Fundulus heteroclitus*) to treatment for *Phragmites* removal. Estuaries 26:484-494
Able KW, Hagan SM, Brown SA (2005) Production of fishes in restored salt marshes in Delaware Bay: Progress towards estimates for mummichogs in treated *Phragmites* marshes. Report to PSEG Estuary Enhancement Program, Salem, NJ
Able KW, Hagan SM, Brown SA (2006) Habitat use, movement and growth of young-of-the-year *Fundulus* spp. in southern New Jersey salt marshes: comparisons based on tag/recapture. J Exp Mar Biol Ecol 335: 177-187
Able KW, Balletto JH, Hagan SM, Jivoff PR, Strait KA (2007) Linkages between salt marshes and other nekton habitats in Delaware USA. Rev Fish Sci 15:161
Able KW, Hagan SM, Brown SA (2006) Habitat use, movement and growth of young-of-the year *Fundulus* spp. in southern New Jersey salt marshes: comparisons based on tag/recapture. J Exp Mar Biol Ecol 335:177-187
Able KW, Hagan SM, McLellan J, Witting DA (2002) Characterization and comparison of benthic gears for sampling estuarine fishes and crustaceans. Rutgers University Marine Field Station Jacques Cousteau Technical Report #100-18
Angradi TR, Hagan SM, Able KW (2001) Vegetation type and the intertidal macroinvertebrate fauna of a brackish marsh: *Phragmites* vs. *Spartina*. Wetl 21: 75-92
Aubrey DG, Weishar LL (1998) Hydrodynamic controls on hydroperiods in marshes. Proceedings of the ASCE Wetlands Journal Engineering and River Restoration Conference; Denver, CO. 62
Balletto JH, Teal JM, Weishar LL (1997) Restoration of a Salt Hay Farm on Delaware Bay: A Progress Report in the First Year after Restoring Tidal Circulation. Proceedings of the 24th Annual Conference on Ecosystems Restoration and Creation; Tampa, FL
Balletto JH, Vaskis Heimbuch M, Mahoney HJ (2005) Delaware Bay salt marsh restoration: Mitigation for a power plant cooling water system in New Jersey, USA. Ecol Eng 25:204-213
Chitty JD (2000) The Response of a Resident Marsh Killifish, *Cyprinodon variegatus*, to Marsh Restoration in southern New Jersey. M.S. Thesis New Brunswick, NJ: Rutgers University

(continued)

Table 10.5 (continued)

Currin CA, Wainright SC, Able KW, Weinstein MP, Fuller CM (2003) Determination of food web support and trophic position of the mummichog, *Fundulus heteroclites*, in New Jersey Smooth Cordgrass (*Spartina alterniflora*), Common Reed (*Phragmites australis*), and restored salt marshes. Estuaries 26:495-511
Gratton C, Denno RF (2005) Restoration of arthropod assemblages in a *Spartina* salt marsh following removal of the invasive plant *Phragmites australis*. Restor Ecol 13:358-372
Grothues TM, Able KW (2003a) Discerning vegetation and environmental correlates with subtidal marsh fish assemblage dynamics during *Phragmites* eradication efforts: Interannual trend measures. Estuaries 26:574-586
Grothues TM, Able KW (2003) Response of juvenile fish assemblages in tidal salt marsh creeks treated for *Phragmites* removal. Estuaries 26:563-573
Grothues TM, Nemerson DM, Able KW (2004) Evaluating salt marsh restoration in Delaware Bay: analysis of fish response at former salt hay farms. Estuaries 27:58-69
Grothues TM, Able KW (2003a) Response of juvenile fish assemblages in tidal salt marsh creeks treated for *Phragmites* removal. Estuaries. 26:563-573
Hagan SM, Brown SA, Able KW (2007) Production of mummichog (*Fundulus heteroclitus*): response in marshes treated for common reed (*Phragmites australis*) removal. Wetl 271:54-67
Hinkle RL, Mitsch WJ (2005) Salt marsh vegetation recovery at salt hay farm wetland restoration sites on Delaware Bay. Ecol Eng 25:240-251
Howes BL, Teal JM, Peterson S (2005) Experimental *Phragmites* control through enhanced sediment sulfur cycling. Ecological Engineering 25:292-303
Hunter KL, Fox M, Able KW (2007) Habitat influences on reproductive allocation and growth of the mummichog (*Fundulus heteroclitus*) in a coastal salt marsh. Mar Biol 151: 617-627
Jivoff PJ, Able KW, Martino EJ. (*In review*) Identifying essential habitats of blue crabs (*Callinectes sapidus*) from inter- and intra-estuarine variability in southern New Jersey
Jivoff PR, Able KW (2003) Blue Crab, *Callinectes sapidus*, response to the invasive Common Reed, *Phragmites australis*: Abundance, size, sex ratio, and molting frequency. Estuaries 26:587-595
Jivoff PR, Able KW (2003) Evaluating salt marsh restoration in Delaware Bay: The response of Blue Crabs, *Callinectes sapidus*, at former salt hay farms. Estuaries 26:709-719
Jivoff PR, Able KW (*In review*) Evaluating Salt Marsh Restoration in Upper Delaware Bay: The Response of Blue Crabs, Callinectes sapidus, to the treatment for the eradication of Phragmites australis and reestablishment of Spartina alterniflora
Kimball ME, Able KW (2007) Nekton utilization of intertidal salt marsh creeks: tidal influences in natural *Spartina*, invasive *Phragmites*, and marshes treated for *Phragmites* removal. J Exp Mar Biol Ecol 346:87-101
Lathrop RG, Cole MB, Showalter RD (2000) Quantifying the habitat structure and spatial pattern of New Jersey (U.S.A.) salt marshes under different management regimes. Wetl Ecol Manag 8:163-172
Litvin SY, Weinstein MP (2003) Life history strategies of estuaries nekton: The role of marsh macrophytes, benthic microalgae, and phytoplankton in the trophic spectrum. Estuar Res Fed 26:552-562
Litvin SY, Weinstein MP, Currin CA (*In review*) Marine transients and the coastal conveyor belt: nutrient flow from primary producers to juvenile weakfish (Cynosion regalis) Ecology
Miller MJ, Nemerson DN, Able KW (2003) Seasonal distribution, abundance and growth of young-of-the-year Atlantic croaker, *Micropogonias undulatus*, in Delaware Bay and adjacent marshes. Fish Bull 101:100-115
Miller MJ, Able KW (2002) Movements and growth of tagged young-of-the-year Atlantic croaker, *Micropogonias undulatus*, in restored and reference marsh creeks in Delaware Bay. J Exper Mar Biol Ecol 267:15-38

(continued)

Table 10.5 (continued)

Minello TJ, Able KW, Weinstein MP, Hays CG (2003) Salt marshes as nurseries for nekton: testing hypotheses on density, growth, and survival through meta-analysis. Mar Ecol Prog Ser 246:39-59
Myers R A (2001) Inferring the Bayes Priors with Limited Direct Data: Applications to Risk Analysis. N Am J Fish Manag 1-29
Nemerson DM (2001) Abstract of the Dissertation: Trophic Dynamics and Habitat Ecology of the Dominate Fish of Delaware Bay (USA) Marsh Creeks. New Brunswick, NJ: Rutgers
Nemerson DM, Able KW (*In review*) Diel and tidal influences on the abundance and food habits of four young-of-the-year fish in Delaware Bay, USA, marsh creeks
Nemerson DM, Able KW (2005) Juvenile sciaenid fishes respond favorably to Delaware Bay marsh restoration. Ecol Eng 25:260-274
Nemerson DM, Able KW (*In review*) Shallow estuarine fish nursery habitats exhibit strong trophic seasonality and resource partitioning
Nemerson DM, Able KW (2003) Spatial and temporal patterns in the distribution and food habits of *Morone saxatilis* (Walbaum), striped bass, in marsh creeks of Delaware Bay, USA. Fish Manag Ecol 10:337-348
Nemerson DM, Able KW (2004) Spatial patterns in diet and distribution of juveniles of four fish species in Delaware Bay, USA marsh creeks: Factors influencing fish abundance. Mar Ecol Prog Ser 276:249-262
Philipp KR (2005) History of Delaware and New Jersey salt marsh restoration sites. Ecol Eng 25:214-230
Philipp KR, Field RT (2005) *Phragmites australis* expansion in Delaware Bay salt marshes. Ecol Eng 25:275-291
Popper AN, Balletto JH, Strait KA, Winchell FK, Wells AW, Vaskis MF (2002) Preliminary evidence for the use of sound to decrease losses of aquatic organisms at a power plant cooling water intake. Bioacoustics 12:306-307
Raichel DL, Able KW, Hartman JM (2003) The influence of *Phragmites* (Common Reed) on the distribution, abundance, and potential prey of a resident marsh fish in the Hackensack Meadowlands, New Jersey. Estuaries 26:511-521
Ramcharitar JM, Deng X, Ketten D, Popper AN (2004) Form and function in the unique inner ear of a teleost: The Silver Perch (*Bairdiella chrysoura*). J Comp Neurol. 475:531-539
Ramcharitar, Popper AN (2004) Masked auditory thresholds in sciaenid fishes: A comparative study. J Am Acoustical Soc 116:1687-1691
Rose KA, Cowan JH, Winemiller KO, Myers RA, Hilborn R (2001) Compensatory density dependence in fish populations: importance, controversy, understanding and prognosis. Fish Fish 2001:293-327
Rountree RA, Able KW (2007) Spatial and temporal habitat use patterns for salt marsh nekton: implications for ecological functions. Aquat Ecol 41: 25-45
Smith KJ, Taghon GL, Able KW (2000) Trophic linkages in marshes: ontogenetic changes in diet for young-of-the-year mummichog, *Fundulus heteroclitus*. In: Weinstein, MP, Kreeger DA (eds.) Concepts and Controversies in Tidal Marsh Ecology. Kluwer Academic Press, The Netherlands
Teal JM, Weishar LL (1998) Salt Hay Farm Restoration Design: Combining Biological and Engineering Objectives. Proceedings of the American Society of Civil Engineers (ASCE) Wetlands Engineering and River Restoration Conference; Denver, CO
Teal JM, Weishar LL (2005) Ecological engineering, adaptive management, and restoration management in Delaware Bay salt marsh restoration. Ecol Eng 25:304-314
Teal JM, Weinstein MP (2002) Ecological engineering, design, and construction considerations for marsh restorations in Delaware Bay, USA. Ecol Eng 18:607-618

(continued)

Table 10.5 (continued)

Teal JM, Peterson SB (2005) Introduction to the Delaware Bay salt marsh restoration. Ecoll Eng 25:199-203

Teo SLH (1999) Population dynamics of the mummichog, *Fundulus heteroclitus*, in a restored salt marsh. Master's thesis. Rutgers University, New Brunswick, NJ

Teo SLH, Able KW (2003a) Growth and production of the mummichog (*Fundulus heteroclitus*) in a restored salt marsh. Estuaries 26:51-63

Teo SLH, Able KW (2003b) Habitat use and movement of the mummichog (*Fundulus heteroclitus*) in a restored salt marsh. Estuaries 26:720-730

Tupper M, Able KW (2000) Movements and food habits of striped bass (*Morone saxatilis*) in Delaware Bay (USA) salt marshes: Comparison of a restored and a reference marsh. Mar Biol 137:1049-1058

Wainright SA, Weinstein MP, Able KW, Currin CA (2000) Relative importance of benthic microalgae, phytoplankton and detritus of smooth cordgrass (*Spartina*) and the common reed (*Phragmites*) to brackish marsh food webs. Mar Ecol Prog 200:77-91

Weinstein MP (1998) What begets success? In: Hayes DF (ed) Wetlands Engineering and River Restoration. Proc Amer Soc Civil Engin Conference, Colorado

Weinstein MP, Balletto JH (1999) Does the Common Reed, *Phragmites australis* reduce essential habitat for fishes? NJ Mar Sci Consort Estuaries 22:793-802

Weinstein MP, Balletto JH, Teal JM, Ludwig DF (1997) Success criteria and adaptive management for a large-scale wetland restoration project. Wetl Ecol Manag 4:111-127

Weinstein MP, Teal JM, Balletto JH, Strait KA (2001) Restoration principles emerging from one of the world's largest tidal marsh restoration projects. Wetl Ecol Manag. 9:387-407

Weinstein MP, Philipp KR, Goodwin P (2000) Catastrophes, near-catastrophes and the bounds of expectation: Success criteria for macroscale marsh restoration. In: Weinstein MP, Kreeger DA (eds.) Concepts and Controversies in Tidal Marsh Ecology. Kluwer Academic Publication, The Netherlands

Weinstein MP, Weishar LL (2002) Beneficial use of dredged material to enhance the restoration trajectories of formerly diked lands. Ecol Eng 19:187-201

Weinstein MP, Litvin SY, Currin CA, Wainright SC, Able KW. (in preparation) Marine transients and the coastal conveyor belt: nutrient flow from primary producers to juvenile weakfish (Cynoscion regalis)

Weinstein MP, Litvin SY, Bosley KI, Fuller CM, Wainright Sc (2000) The Role of Tidal Salt Marsh as an Energy Source for Marine Transient and Resident Finfishes: A Stable Isotope Approach. Trans Am Fish Soc 129:797-810

Weishar LL, Teal JM (1998) The Role of Adaptive Management in the Restoration of Degraded Diked Salt Hay Farm Wetlands. Proceedings of the ASCE Wetlands Engineering and River Restoration Conference; Denver, CO

Weishar LL, Teal JM, Balletto JH (1996) The Design Process Utilized to Restore Diked Salt Hay Farms to Natural Marshes. Proceedings of Ecosystems Restoration and Creation

Weishar LL, Teal JM, Hinkle R (2005) Designing large-scale wetland restoration for Delaware Bay. Ecol Eng 25:231-239

Weishar LL, Teal JM, Hinkle R (2000) Development of Marsh Hydrogeomorphology and Marsh Vegetation within a Salt Hay Farm Wetland Restoration Site. 27th Annual Conference on Ecosystems Restoration and Creation.; Tampa, FL

Weishar LL, Teal JM, Hinkle R (2005) Stream order analysis in marsh restoration on Delaware Bay. Ecol Eng 25:252-259

Weishar LL, Teal JM, Hinkle R, Philipp KR (1998) A Comparison of Two Restoration Designs, for Degraded New Jersey. Proceedings of the 25th Annual Conference on Ecosystems Restoration and Creation; Tampa, FL

10.10 Lessons Learned and the Use of the EEP Model

There are a number of strategies used in EEP that are transferrable to other complex environmental projects.

10.10.1 Development of Win–Win Solutions

In a complex project such as EEP, problems present themselves on an ongoing basis. These problems must be viewed as an opportunity and potentially developed into a win–win situation. For example, when PSEG had to decide on an approach to responding to the 1990 draft permit requiring cooling towers, PSEG had three options: litigation, cooling tower construction, or an ecologically beneficial program. PSEG decided on an innovative approach that was a win for the environment and a win for PSEG since the EEP option did not impact plant operations and was not the most expensive approach. Although not the most expensive, EEP was an order of magnitude more expensive than litigation, the least expensive approach.

Another example of a win–win approach was when NJDEP expressed concern that EEP's implementation would reduce the amount of high marsh habitat used by the threatened and endangered species. PSEG could have decided that it was too late in the process for NJDEP to raise the issue and balked at a new requirement that had the potential of creating substantial delays in implementation. Alternatively, NJDEP could have issued land use permits requiring large areas of high marsh habitat be created (at a high cost). PSEG's science advisors reviewed the literature and learned that there was no scientific understanding of the minimum habitat areas for those species, a scientific experiment was proposed to develop the scientific basis for the minimum acreage required. This provided an agreeable solution to NJDEP and did not result in an implementation delay. Also, the requirement for creating large areas of high marsh habitat was averted.

10.10.2 Use of Sound Science

There are many components to a decision for project permitting besides environmental issues such as legal, political, economics, and socioeconomics; however, the foundation for resolution should be sound science. If restored salt marshes did not provide habitat for fish, the concept of EEP would have been untenable. Existing data on salt marsh functions ensured the likelihood that restored marshes would meet the regulatory requirement for fish habitat, but PSEG funded research to expand the knowledge of fish production in marshes to confirm the process for Delaware Bay. In addition, PSEG helped fund research to demonstrate secondary marsh production benefits to key species such as weakfish (*Cynoscion regalis*) that spend limited time in the salt marshes (Weinstein et al. 2000, 2005, 2009a, b; Litvin and Weinstein 2003). With sound scientific information, the regulators then must address the legal, political, and socioeconomic issues.

10.10.3 Early and Open Communication with Stakeholders

Not only does the implementation of a complex project such as EEP require many permits and approvals prior to the first spade going into the ground but ongoing operations also involve regulatory review since the EEP spans a number of NPDES permit renewals. All of these permits (over 200) require (or provide) the opportunity for public input with numerous opportunities for public intervention. By informing the stakeholders and soliciting their input very early in the planning process, many of their concerns were addressed by revisions to the design. The use of the newsletter, the hotline, and the creation of the website were all tools that kept interested parties informed of EEP progress. The implementation of the public advisory committee process also alleviated many of the technical concerns raised by the stakeholders. Many of the technical issues raised were difficult for the nonscientific members of the committee to understand. By observing the debate amongst the technical experts on the advisory committee, stakeholders became more comfortable or knowledgeable about the issues. With more concerns alleviated, there were fewer challenges to the project. All of these activities served to reduce the amount of opposition observed early in the EEP process and resulted in some opponents become staunch advocates.

10.10.4 Inclusion of Public Access on Sites

All restoration sites had public access features (boardwalks, small boat launch ramps, bird observation platforms, and blinds) encouraging passive and active recreation (e.g., hunting, fishing, bird watching). This public access allowed people to experience the restored marsh sites firsthand, including the biological richness of the marshes, and the contrast to the ecological diversity of the upland buffer areas. Virtually all public access features were in compliance with the American with Disabilities Act and allowed people with disabilities to safely get out into the marsh and experience the wonders of the salt marsh ecosystem. The designs developed for these public access points have since been incorporated into other restorations in the region by members of the stakeholder groups initially introduced to each other as part of the PSEG restoration project.

References

Able KW, Balletto JH, Hagan SM, Jivoff PR, Strait KA (2007) Linkages between salt marshes and other nekton habitats in Delaware Bay, USA. Rev Fish Sci 15:1–61

Able KW, Hagan SM, Brown SA (2006) Habitat use, movement and growth of young-of-the-year *Fundulus* spp. in southern New Jersey salt marshes: comparisons based on tag/recapture. J Exp Mar Biol Ecol 335: 177–187

Balletto JH, Vaskis-Heimbuch MF, Mahoney HJ (2005) Delaware Bay salt marsh restoration: A model solution for an environmental problem. Ecol Eng 25:204–213

Barnthouse LW, Klauda RJ, Vaughan DS, Kendall RL (1988) Science, Law and Hudson River Power Plants: A Case Study in Environmental Impact Assessment. Am Fish Soc Monogr 4:347

Chitty JD, Able KW (2004) Habitat use, movements and growth of the sheepshead minnow, *Cyprinodon variegatus*, in a restored salt marsh. The Bull N.J. Acad Sci 49:1–8

Currin CA, Wainright SC, Able KW, Weinstein MP, Fuller CM (2003) Determination of food web support and trophic position of the mummichog, *Fundulus heteroclitus*, in New Jersey smooth cordgrass (*Spartina alterniflora*), common reed (*Phragmites australis*) and restored salt marshes. Estuaries 26: 495–510

Haney A, Power RL (1996) Adaptive management for sound ecosystem management. Environ Manag 20:879–886

Holling CS (1978) Adaptive environmental assessment and management. In: Applied Systems Analysis and Management (International Series on Applied Systems Analysis). Edited by CS Holling. International Institute for Applied Systems Analysis, and Wiley, Toronto p. 377

Howes BL, Teal JM, Peterson SB (2005) Experimental Phragmites control through enhanced sediment sulfur cycling. Ecol.Eng 25:292–303

Kentula, ME, Brooks RP, Gwin SE, Holland CC, Sherman AD, Sifneos JC (1993) Chapter 1, Introduction. In: Hairston AJ (ed) An Approach to Improving Decision Making in Wetland Restoration and Creation. Boca Raton, FL: Smoley CK, Inc.

Litvin SY, Weinstein MP (2003) Life history strategies of estuarine nekton: the role of marsh macrophytes, microphytobenthos and phytoplankton in the trophic spectrum. Estuaries 26: 553–563

Makansi J (2007) Lights Out: The electricity crisis, the global economy, and what it means to you. Wiley, New Jersey

Mitsch WJ, Jørgensen SE (1989) Introduction to ecological engineering. In: Mitsch WJ, Jørgensen SE (eds.) Ecological Engineering: An Introduction to Ecotechnology. Wiley, New York

New Jersey Department of Environmental Protection (NJDEP) (1990) Fact Sheet for Draft NJPDES Permit Renewal Including Section 316(a) Variance Determination and Section 316(b) "BTA" Decision. NJPDES Permit No. NJ0005622

Price KS, Beck RA (1988) Fisheries, in. The Delaware Estuary: Rediscovering a Forgotten Resource. Edited by TL Bryant and JR Pennock. University of Delaware Sea Grant College Program. Delaware

PSEG (2009) Applicant's Environmental Report – Operating License Renewal Stage: Salem Nuclear Generating Station. http://www.nrc.gov/reactors/operating/licensing/renewal/applications/salem/salem-envir-rpt.pdf Accessed 1 January 2010

PSE&G (1993) September 16, 1993 Comments on Draft NJPDES Permit Appendix H. PSEG; Newark, NJ.

Teal JM, Peterson S (2005) The Interaction between science and policy in the control of *Phragmites* in oligohaline marshes of Delaware Bay. Restor Ecol 13:223–227

Teal JM, Weishar L (2005) Ecological engineering, adaptive management, and restoration management in Delaware Bay salt marsh restoration. Ecol Eng 25:304–314

Teo SLH, Able KW (2003) Habitat use and movement of the mummichog (*Fundulus heteroclitus*) in a restored salt marsh. Estuaries 26:720–730

Thom R (1996) Goal setting and adaptive management, Chapter 4. In: Anonymous. Planning and Evaluating Restoration of Aquatic Habitats from an Ecological Perspective. IWR Report 96-EL-4, September 1996. Alexandria: Institute for Water Resources. p. 20

Walters C, Holling CS (1990) Large-scale management experiments and learning by doing. Ecology 71:2060–2068

Weinstein MP, Litvin SY, Bosley KL, Fuller CM, Wainright SC (2000) The role of tidal salt marsh as an energy source for marine transient and resident finfishes: A stable isotope approach. Tran Amer Fish Soc 129:797–810

Weinstein MP, Keough JR, Guntenspergen GR, Litvin SY (2003) Dedicated issue Phragmites australis: A sheep in wolf's clothing? Estuaries 26(2B):397

Weinstein MP, Litvin SY, Guida VG, Chambers RC (2005) Considerations of habitat linkages, estuarine landscapes, and the trophic spectrum in wetland restoration design. J Coast Res 40:51–63

Weinstein MP, Litvin SY, Guida VG, Chambers RC (2009a) Is global climate change influencing the overwintering distribution of weakfish *Cynoscion regalis*? J Fish Biol 75:693–698

Weinstein MP, Litvin SY, Guida VG, Chambers RC (2009b) Essential fish habitat and wetland restoration success: A tier III approach to the biochemical condition of common Mummichog *Fundulus heteroclitus* in Common Reed *Phragmites australis*- and Smooth Cordgrass *Spartina alterniflora*-dominated salt marshes. Estuar Coasts 32:1011–1022

Weston KC (1992) Energy Conversion – The EBook. Originally published by Brooks/Cole. http://www.personal.utulsa.edu/~kenneth-weston/. Accessed 30 December 2009

Chapter 11
Joint Fact Finding and Stakeholder Consensus Building at the Altamont Wind Resource Area in California

Gina Bartlett

Contents

G. Bartlett (✉)
Center for Collaborative Policy, California State University, Sacramento, 160 Delmar Street, San Francisco, CA 94117, USA
e-mail: gina@ccp.csus.edu

J. Burger (ed.), *Stakeholders and Scientists: Achieving Implementable Solutions to Energy and Environmental Issues*, DOI 10.1007/978-1-4419-8813-3_11,

Abstract This chapter provides a brief overview of conflict in the Altamont Pass Wind Resource Area and an in-depth description of the joint fact-finding process designed to improve stakeholder consensus on the complex scientific issues associated with avian fatality and wind power. High political and economic stakes and contentious, uncertain science has created a highly complex environment for crafting policy. The chapter provides a brief history of research and avian mortality issues, the policy and political environment, and key stakeholders and interests. The joint fact-finding process and the consensus-building structure and collaborative outcomes are illustrated. The author concludes that the Scientific Review Committee has successfully fulfilled its charge, yet the larger policy question associated with minimizing avian fatalities and maximizing wind energy has remained largely unresolved. As the wind companies and other interested parties move toward replacing outdated turbines, a different consensus-seeking approach will likely be necessary to grapple with the scientific, political, economic and policy issues necessary to affect change and realize consensus.

11.1 Introduction

In Northern California, just east of the San Francisco Bay Area, lies a broad expanse of rolling hills, home to more than 4,000 wind turbines. This area is known as the Altamont Pass Wind Resource Area (Altamont Pass). These turbines, largely installed in the 1980s, have provided approximately 1% of California's annual energy needs. As the public has increased its demand for wind as an energy source, concerns about avian fatalities in the Altamont Pass and other wind farms have become more apparent. While most stakeholders involved in the Altamont say that wind energy and wildlife protection are not mutually exclusive, much debate and tension have arisen over determining wind energy's wildlife impacts and reaching agreement on ways to shift operations to reduce impacts. High stakes, both policy and economic, have spawned a litigious and politically charged atmosphere. Contentious, uncertain science has created a highly complex environment for crafting policy. Improving science and managing uncertainty, however, are critical to resolving the resource management concerns that stakeholders have raised.

A joint fact-finding process through a Scientific Review Committee is the primary vehicle for stakeholder involvement on issues related to Altamont avian mortality. A local governmental entity responsible for permitting the turbines convened the Scientific Review Committee to provide insight and guidance on scientific issues associated with Altamont Pass avian fatalities. Stakeholders have engaged actively with the Scientific Review Committee. The term stakeholders refers to all parties who have an interest in the Altamont Pass, including wind operators, environmentalists, landowners, scientists, and county and state government.

This chapter provides a brief overview of conflict in the Altamont Pass and an in-depth description of the joint fact-finding process designed to improve stakeholder consensus on these complex scientific issues. While the Altamont Pass Wind

Resource Area is perhaps the most controversial wind farm in the United States because of its high avian mortality, the process used there to address its environmental problems could provide insights for other wind facilities and energy-related developments and the joint fact-finding process.

The chapter first provides a brief history of research and avian mortality issues at the Altamont Pass Wind Resource Area. Second, the policy and political environment and controversy associated with the project supplement the history. The key stakeholders and interests follow since this is necessary to understand the consensus-building process. The joint fact-finding process and the consensus-building structure for stakeholders and scientists designed by the facilitators are illustrated. The Scientific Review Committee's collaborative outcomes are detailed in the fourth section. The challenges and lessons learned conclude the chapter.

The Scientific Review Committee has successfully fulfilled its charge, yet the larger policy question associated with minimizing avian fatalities and maximizing wind energy has remained largely unresolved. As the wind companies and other interested parties move toward replacing outdated turbines, a different consensus-seeking approach will likely be necessary to grapple with the scientific, political, economic, and policy issues necessary to affect change and realize consensus.

11.2 Brief History

A broad range of stakeholders have invested substantial human and financial resources into the challenging issue of avian mortality at Altamont Pass. In this section, a brief history of events provides context for establishing the joint fact-finding scientific forum.

Most Altamont wind turbines were installed in the early-to-mid 1980s. Alameda County issued the majority of the permits between 1981 and 1988 for 20 years, on approximately 40,000 acres in the Alameda County portion of the Altamont Pass Wind Resource Area. In the early 1990s, more than 7,000 turbines operated with an average turbine rating of 101 kW. Altamont Pass reportedly has provided enough energy to supply about 150,000 households per year (Board Resolution, Adopted September 22 2005, Number R-2005-453). Almeda County purports that Altamont has less variable costs than other energy sources, can assist in meeting peak summer energy demand, is nonpolluting, and does not use fossil fuels or large amounts of water. Multiple companies have operated, buying and selling turbine fields in the Altamont over the years.

The State of California began to investigate the issue of avian mortality as bird deaths associated with wind turbines was being recognized. In 1992, the California Energy Commission issued a report beginning to outline the unexpected impact of wind turbine development in California on the death of birds from collisions with wind turbines (Orloff and Flannery 1992). Orloff and Flannery studied the significance of the impact of wind turbines on birds, identifying causes of death and recommending mitigation factors. Searching the ground for dead birds for six seasons, the authors documented a number of dead birds, the majority of which were raptors (red-tailed hawks, American kestrels, and golden eagles, among others). They attributed approximately 74% of the deaths to turbine collisions, electrocution, or wire collisions. Significantly, they concluded that golden eagles, red-tailed hawks, and American kestrels were 3–9 times more likely to be killed than were turkey vultures (Orloff and Flannery 1992).

As Orloff and Flannery's (1992) paper and the challenge of avian fatalities associated with wind farms became more widely discussed, environmental and conservation organizations' concern about avian fatalities heightened. Also, the wind companies' 20-year permits were coming up for review. The County of Alameda East County Board of Zoning Adjustments in November 2003 held a hearing on the permits for wind farm operations and exempted those permits from the California Environmental Quality Act, which sets regulations for environmental impact analysis. This exemption decision triggered environmentalists to sue, demanding that an environmental review occur and better mitigation measures be identified to prevent a "take" of protected avian species under the Endangered Species Act. Local Audubon Society chapters and a nonprofit organization, Californians for Renewable Energy (CARE), petitioned Alameda County Superior Court to set aside the Alameda County Board's issuance of the permits, successfully arguing that such action violated the county's general code and the California Environmental Quality Act.

Amid this controversy, another major Altamont Pass avian mortality study was released (Smallwood and Thelander 2004). In 1998, the National Renewable Energy Laboratory (NREL) initiated research to address the causal relationships between wind turbines and bird mortality. Then, the Public Interest Energy Research Environmental Area (PIER) of the California Energy Commission funded a project to expand upon the NREL efforts. The resulting study attempted to improve understanding of wind-turbine-related fatalities. The researchers, however, concluded that "many bird collisions with wind turbines are associated with factors that could not be understood within the scope of the project." The researchers identified and evaluated possible measures to mitigate bird mortality, including some recommendations for management measures, including careful repowering or replacing small turbines with larger turbines on taller towers (Smallwood and Thelander 2004).

Political controversy ensued. Wind industry representatives questioned the validity of the 2004 study. Environmental organizations continued actively pursuing legal options and urging environmental review. One contention was that a lack of data would prevent an environmental review under the California Environment Quality Act from occurring. Through negotiation among the parties and in consultation with a working group of professionals knowledgeable about the Altamont Pass, the county developed a set of conditional use permits. The Board of Supervisors issued the permits with conditions that scientific research occur, management strategies be introduced and monitored, and efforts to explore replacing older generation turbines with new, larger turbines (repowering) proceed. The conditional-use permits also stipulated that an environmental impact report occur consistent with the California Environmental Quality Act once 3 years of data were collected.

In 2006, the county convened the Scientific Review Committee, as stipulated in the permits, as the primary body seeking consensus on the best methods and tools to improve scientific understanding in Altamont Pass. The Alameda County Conditional Use Permit (Exhibits G-1 and G-2) stipulate that throughout *all years of the [Avian Wildlife Protection and Settling Parties Avian Wildlife Protection] Program, the Scientific Review Committee shall investigate, monitor, and evaluate the effectiveness of the Program, using input from the Permittee(s), the County consultant and state-sponsored research, and subsequently recommend adjustments, and design and implementation of alternative strategies.*

At the same time that the county was appointing scientists to the committee, the county hired a team of consultants, known as the Monitoring Team, to begin conducting field work in the Altamont to search for birds injured or killed by wind turbines. Building on the work of the wind companies' previous consulting team, the Monitoring Team, in consultation with the Committee, developed field protocols and initiated field work. The county also retained the services of an independent, impartial facilitation team to facilitate the Scientific Review Committee.

As the field work progressed, the Scientific Review Committee grappled with issues that came before it. The Committee reviewed compliance with permits and the potential benefits of various management actions. The parties to the lawsuit continued deliberating and in 2007 announced that a settlement had been reached. The settlement agreement stipulated that the wind companies would have to demonstrate a 50% reduction in avian mortalities within 3 years. The settlement agreement picked a baseline number of annual fatalities from a range in the 2004 California Energy Commission study (Smallwood and Thelander 2004). The agreement also required that the wind companies implement particular management actions, contingent on progress made, to reach the 50% reduction. These included seasonal shutdown, removing or relocating high-risk turbines, and studying blade painting.

11.3 Policy and Political Context

Resolving the issues associated with avian fatalities and wind energy in the Altamont Pass has implications on various levels. First, the birds killed are protected by federal and state laws. Second, turbine operation and energy production are central to expanding renewable energy to address climate change and meet California's clean energy mandates. The growing unease around avian fatality has raised challenges at other wind power facilities. Lastly, stakeholders at the local level have reached consensus that aging turbines of the Altamont Pass need to be replaced with newer turbines. The wind companies have hesitated to proceed with plans to repower given the legal uncertainty associated with wind power in this context. Finally, tensions have run high. Scientific uncertainties, political demand for wind energy, and hostile stakeholder activities have yielded a contentious, controversial environment.

11.3.1 Avian Fatalities

Disagreement among stakeholders exists both over the extent of avian mortality and which measures reduce avian mortality most effectively. Most bird species killed in the Altamont Pass are protected by federal or state law, including species protected by the Bald and Golden Eagle Protection Act, the Migratory Bird Treaty Act, and the federal and state endangered species acts. Key species of concern include golden eagles, American kestrels, red-tailed hawks, and burrowing owls. Estimates of the number of birds killed in the past 20 years range from 10,000 to 44,000 (Board resolution). During the current 3-year monitoring study, field crews found 1,117 raptors (43%) and 1,472 (57%) nonraptors for a total of 2,588 birds (ICF Jones and Stokes 2009). These raw numbers do not correct for searcher error or birds missed due to scavenging.

11.3.2 Climate Change and Clean Energy Mandates

In recent years, climate change has gained prominence as a major public policy issue. Reducing carbon emissions associated with energy production is one strategy to contribute to slowing climate change. A central component of this strategy is to increase the proportion of carbon-free energy production sources, such as wind (Pachauri and Reisinger 2007).

In California, the state has mandated the development of renewable energy. California's Renewable Portfolio Standard requires that the state achieve 20% renewable energy production in 2010 (2002 Senate Bill 1078 and 2006 Senate Bill 107) and 33% by 2020 (California Executive Order S-14-08). For comparison, the California Energy Commission estimates that in 2008, renewables (geothermal (4.46%), wind (2.39%), biomass (2.08%), small hydro (1.44%), and solar (0.24%)) made up 10.61% of California's power supply as measured in gigawatt hours (CEC 2008). In short, California has to transition approximately 10% more of its energy production to renewables between 2010 and 2020.

Concerns about wind energy's toll on birds and bats have complicated support. The National Audubon Society in 2007 publicized its support for properly sited wind energy as a renewable resource over its standing concerns about wind energy effects on avian species. The Society concluded that the effects of climate change on avian species would be greater than those caused by wind turbines. While continuing to seek to reduce avian fatalities associated with wind turbines, the Society has said it will support properly sited wind turbine projects.

11.3.3 Mortality Impedes Wind Farm Development

Demand for renewable energy continues to rise, and wind energy is a prominent element of renewable portfolios (Brown 2006). The impartial facilitator's issue assessment of 2006 reported that while wind operators, environmentalists, and other stakeholders were generally optimistic that birds and wind energy production could safely coexist, the industry had yet to implement a proven strategy to reduce avian fatalities, and concern around avian and bat mortality has continued (Bartlett 2006). During interviews and meetings, Altamont stakeholders generally articulate support for wind energy and its necessity when faced with challenges of climate change and air quality.

Altamont stakeholders continue to express concern that avian mortality is holding back the potential of wind power in California and beyond. Scientists and consultants working on issues related to Altamont Pass report that they are called upon to consult with other potential wind farms, and that stakeholders are hesitant to support wind farms because of concerns tied back to the Altamont Pass. Further, Alameda County has placed a moratorium on issuing permits to increase permitted Altamont Pass electrical production capacity until bird mortality is significantly reduced (Daulton 2007).

11.3.4 Need to Replace Aging Turbines (Repowering)

In the Altamont Pass, many of the turbines, installed in the early 1980s, are reaching or surpassing the typical 20-year lifespan of a turbine. New turbines are typically much larger and more productive (1 MW) than those currently in the Altamont. One turbine can replace every 8–10 existing turbines. This has a significant possibility to reduce avian fatalities (Smallwood and Neher 2004). However, the wind companies have been reluctant to replace aging turbines in the uncertain legal climate attributed to stakeholder concerns with avian wildlife fatalities.

The science is starting to demonstrate that bird fatalities would be reduced by replacing older small turbines with fewer large new turbines. In December 2009, the Altamont Pass Wind Resources Area Monitoring Team concluded "a marked reduction in the average annual mortality rates and estimated number of fatalities for all four target species – with the exception of burrowing owls" in the repowered project known as Diablo Winds as compared to other parts of the Altamont operating with older generation turbines (ICF Jones and Stokes 2009). Further, the Scientific Review Committee recommends repowering as a management action that could significantly reduce mortality (See January 2010 SRC Meeting Summary at http://www.altamontsrc.org).

11.3.5 The Controversy

Neither the brief history nor the policy context really details the hostile, adversarial relations of Altamont Pass stakeholders. Environmental organizations, the county and the wind companies have been involved in lawsuits. Scientists engaged in studying the issues have often felt that their credibility has been attacked publicly. The wind companies operating in the Altamont have disagreements about how to approach the issues. Further, scientific understanding of the vast complex area has been rife with uncertainty and need for further inquiry. Karl et al. (2007) discuss how incomplete understanding is used to delay decisions opposed by one group or another in adversarial processes. In the Altamont, uncertainty has delayed studies and implementation of management strategies. The need for more data, as well as the controversy associated with previous studies has delayed environmental analyses. In some cases, scientists were hired to attack or create questions about other scientific work, thereby canceling out or calling into question the basic findings that avian fatalities were occurring and being attributed to the wind turbines. The controversy and actions taken have created an environment of distrust and hostility, making progress on any issues challenging. Amid this controversy, the parties agreed to the concept of the Scientific Review Committee, as a tool to assist in reaching consensus on the science and policy issues in the Altamont Pass Wind Resource Area. The next section details who the key parties are, as well as their major interests.

11.4 Stakeholder Identification

The primary stakeholders in the conflict over wind and avian mortality are highlighted in Table 11.1. The stakeholders include the wind companies, environmental organizations, Alameda County, state and federal natural resource agencies, and other interested scientists. Most of the stakeholders identified are active in the work of the Scientific Review Committee.

The wind companies are a central stakeholder in that they install and operate the wind turbines in the Altamont Pass. Their primary interest is to maximize wind energy production and revenue for the companies. Multiple companies operate turbines in the Altamont Pass (approximately six companies). The companies vary significantly in size of operations. Most of the companies operate wind farms in areas outside of the Alameda County in the adjacent county and in other parts of California and the country. Also, the wind companies have differing perspectives with regard to how significant the problem of avian mortality is and how it should be managed. Due to the differing size of their operations, they also differ in the amount of resources they can or will contribute to addressing issues related to monitoring. The companies' common interest, regardless, is to maximize wind production now and into the future. All of the wind companies lease the lands on which the turbines operate, and many landowners graze cattle around the turbines.

Alameda County issues the permits to the wind companies for operations. The county's relationship to the wind farm is complex, as Table 11.1 suggests. First, two boards and several departments are tasked with the responsibility of the wind farm. Second, the county relies upon the income generated through the wind company permits as part of its annual revenue base. The county's interests are to maximize power output (and revenue) to meet statewide renewable goals and to reduce avian mortality so that the turbines can run unfettered. Also, County staff spend substantial staff time dealing with the situation at the Altamont Pass. County Counsel also participates in the legal proceedings, representing the county.

Table 11.1 Key stakeholders and interests involved in the Altamont pass wind resources area

Stakeholders	Involvement	Interests
Wind Power Companies	Operate turbines	Maximize wind power production Maximize profit and minimize costs associated with wind production Minimize permitting and other regulatory disruptions
Landowners	Lease land to wind power companies for wind turbine placement	Receive fees associated with land leases Continue landowner operations, such as grazing on lands, uninterrupted
Alameda County Board of Supervisors	Approve conditional-use permits Approve county contracts for Scientific Review Committee members and consultants who perform monitoring	Maximize wind power production to ensure income generation for county through permits and contribute to state goals/target for renewables Be responsive to constituents (including wind companies) and those demanding renewable energy
Alameda County Board of Zoning	Approve permits for turbines to operate	Maximize wind power production Minimize impacts on wildlife
Alameda County Department of Community Develop and Planning	Coordinate with wind companies Monitor compliance Convene the Scientific Review Committee Participate in legal negotiations with settling parties	Maximize wind power production Minimize impacts on wildlife Reduce conflict associated with wind farm to save staff time and meet other goals Ensure that income from permits continues
Alameda County Counsel	Interpret conditional-use permits and negotiate with settling parties	
California Department of Fish and Game		Protect wildlife
California Energy Commission	Funds research associated with the wind farm	Promote renewables Improve understanding of the issues
California Attorney General	Observes wind farm activities and monitors science	Minimize impacts on wildlife Maximize wind power production
U.S. Fish and Wildlife Service	Regulate Migratory Bird Act	Protect wildlife
Environmental/ conservation organizations	Minimize impacts to wildlife Support renewable energy	Minimize impacts on wildlife Maximize wind power production Reduce emissions contributing to climate change
Scientific Review Committee	Provide independent guidance on scientific issues	Establish credible scientific program
Consultants	Provide scientific services for monitoring and data collection and interpretation	Establish credible scientific program Maintain professional credibility

The Alameda County Community Development Planning Department oversees the permits that authorize the wind companies to operate turbines in the project area. The Planning Director is authorized through the conditional use permits on wind farm operations to make decisions related to the permits arising from Committee recommendations. The Planning Director reports to the Alameda County Board of Supervisors.

The California Department of Fish and Game is charged with protecting wildlife. Its staff are minimally involved in the Scientific Review Committee. They participate in meetings, contributing agency staff expertise on wildlife and state policy. The California Energy Commission monitors the work of the Scientific Review Committee. Commission staff also contract with some of the scientists on the committee to perform research. The Commission also grants funds to perform research which the county has occasionally pursued to fund some of the Committee-recommended research. The California Attorney General's office monitors activities of the wind farm and the Committee. Staff attorneys attend meetings in which major research findings are presented or significant policy issues discussed. The U.S. Fish and Wildlife Service has been less active in the Scientific Review Committee.

Environmental and conservation organizations participate in Committee meetings. Several Audubon Chapters and Californians for Renewable Energy are active in the settlement negotiations with the wind companies. The Center for Biological Diversity also participates occasionally. The main interest of these organizations is to minimize the impacts on wildlife while maximizing wind energy production.

11.5 Structure of Scientist and Stakeholder Involvement

One of the primary vehicles for stakeholder involvement in issues related to avian mortality for the Altamont is a joint fact-finding process through the Scientific Review Committee. The primary role of the Scientific Review Committee is to provide scientific interpretation and recommendation to the county (Fig. 11.1). For very complex environmental problems, decisions based on sound science must integrate social science, natural science, and stakeholder concerns (Karl et al. 2007). The Committee is structured to build consensus and understanding on scientific and policy issues in the broader stakeholder community. This section describes the structure of stakeholder involvement, including the roles and responsibilities of the parties in the joint fact-finding process.

11.5.1 County of Alameda

As discussed previously, Alameda County issues and oversees wind company permits. The county appointed the Scientific Review Committee to guide its efforts on the wind farm, and actively seeks the Committee's advice. The county frames the

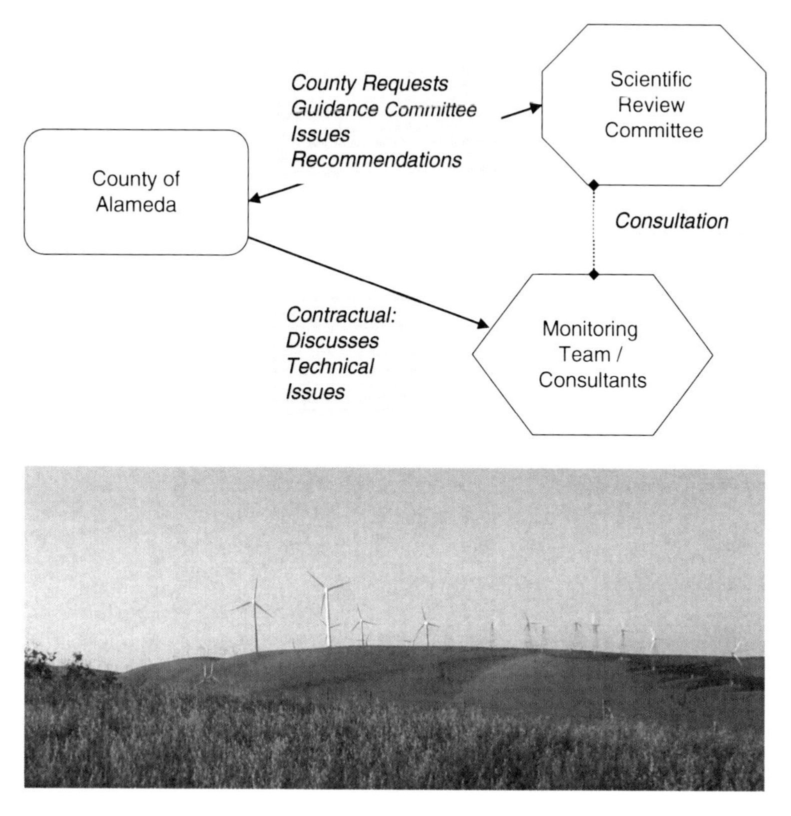

Fig. 11.1 The scientific review committee provides scientific interpretation and recommendation to the county

Committee's agenda, calling upon it to assist with particular tasks or research efforts. For example, the county relies upon the Committee to review and consult on the monitoring protocols and results.

11.5.2 Scientific Review Committee

Convened by Alameda County, the Scientific Review Committee is composed of five independent scientists. Its primary role is to provide independent review and expertise on monitoring and research related to wind energy production and avian behavior and safety. To this end, the goals of the committee are to reach agreement on analysis and interpretation of data and ensure sound and objective scientific review of avian safety strategies (SRC Charter – http://www.altamontsrc.org). The Committee has a collegial working relationship with the monitoring team who performs avian monitoring and advises the monitoring team as to their protocols, activities, and analyses.

Each of the five major interest groups nominated a scientist to participate in the Committee. The interest groups were:

- Permittees (i.e., wind farm companies or turbine operators and their personnel)
- Environmental community
- County Planning Department
- California state agency (California Energy Commission and Department of Fish and Game)
- Federal resources agency (U.S. Fish and Wildlife Service)

The Board of Supervisors then appointed each committee member. This appointment has subjected the Committee's meetings and deliberations to the Brown Act, California's public sunshine law. The Brown Act dictates that the Committee's deliberations must occur in public. Meeting agenda are required to be posted at least 72 hours in advance; the Committee cannot reach a decision on anything not included in the agenda. Members are unable to build consensus through serial conversations via telephone or e-mail or unscheduled meetings.

While appointed by these interests, committee members strive to be objective in reviewing and providing guidance on science related to its charge. They do not represent the interests of those who nominated them, and neither consult nor discuss perspectives with any interest group preferentially. The goals of the committee are to provide a neutral forum for open dialogue among experts in the field with different perspectives, reach agreement on analysis and interpretation of data, and ensure sound and objective scientific review of avian safety strategies (See Conditional Use Permits Attachment D, Altamont Pass Wind Resource Area Scientific Review Committee, September 22, 2005, http://www.altamontsrc.org).

The scientists on the committee are under contract with Alameda County. Scientists are paid for their time preparing for and attending meetings. Funds from the wind turbine permits pay for the Scientific Review Committee contracts.

11.5.3 Stakeholder Interest Groups (See Fig. 11.2)

Stakeholder interest groups participate in scientific deliberations on two levels. On one level, a small group of highly informed individual stakeholders participate in most Committee meetings. This core stakeholder group includes a wind company biologist, several wind company officials, and representatives of environmental organizations involved in the settlement agreement. At most meetings, these stakeholders help to frame research questions and actively engage with the scientists and Monitoring Team on data collection and analytical issues. They often provide a clarifying role with regard to turbine operations. These stakeholders have a very deep stake in the outcomes and analyses. At times, the county may request, or the settlement agreement may require, the Committee to evaluate study proposals for particular management actions. These requests have involved a great deal of interaction between the scientists, the companies, and the other interested parties engaged through the core stakeholder group.

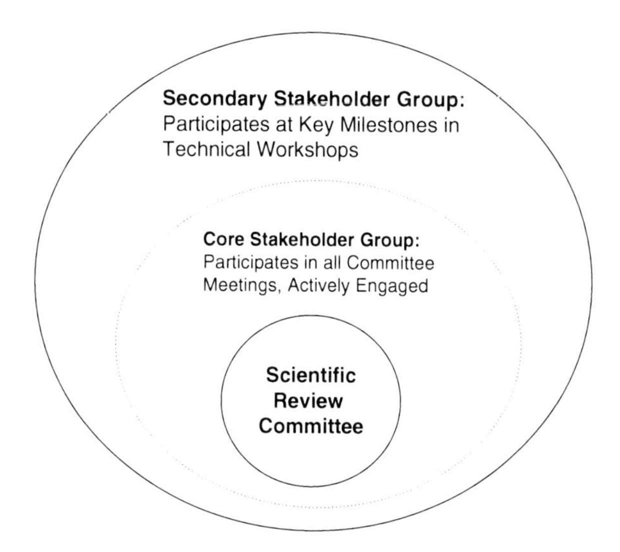

Fig. 11.2 Stakeholder interest groups

The secondary stakeholder group is interested in the wind farm, but not as intensively as the highly informed or core stakeholders identified above. The goal for this secondary group of stakeholders is to provide a public forum in which the Committee can seek understanding of public concerns and share its recommendations. At major milestones, such as the release of a major report, the Scientific Review Committee has held technical workshops geared specifically to this group. The goal is to share information and understanding gained through Committee deliberations and Monitoring Team analyses, but information is presented in a way that the secondary stakeholders can digest and understand so that they may comment and provide meaningful input.

11.5.4 Consultants Who Perform Monitoring (Monitoring Team)

A team of consultants, known as the Monitoring Team, carry out the monitoring program. Although they are employed directly by the county, they are also stakeholders in that they have an interest in the science and outcomes. The Monitoring Team developed the monitoring protocol in consultation with the Scientific Review Committee. Staff and subcontractors from the Monitoring Team walk the fields of the wind farm documenting the number of fatalities that occur. Other members of this Monitoring Team analyze the data and write the findings report. Alameda County calls upon the Scientific Review Committee to help it make decisions regarding the scope of work and study design that the Monitoring Team implements. The Monitoring Team contracts with Alameda County and is paid through permitting fees that the county collects from the wind companies.

11.5.5 Facilitators

Alameda County also contracts with independent third-party facilitators to facilitate the Scientific Review Committee and handle the broader public outreach associated with the Committee. A program of California State University, Sacramento, the facilitators remain impartial toward the content of the discussions. Facilitators work with the Scientific Review Committee, the county, and the Monitoring Team to develop the Committee's agendas and work program. They also work closely with members of the Committee to document its agreements and recommendations. The facilitators run the meetings of the Committee in a fair and objective manner, keeping the issues on track and allowing the committee to conduct its business while allowing sufficient and meaningful input by other stakeholders. Alameda County pays the facilitator through fees collected from the wind company permits.

11.6 Facilitator's Role: Effecting Successful Collaboration and Joint Fact Finding

The facilitation team structured the committee's deliberation using a phased collaborative joint fact-finding model. High-quality joint fact finding ensures that the best-quality science is used in the process (Ehrman and Stinson 1999). Given the scientists' commitments to their work and the anticipated scrutiny in this process, high-quality science has been paramount. This section discusses a phased model of joint fact finding that was employed at the Altamont. This model includes organization, education, negotiation, resolution, and implementation phases. The section also details the Committee's collaborative tasks and outcomes.

Joint fact finding refers to best practices or procedures that ensure that science and politics are balanced appropriately in environmental decision making (Karl et al. 2007). Joint fact finding engages scientists and stakeholders in framing research questions, discussing assumptions, and analyzing and interpreting results, thus making the research more relevant and generally better accepted across divergent stakeholder groups.

The Scientific Review Committee employs a process of deliberative interaction with stakeholders and the Monitoring Team at its public meetings. The Committee typically prefers to deliberate internally, interspersed with public interaction. Initially, the core stakeholder group frames questions of interest regarding the research. The Committee then integrates these questions into its deliberations. As the Committee deliberates, it periodically invites discussion with the core stakeholder group. The Monitoring Team is also integral to this process as it has first-hand knowledge about field protocols and data. The Monitoring Team interjects as necessary to keep the conversation consistent with the reality of the monitoring program. As the Committee narrows to reach an agreement, the core stakeholder group asks questions and contributes information for the Committee's consideration. In this way, the parties are able to create shared knowledge that is technically credible, publicly legitimate, and especially relevant to policy and management decisions (Karl et al. 2007).

During the initial assessment, one of the critical roles envisioned for the Committee was to build trust among the parties in conflict. Parties that do not trust each other are more likely to criticize each other's interpretations of scientific findings. Through the Committee's joint fact-finding activities, deliberations, and public workshops, the goal is for all the parties to invest in the Committee's recommendations and help build trust among the parties over time. This process of deliberation and interaction with stakeholders has occurred throughout the phases of the process designed to support collaboration and build confidence and trust in the collaborative outcomes.

11.6.1 Phased Model to Support Collaboration

To successfully implement the joint fact-finding process, the facilitation team has used a phased model designed to support collaboration and move toward consensus on scientific research methods and results. The model is based on a reflective practice framework used by the facilitation consultant, the Center for Collaborative Policy (Five Stages, 2010).

Phase 1 Organization: In this phase, an assessment occurs in which an independent third party can interview stakeholders and determine the interests of different groups. Through the assessment, the third party can recommend a process structure that is appropriate to the situation. The key parties are identified. Scientists are vetted with the interested parties so that a credible group of experts are identified to guide the effort. A charter that defines the charge, decision-making rules, roles, responsibilities, and communication protocols is developed, discussed, and approved by the stakeholder group or committee at its convening meeting. All these process guidelines are decided at the beginning, before the parties commence substantive discussions.

The Committee's impartial facilitator conducted an initial assessment with the scientists and with all the major stakeholders. The facilitator used the information gained in these interviews and best practices to clarify the role and charge of the Committee and other interested parties, to introduce a consensus-seeking process for the scientists' decision making, and to design a process with robust stakeholder involvement.

Typically, the impartial third party would make recommendations on the composition of the stakeholder group. In this case, the scientific committee had already been designated, so the facilitators confirmed the other interested stakeholder groups and factored their involvement into the Committee's work plan and meetings.

Phase 2 Education: During the education step, everyone involved improves their understanding of the existing science and each party's issues and concerns. The scientists, with input from the interested parties, agree on background materials and

Table 11.2 Key steps to collaborating in a joint fact-finding approach

	Steps
Phase 1 organization	Interview and assess interests in situation
	Make recommendations on appropriate collaborative process
	Identify key parties
	Identify credible scientists
	Develop a charter defining charge, decision-making rule, roles, responsibilities, and communication protocol
	Convene collaborative group(s)
Phase 2 education	Review background materials and research
	Scientists and stakeholders educate one another about issues and concerns
	Discuss existing research and future research needs
Phase 3 negotiation on research	Stakeholders and scientists define the questions to be analyzed
	Design research studies
	Articulate analytical limitations
	Incorporate management actions into design if appropriate
Phase 4 resolution and implementation	Commence scientific research (data collection)
	Conduct analyses
	Review analytical results
	Modify studies as appropriate, continue scientific work
	Recommend agreed-upon management strategies
	Continue monitoring and evaluation
	Adjust management strategies periodically
	Return to Phase 3 to modify studies and identify additional research based on analytical results

Source: Gina Bartlett, Center for Collaborative Policy, California State University

research for review. The parties discuss existing research and future research needs. Another important element of this step is that all the parties educate one another about their issues and concerns. In this way, the central research questions will be informed by stakeholder issues and existing research.

In the Altamont, Committee members identified studies that would inform its work. Of interest, two of the scientists did not have prior experience with the Altamont. The other three scientists had extensive direct research experience there. Two of three had been primary authors on separate Altamont Pass avian mortality studies that the California Energy Commission published. Sue Orloff coauthored the first significant study on avian mortality in the Altamont (Orloff and Flannery 1992). Shawn Smallwood coauthored the next substantial study that studied mortality (Smallwood and Thelander 2004).

Another unusual element is that much of the Committee's work is stipulated by conditional use permits. The Committee commenced in 2006 prior to the settlement agreement, so the Committee's work plan and efforts are tied to the tasks and milestones associated with the permits. The Committee solicits and benefits from continual interaction with the interested parties and the Monitoring Team. These interactions help to educate everyone.

Phase 3 Negotiation on Research: Once the group has convened and educated itself about both the science and the issues and concerns, the research study is designed. Scientists along with stakeholders negotiate the central research questions. In this way, the research is designed to be responsive to interested parties.

The first task in the Scientific Review Committee was to frame the central research questions that the monitoring program should address, and then to design a monitoring program to gather data to address those research questions. Since the monitoring had begun in November 2005 before the Scientific Review Committee convened, the Scientific Review Committee modified the existing protocol. As part of the research design, the scientists considered what management actions would be implemented. Some of the management actions were stipulated in the permits although the scientists could recommend others. Interested parties participated in these conversations and contributed to the substance of the discussions. Initially, the Monitoring Team was large and included many scientists with a long history in the Altamont. So, team members have also engaged in a lively exchange with the Scientific Review Committee on research design issues. Once the draft protocols had been crafted, the Scientific Review Committee held a technical workshop for the Monitoring Team to present and discuss the protocols with the broader stakeholder community. Once this process moved forward, the Monitoring Team began field work and data collection under guidance from the protocols.

The Committee repeats a similar process to develop other research proposals and study designs. The core stakeholder group helps frame the questions. The Committee deliberates. The Monitoring Team contributes information about the monitoring program. The Committee entertains questions from the core stakeholder group and then, as it narrows toward agreement, engages stakeholders in discussion before reaching agreement. At key milestones, technical workshops engage the secondary stakeholders, keeping them abreast of research proposals and findings.

Phase 4 Resolution and Implementation: In this phase, research or data collection is under way. The scientists can also agree on management strategies and implementation methods. At periodic milestones, data can be reviewed and analyzed so that management strategies can be adjusted. Periodically, the scientists or the stakeholders may recommend modifications to studies or management strategies. In this case, the parties return to phase 3 to negotiate specific research to address these modifications or, in some cases, adapt the existing research design or management strategies.

Data collection and analysis have continued over a 3-year period. Due to various complications during the first 2 years, the annual analyses tended to be tardy or incomplete. These delays created challenges for modifying the research approach and agreeing on the best methods of data analysis. One significant breakthrough occurred in the third year when the Monitoring Team was able to stabilize the project database and make it available to the public. This data transparency has significantly shifted the dynamics since all the parties are able to review and even analyze the data. Prior to this, some stakeholders had expressed a great degree of suspicion about the work of the Committee and Monitoring Team. Access to all of the information has enabled stakeholders to have a far greater understanding of the complexities revolving around the data and the science.

When conditions change, the Scientific Review Committee returns to Phase 3 and reevaluates the study. For example, when the Settlement Agreement emerged in 2007 with a stipulated baseline number to compare against newly collected data, the Committee had to evaluate how, if at all, the research design would change.

11.6.2 Collaborative Tasks

There are a number of tasks required of the Committee and Monitoring Team to move toward evaluating avian mortality in the Altamont, suggesting management measures to reduce mortality, and evaluating those management measures.

11.6.2.1 Monitoring Program

One of the first major joint fact-finding tasks was to develop monitoring protocols to perform biological monitoring to improve understanding and evaluate fatalities associated with turbine operations. This meant going out in the field and finding dead birds, using a systematic protocol and a stratified sampling approach. The objective was to determine overall trends to reduce bird mortality at Altamont and to be able to evaluate the effects of different management strategies and other factors.

The Committee, in consultation with the Monitoring Team and with input from other interested parties, crafted the protocol that has been used for the 3-year monitoring program. The Committee felt that the monitoring program should meet standards credible in scientific studies with 15% data precision, especially given the heightened political sensitivities of this program. They discussed many other issues and considerations for the monitoring program, such as bird abundance, behavior studies, turbine function, and field methods. All these robust discussions have contributed to the protocol that was adopted in the Altamont Pass.

11.6.2.2 Interpreting Monitoring Results

Another major activity of the Committee is to assist in interpreting the monitoring results. The Monitoring Team presents results, and then the Committee comments in a public forum (announced publicly usually several weeks before the meeting), discusses the results, and often submits written comments to the team. These efforts have served to improve the subsequent analyses and monitoring methods. This process also assists the stakeholders in understanding and contemplating the results. Interested parties, especially those involved in the settlement agreement, also call upon the Committee during its public meetings to assess the meaning of the results in relationship to goals of the settlement agreement. Sometimes settling parties pose specific questions to the Committee for discussion and response.

One example of the Committee modifying the research and management approach concerned the annual winter shutdown of turbines to reduce bird fatalities. The Committee recommended modifying the proposed design to sequence the shutdown to allow the Monitoring Team to conduct searches prior to shutting down the turbines. Upon further data review and consultation with field crews, the Committee recommended a different shutdown sequence to reduce the possibility of bird confusion (and more fatalities). This is an example of the Committee using Monitoring Team data, field expertise, and the operational realities of a wind farm to make recommendations on management actions. The goals of these management actions are to try to meet policy goals of maximizing wind power while minimizing avian fatalities.

The monitoring analyses have been inconclusive. From the beginning, the Monitoring Team and Scientific Review Committee determined that the scientific program would be unable to assign a reduction in mortality to any one management strategy. Because the area is so vast and so many variables are present, there is no way to control the environment to isolate any one action. On another level, the Monitoring Team has found that its analyses have not met the scientific rigor that stakeholders expected. The team has been unable to secure the resources necessary to complete bird use data entry and analysis. The lack of bird use or abundance data has been a major impediment to understanding the implications of the data. For these reasons, the overall trend results were often inconclusive. The Monitoring Team and Committee have been able to confirm that the 50% reduction was not reached after 3 years, but the results of the monitoring program are far from conclusive and have left many stakeholders with a feeling of dissatisfaction.

11.6.2.3 Future Research

The Scientific Review Committee has also recommended research to improve understanding of how to reduce fatalities. They have crafted research proposals for which the county could seek funding. One research design is to better understand a specific species (burrowing owls) that scientists believe significantly inflates the number of fatalities. However, the costs of conducting the study have exceeded the resources available. So, the Committee developed a research study that the county could then share with other potential funders.

The Scientific Review Committee has also recommended studies to improve adjustment factors. Raw fatality numbers are typically adjusted as part of the analysis to manage for the number of birds that were likely scavenged between the time the bird was killed and the bird was discovered by a Monitoring Team field member. The numbers are also adjusted by a searcher detection error, based on an estimate of how likely a field member is to miss a carcass. These correction factors are always subject to great debate and a high error factor, so improving the research on these factors has been deemed significant for the Altamont Pass and potentially other wind farms.

11.6.2.4 Management Actions

The Scientific Review Committee has also issued recommendations, at the request of the county and the other settling parties, on management actions that the wind companies can take to potentially reduce fatalities. The conditional use permits stipulate some of the management actions, but not all. Also, the parties sometimes ask the Scientific Review Committee to prioritize management actions to inform the settlement discussions and negotiations on what to do in response to mortality that has not achieved a 50% reduction as required by the settlement agreement. After providing recommendations, the county and settling parties have then negotiated to what extent the wind companies would implement the strategies.

11.7 Collaborative Outcomes from the Altamont Process

There are a number of positive outcomes from the process employed at the Altamont, which can serve as models for resolution at other energy facilities. In addition to the collaborative tasks, the joint fact-finding process has successfully created transparency around the data and reaching consensus on turbine locations.

Open access to data is one challenge cited in reaching agreement involving scientific and technical information (Adler et al. 2000). In 2009, the Monitoring Team made a strong effort to make data transparent. After seeking consensus on filters to determine data quality, the Monitoring Team made the database publicly accessible on the Internet. All the interested parties could review and analyze all the fatality records collected by the Monitoring Team. Over time, the public database has been modified to include additional data and records from earlier studies that form the baseline study (NREL and CEC data). All interested parties can access the data, download the data, and run analyses. The "stability" of the data has improved substantially since this occurred. Further, all the parties conducting analyses on their own are using the same data. Historically, each analyst would develop his or her own data inclusion/exclusion rules. Consensus on the data filters and the public database are a substantial improvement for the Altamont. This transparency supports a stronger consensus on the data and the analytical results. This transparency and consensus has translated into increased trust among the stakeholders.

During the course of its work, the Committee has made two site visits to rate turbines for their potential hazard. Smallwood and Spiegel had previously tiered high-risk turbines (2005), and county permits stipulated a staged removal process using this tiering system. The wind companies are constantly working on, moving, or dismantling turbines for maintenance. As a result, turbine configuration is constantly shifting. Because of this shifting turbine landscape, the Committee and settling parties agreed that the Committee would go on site visits to make field observations and establish a risk rating system for turbines. The Committee has been able to reach consensus on turbine siting and location to reduce mortality that could be applied elsewhere or during repowering (P70 SRC Hazardous Turbine Relocation Guidelines).

11.8 Evaluating Success, Lessons Learned, and Future Challenges

Three demonstrated measures evaluate collaborative outcomes: agreements reached, quality of agreements, and improved working relationships (Emerson et al. 2009; Orr et al. 2008). One important performance outcome cited in the literature and practice is reaching agreement or stated outcome (Bingham 1986; Buckle and Thomas-Buckle 1986; Dukes 2004). The Scientific Review Committee has been able to reach many agreements and achieve consensus on its recommendations for study and management actions. The Committee has reached agreements with the Monitoring Team and others to support a common, credible database accessible to all. While the results of the monitoring study have been uncertain, the Committee and stakeholders have been able to reach consensus that repowering is necessary to reduce mortality significantly while maximizing power production. The Scientific Review Committee has been able to reach consensus on its recommendations even though not all of the recommendations have been implemented, which raises questions about the quality of the agreements.

Agreement quality can be considered through the (1) durability of an agreement to last over time; (see Table 11.3) (2) the practicality or ability of the parties to implement the agreement; (3) the agreement's flexibility or responsiveness to changed conditions; and (4) the agreement's provisions to be modified in response to changed conditions with monitoring and evaluation components incorporated (Susskind and Cruikshank 1987; Hamilton 1991; Sipe and Stiftel 1995; Innes and Booher 1999a, b; Susskind et al. 2000; Todd 2001; Kloppenburg 2002; D'Estree and Colby 2004). The management strategies that the Committee has recommended have often not been implemented fully consistent with its recommendations. Further, the Committee's recommended research proposals, as well as the monitoring program itself, have been quite costly. Often, the scientific deliberations have been lengthy. The length of time and the level of detail necessary to make decisions made interested parties sometimes hesitant to rely on the Committee. The Committee has been effective in modifying its recommendations in response to new information and data. In this manner, the monitoring and evaluation have been quite effective.

The parties developed shared intellectual capital. Innes and Booher (1999a, b) describe this as mutual understanding of each others' interests, shared definitions of the problem, and agreement on data, models, and quantitative or scientific descriptions of the issues. The Scientific Review Committee members, through the process of joint fact finding, work well together and have developed shared knowledge and understanding of the avian mortality issue and associated scientific complexities. This shared knowledge was extended toward other interested parties.

The Scientific Review Committee has deepened its knowledge and developed a productive working rapport. All of the scientists on the Committee have demonstrated enhanced collaborative capacity and skills as time went on. Their ability to engage in problem solving and deliberate on issues using the collaborative staged model has been evident as time progressed. They have become more sensitive to each

Table 11.3 Evaluating outcomes of the collaborative effort

Evaluative element	Altamont pass
Agreements reached	Monitoring protocols established Management strategies recommended Research studies designed to seek funding for additional research questions Common, credible data in accessible database Repowering is necessary to address the central problem (reducing mortality and maximizing power production)
Quality of agreements	Management strategies not always implemented
Agreement durability	Monitoring and periodic evaluation incorporated
Practicality	Agreements not always practical (excessive labor requirements or costs)
Flexibility, responsive to changing conditions	
Monitoring and evaluation incorporated	
Improved working relationships	Scientific Review Committee deepened knowledge and developed productive working rapport Intellectual capital enhanced Capacity to collaborate improved Stakeholders engaged with scientists on issues; relevant information shared in the process

other's expertise and found ways to integrate and rely on each other's disciplines and analytical strengths. This is consistent with Innes's (1998) observations that once this shared intellectual capital exists, parties reduce or better manage conflict.

The Scientific Review Committee has met its stated goals and achieved its mandate. It has improved its working relationship as an internal body. Once consensus was reached on common data and the data became transparent, many stakeholder relationships have improved as well. However, this forum has not succeeded in resolving the problem in the Altamont: how to maximize power production and reduce avian fatalities. This is a long, arduous process that necessitates a great deal of lengthy negotiation. At the time of this writing, the parties are proceeding toward repowering in a slow, cautious manner. All parties have concluded that repowering is necessary.

11.8.1 Challenges

The challenges to this project have been many. The monitoring results have had limitations, and the findings have been uncertain. This is complicated by the expense required to implement monitoring, analyses, and studies. The parties have a high degree of distrust, which the joint fact finding has only partially addressed. And, lastly, the settling parties have shifted the central question of how to repower to a different forum.

One of the major challenges to this scientific effort has been the high degree of scientific uncertainty. In the early years of monitoring, analysts have been unable to conclude with certainty whether there has been a reduction in avian mortality. Another frustration for interested parties is that while the monitoring is expensive and exhaustive from a labor standpoint, the geographic area is so vast and complex that the analysts cannot link trends in fatalities to any particular wind farm management action. Although the Monitoring Team and the Scientific Review Committee, during the design phase and since, have been explicit that the study would not be able to attribute the change in mortality to a specific action, the parties are still frustrated by this. For example, the effect of seasonal shutdown (when all turbines are shut down) has not been strong (ICF Jones and Stokes 2009).

The monitoring program costs have exceeded $1 million annually. The Committee and facilitation budget has added another layer of substantial expense. This has exceeded expectations. The elected officials who provide oversight and policy direction find this quite difficult to understand and support, given the inconclusiveness of the results.

Despite the costly budget, the monitoring work has had financial constraints. The budget has limited the ability of the Monitoring Team to analyze bird use data. There is no way to attribute the effect of the regional bird population fluctuations to the monitoring data and avian fatalities. Another related challenge to this effort has been the time and resources necessary to undertake the Committee's work in a difficult political environment. The Committee has met more often than was anticipated. Exploring possible management actions has required the parties to provide extensive materials, maps, and data. Discussions have often raised other questions that necessitated additional thinking and information development. Committee deliberations on a particular issue might stretch over several months due to these issues.

Over time, the Committee's joint fact-finding activities have built transparency and increased public/stakeholder trust in the data and analyses somewhat; however, relationships between stakeholders, and between some stakeholders and the Committee, have remained strained. The parties still question the motives of others, and not all parties view each other as acting in good faith.

Repowering will occur through a formal state and federal conservation planning. The parties have chosen to create a separate scientific panel to move forward with repowering. The reasons behind this decision have never been publicly stated. This could be significant since all parties have reached consensus that repowering, if done correctly, holds the most potential for reducing avian fatalities.

11.8.2 Lessons Learned

Although not its direct charge, the Committee, through its recommendations and research forum, has been unable to affect the policy dilemma of reducing avian fatalities while maximizing wind power production. The Committee has issued recommendations, but they have not always been implemented as planned. While the charge and responsibility of the Committee did not include resolving the key policy issue, the overall focus of the effort was to do so. This lack of resolution is not the fault of the Committee; however, it does call into question how well the scientific review function has contributed to resolving these difficult policy issues.

The monitoring program, faced with a high degree of uncertainty, has been expensive and inconclusive. Analyses have so far been unable to show much in the way of statistically significant avian mortality downward trend. It is not clear whether the cause of this outcome is poor study design, a lack of appreciation for the scientific and statistical complexities of the site, incomplete implementation by the wind companies of recommended management actions, an increase in avian populations that has masked the effect of management actions, a combination of these factors, or other factors entirely. The Committee, Monitoring Team, and parties have been unable to reach strong conclusions.

Occasionally the Committee has fallen into groupthink and become somewhat isolated from the reality of the cost and expenses necessary to implement its suggestions. This may have been a significant weakness of the Committee structure; however, it may have been what was necessary. Designing and making recommendations on scientific studies was one of the Committee's primary functions. The scientists seemed to have occasionally taken a purist attitude, placing the advancement of science above cost considerations. Devoting more extensive Committee meeting time to budgetary and financial issues and making budgetary issues more transparent might have enabled Committee members to agree on more financially pragmatic recommendations. If the scientists were working under a grant, they would have been limited in some way financially. Keeping budget limitations vague and unclear probably undermined the Committee's ability to grapple with resource limitations. On the other hand, maybe having the scientists identify the ideal recommendations served the county and other settlement parties in negotiating the choices, and associated costs, of the research program

The next major endeavor to address the issue of avian mortality and wind power is repowering, i.e., replacing old turbines with fewer, larger turbines. The issues in the Altamont revolve around science, environmental and energy politics, economics, and public policy. Altamont wind companies, the county and environmental organizations have differing objectives, yet they have common interests. Bird kills are theoretically breaking laws that are not being enforced. This has given the environmental interest groups power to make demands on the wind companies. However, environmental organizations support wind power as a form of electrical production. The wind companies, through their permits, are required to work on these issues and have an economic interest in maximizing wind power production and reducing conflict. The county would like wind power to be successful and are striving to create a political and policy environment to support county objectives. Despite these common interests, the existing structure is unlikely to yield consensus on the future of the Altamont.

If the parties chose to rely upon the Scientific Review Committee as a distinct entity, calling upon it for scientific recommendations only, broad-scale consensus on repowering is unlikely to be achieved. The decision-making structure needs to shift to consensus seeking for all the stakeholders and all the issues, including scientists.

While the county, wind companies and environmental organizations can rely on the recommendations of the Committee to inform their scientific discussions, all the stakeholders need to engage and reach consensus on the approach with scientists actively engaged. The issues are not just scientific: they are a complex web of political, scientific, and policy issues that need to be grappled with as a whole. Using the existing structure will likely require significant resources and time since the scientists function and deliberate in a separate scientific forum from where the parties are making the political and policy decisions. The scientists, wind companies, permitting agencies, wildlife protection agencies, landowners, and environmental organizations need to agree on the best approach to repowering Altamont. All of the stakeholder interests and scientific information need to coalesce in one consensus-seeking forum. This will ultimately ensure that the true policy dilemma, maximizing wind power and minimizing avian fatalities, is addressed in a way that all the parties themselves can recognize the achievement.

11.9 Conclusion

The Altamont Pass Scientific Review Committee has functioned effectively during its existence. The body has a clearly defined structure and role. It has performed its tasks, created working relationships among its members, increased its collaborative capacity, and contemplated methods to examine these issues in a manner that has scientific credibility. The Committee has had multiple collaborative outcomes that informed the work at Altamont Pass and could serve as models for resolution at other energy facilities. The Scientific Committee has performed its function of recommending and evaluating methods and the results of analyses regarding avian fatalities. The process has required significant human and financial resources that might be difficult to replicate in other places.

Significantly, during the tenure of the Scientific Review Committee, the key policy question of maximizing wind power while minimizing avian fatalities has remained largely unresolved. The parties have reached consensus that replacing outdated turbines with fewer more powerful new turbines, known as repowering, is the preferable and necessary approach. To address this in the future, a consensus-seeking structure that grapples with the scientific, political, and policy issues in one forum is necessary to truly engage and resolve the central issue.

References

D'Estree TP, Colby BG (2004) Braving the Currents: Evaluating Environmental Conflict Resolution in the River Basins of the American West. Boston: Kluwer Academic

Adler PS, Barrett RC, Bean MC, Birkhoff JE, Ozawa CP, Rudin EB (2000) Managing Scientific and Technical Information in Environmental Cases: Principles and Practices for mediators and Facilitators. Published by RESOLVE, Inc. US Institute for Environmental Conflict Resolution and Western Justice Center Foundation

Bartlett, G (2006) Center for Collaborative Policy, Issue Assessment: Altamont Pass Wind Resource Area and Avian Mortality. Available via AltamontSRC.org. Accessed 31 Mar 2010

Bingham G (1986) Resolving Environmental Disputes: A Decade of Experience, Washington: Conservation Foundation

Brown L (2006) Wind Energy Demand Booming. March 27, 2006. Available via http://www.RenewableEnergyWorld.Com. Accessed 15 Mar 2010

Buckle LG, Thomas-Buckle SR (1986) Placing Environmental Mediation in Context: Lessons from "failed" mediations in Environmental Impact Assessment Review 6(1):55-60

California Energy Commission 2008 Net System Power Report – Staff Report, Publication number CEC-200-2009-010, to be considered for adoption 15 Jul 2009

Center for Collaborative Policy Five States of Collaboration in Decision Making. Available via. http://www.csus.edu/ccp/collaborative/stages.stm, Accessed 31 Mar 2010

County of Alameda Board of Supervisors, Board Resolution, Adopted September 22, 2005, Number R-2005-453

Daulton M (2007) Impacts of Wind Turbines on Birds and Bats. Congressional Testimony of Mike Daulton, Director of Conservation Policy, National Audubon Society, Before the Committee on Natural Resources, Subcommittee on Fisheries, Wildlife and Oceans. Accessed 19 Mar 2010

Dukes EF (2004) What We Know about Environmental Conflict Resolution: An Analysis Based on Research. Conflict Resolution Quarterly 22(1-2):191-221

Ehrman JR, Stinson BL (1999) Joint Fact-Finding and the Use of Technical Experts. In: Susskind L, McKearnan S, Thomas-Larmer J (eds.) The Consensus Building Handbook, Sage Publications, Thousand Oaks

Emerson K, Orr P, Keyes D, McKnight K (2009) Environmental Conflict Resolution: Evaluating Performance Outcomes and Contributing Factors. Conflict Resolution Quarterly. 27(1):27-65

Hamilton MS (1991) Environmental Mediation: Requirements for Successful Institutionalization. In: Mills MK (ed.) Alternative Dispute Resolution in the Public Sector, Nelson-Hall, Chicago

Jones ICF, Stokes (2009) Draft Monitoring Report (M21) Altamont Pass Wind Resource Area Bird Fatality Study. Available via http://www.altamontsrc.org. Accessed 1 April 2010

Innes JE (1998) Information on communicative planning. J Am Plan 64:52-63

Innes JE, Booher DE (1999) Consensus building and complex adaptive systems: a framework for evaluating collaborative planning. J Am Plan, 65:412-422

Innes JE and Booher DE (1999) Reframing Public Participation: Strategies for the 21st Century, Planning Theory and Practice 5(4):419-436

Karl HA, Susskind LE, Wallace KH (2007) A Dialogue, not a Diatribe: Effective Integration of Science and Policy through Joint Fact Finding. Environment 49(1):20-34

Kloppenburg LA (2002) Implementation of Court-Annexed Environmental Mediation: The District of Oregon Pilot Project. Ohio State J Disput Resolution 17(3):559-596

Orloff S, Flannery A (1992) Wind Turbine Effects on Avian Activity, Habitat Use, and Mortality in Altamont Pass and Solano County Wind Resource Area. California Energy Commission: CA

Pachauri RK, Reisinger A (eds.) (2007) Climate Change 2007, Synthesis Report, Intergovernmental Panel on Climate Change, Contribution of Working Groups I, II and III to the Fourth Assessment Report of the Intergovernmental Panel on Climate Change, Accessed 19 Mar 2010

Orr P, Emerson K, Keyes D (2008) Environmental Conflict Resolution Practice and Performance: An Evaluation Framework. Conflict Resolution Quarterly. 25(3):283-302

Sipe N, Stiftel B (1995) Mediation Environmental Enforcement Disputes: How Well Does it Work? Environmental Impact Assessment Review 25:139-156

Smallwood, Neher (2004) Repowering the APWRA: forecasting and minimizing avian mortality without significant loss of power generation, California Energy Commission, CA

Smallwood, Thelander (2004) Developing Methods to Reduce Bird Mortality in the Altamont Pass Wind Resource Area, California Energy Commission, CA

Smallwood, Spiegel (2005) Combining Biology-Based and Policy Based Tiers of Priority for Determining Wind Turbine Relocation/Shutdown to Reduce Bird Fatalities in the APWRA, California Energy Commission, CA

Susskind L, Cruikshank J (1987) Breaking the Impasse: Consensual Approaches to Resolving Public Disputes, New York, Basic Books

Susskind LM, Van der Wansem, Ciccarelli A (2000) Mediating Land Use Disputes, Cambridge: Lincoln Institute of Land Policy

Todd S (2001) Measuring the Effectiveness of Environmental Dispute Settlement Efforts in Environmental Impact Assessment Review, 21(11):97-110

Chapter 12
Wind Energy in Vermont: The Benefits and Limitations of Stakeholder Involvement

Mary R. English

Contents

Abstract Wind energy … what's not to like about it? With growing concerns about climate change and tightened regulation of conventional air pollutants, the United States is climbing on the wind energy band wagon. Wind and other renewable sources of electricity are being promoted at the federal level through production tax credits and at the state level through renewable portfolio standards. But how do

M.R. English (✉)
Institute for a Secure and Sustainable Environment,
University of Tennessee, Knoxville, TN 37996, USA
e-mail: menglish@utk.edu

J. Burger (ed.), *Stakeholders and Scientists: Achieving Implementable Solutions to Energy and Environmental Issues*, DOI 10.1007/978-1-4419-8813-3_12,

utility-scale wind energy projects play at the local level? Not well. Focusing on a proposed 80-MW project in southwestern Vermont, this chapter examines both the possibilities and the limitations of stakeholder involvement in large-scale wind turbine projects.

12.1 Introduction

As concerns about climate change due to anthropogenic carbon emissions soar and as conventional air pollutants become more tightly regulated, we are turning to renewable sources of energy, including wind energy, for electricity generation. National and state policies are being crafted to help promote wind energy. Nevertheless, wind energy on a grand scale is encountering both technical impediments and local resistance. The two are intertwined: If wind energy is not a reliable, significant source of electricity generation, are the sacrifices that may be required from locally affected people worth it? What role can meaningful local involvement play in helping to make wind energy projects viable?

12.1.1 Objectives for this Chapter

This chapter is meant to prompt reflection about the possibilities and limitations of stakeholder involvement in proposed projects that, while potentially beneficial to society, may have negative local impacts. The focus here is on utility-scale wind energy projects in general and on one case in particular: a proposed 40-turbine project in southwestern Vermont. To understand such projects, the technical and regulatory context must be understood. Only then can one assess the possible roles of stakeholder involvement.

12.1.2 Wind Energy for Electricity: A Viable Technology?

Electricity is produced from wind turbines that typically have a tubular tower supporting three blades attached to a nacelle. Figure 12.1 shows a 1.5-megawatt (MW) wind turbine.

The nacelle houses a blade-driven shaft, a gearbox, and a generator. The tower is usually made of steel; the blades, of a lightweight composite material. In 1995 the total height of a typical land-based utility-scale wind turbine (tower with blades) was about 50 m (164 ft), but by 2005 the total height was about 126 m (413 ft) – roughly comparable to a 40-story building (European Wind Energy Association [EWEA] 2009). The span of the blades grew, accordingly, to as much as 40–80 m (131–262 ft) for a land-based turbine (Ueda and Shibata 2004). The installed capacity of a land-based utility-scale wind turbine – i.e., the theoretical amount of electricity

Fig. 12.1 A 2-MW wind turbine

that the turbine could generate if it operated continually and with optimum wind speed – is now usually in the 1.5–2 MW range.

Offshore wind turbines – common in Europe but only now being introduced in the United States – are expensive and technically challenging to develop and, thus, because of economies of scale with greater size, are usually larger than land-based turbines. For example, a 4.5-MW offshore wind turbine being developed by Vestas Wind Systems has a blade diameter of 120 m (393 ft, e.g., longer than a football field) and a total height of over 160 m (525 ft, e.g., as tall as a 50-story building) (Randolph and Masters 2008). This chapter will discuss only land-based wind turbines.

Wind energy offers several advantages compared with conventional sources of electricity. Wind energy does not contribute to global climate change or to air pollution, as do coal-fired plants and, to a lesser extent, natural gas-fired plants. Wind energy does not require resource extraction, as does electricity from coal, natural gas, and nuclear power, nor does it have the waste problems of coal and nuclear power.

Wind energy is not universally regarded as an unqualified good, however. Wind turbine projects and their associated infrastructure (e.g., access roads and transmission lines) may disturb wildlife habitat. The turbines may be visually intrusive; in addition, their moving blades may result in bird and bat kills (National Research Council 2007, see Bartlett 2011) and may produce sound that annoys and perhaps compromises the health of nearby residents (Nissenbaum 2010). These downsides depend on the location and scale of the wind turbine project: e.g., whether they are located on rolling plains or on mountain ridgelines, and whether a single turbine or numerous turbines are to be installed.

Intermittency can be a significant limitation of wind energy as an electricity source: The more variable the wind, the less a wind turbine project can be relied on to generate electricity when it is needed. The "capacity factor" (i.e., the percentage of installed capacity that is actually generated) for utility-scale wind turbines in U.S. areas with good wind regimes ranges from about 25 to 35% (Randolph and Masters 2008). In contrast, nuclear and coal-fired plants have average capacity factors of 91 and 72%, respectively (Energy Information Administration [EIA] 2010a).

Intermittency would diminish as a problem if the electricity generated could be stored and then dispatched at a later point, but an affordable means of large-scale electricity storage is not yet available. The main method of providing energy storage for electric utilities has been hydroelectric pumped storage, where water is pumped into an uphill reservoir when demand for electricity is low and released downhill through electricity-generating turbines when demand is high. Pumped storage systems, however, are not only difficult to site but also inefficient and costly (Randolph and Masters 2008). Various alternatives are being developed, notably including batteries. Large-scale batteries for stationary electricity sources, while promising, are not yet cost-effective (American Wind Energy Association n.d.).

It has been argued that electricity storage may not be needed to make wind a reliable source: Aggregating the electricity generated by a number of wind turbine projects located over a large geographic area will help to remedy the intermittency problem (EWEA 2009; U.S. Department of Energy 2008). The wind is always blowing somewhere. But large-scale integration of wind energy requires a commensurately large-scale grid; this poses a challenge for modernizing and expanding transmission line infrastructure. Integration of wind energy also requires balancing with other, more readily dispatchable electricity sources that can be ramped up to meet changing demands.

12.2 Background on Utility-Scale Electricity from Wind Power

Total global wind energy capacity has grown dramatically in the past two decades, from 1,743 MW in 1990 to 94,122 MW in 2007 (EWEA 2009). This translates to over 94 gigawatts (GW) of global installed capacity in 2007. (By comparison, the installed capacity of a large coal-fired plant or nuclear power plant is roughly 1 GW).

The European Union (EU) is responsible for a large percentage of the recent growth in wind energy capacity: In 1990, the EU accounted for 25% of the total global installed capacity; by 2007, it accounted for 60% (EWEA 2009). The growth of wind energy in the EU is likely to continue. In 2001 the EU passed a directive promoting renewables in its internal electricity market, and in March 2007 the EU heads of state adopted a binding target of 20% of electricity from renewables by 2020. (In 2005, renewable energy in the EU accounted for 8.5% of electricity generated).

The path to wind energy penetration of the U.S. electricity market has been more erratic.

12.3 Wind Energy in the United States

In the United States, the oil shocks of the 1970s triggered interest in alternative sources of energy. A combination of federal and state tax credits set off the expansion of wind capacity in California in the 1980s, with a proliferation of wind turbines (individually small by today's standards) in areas such as Altamont Pass, San Gorgonio, and Tehachapi (EWEA 2009). Following this growth spurt, wind energy development in the United States flagged during the 1990s, due perhaps to deregulation of electric utilities and the resulting restructuring of the industry.

Since 1999, with monetary incentives – notably a federal renewable electricity production tax credit (PTC) – and with state-mandated renewable energy requirements – notably through state renewable portfolio standards (RPSs) – wind energy has grown rapidly. The PTC acts as a carrot, fostering investment in wind turbine projects. The RPS acts as a stick, mandating that utilities expand beyond conventional, nonrenewable sources of electricity. Figure 12.2 shows the cumulative U.S. wind capacity, by year.

The PTC is a federal tax credit for electricity generated and sold by qualified renewable sources. Originally enacted in 1992, the PTC provision expired at the end of 2001 but – with some lapses – has since been renewed and expanded several times, most recently in February 2009 through the American Recovery and Reinvestment Act. Wind turbine projects that are in service by the end of 2012 are eligible for a tax credit of 2.2 cents per kilowatt-hour; the duration of the tax credit is generally 10 years (Database of State Incentives for Renewables and Efficiency 2010).

A federal RPS has been proposed but as of mid-2010 had not been enacted. As of the end of 2009, however, 30 states had enacted RPSs with enforceable mandates (EIA 2010b). Most were adopted within the past 10 years. State RPS requirements range between 15 and 30% of electricity sales or, in some states, of installed capacity. Requirements typically are set for a future year (e.g., 2020), with interim requirements prior to that year. In some states, more stringent requirements are imposed on investor-owned utilities (IOUs) than on the typically much smaller cooperative and municipal utilities.

As of the end of 2009, the U.S. as a whole had nearly 35 GW of installed wind energy capacity – up from fewer than 3 GW at the end of 1999. Texas had 9.4 GW, Iowa, 3.6 GW; and CA, 2.8 GW. Nine other states – Colorado, Illinois, Kansas, Minnesota, New York, Oklahoma, Oregon, Washington, and Wyoming – had

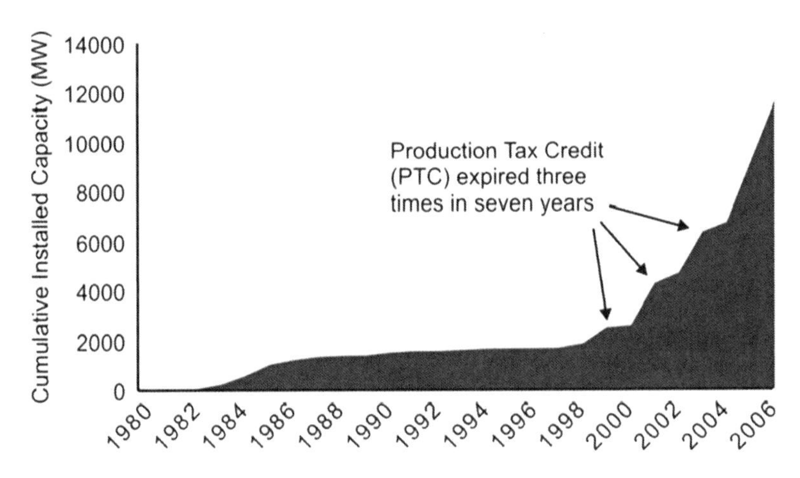

Fig. 12.2 Cumulative U.S. wind capacity, by year (in megawatts)

between 1 and 2 GW. Most other states had at least a nominal amount; only eight states – Connecticut, Delaware, Florida, Georgia, Maryland, Nevada, North Carolina, and Virginia – had none as of the end of 2009 (Wind Powering America n.d.). In 2008, electricity from wind energy represented 2.3% of total installed capacity (EIA 2010a).

The rate of growth in demand for electricity has slowed dramatically in recent decades. Between 2000 and 2008, the growth rate was 0.9% per year, down from 9.8% per year in the 1950s (EIA 2010b). This change is due largely to increased energy efficiency in buildings, industrial processes, lighting, and appliances. Nevertheless, population growth and related economic expansion continue to push total electricity demand upward. Electricity demand in the United States is expected to increase 30% between 2008 and 2035, assuming a "Reference" case in which current trends continue (EIA 2010b). With 45 GW of generating capacity expected to be retired by 2035 (due, e.g., to aging plants), an additional 250 GW of capacity will be needed to meet demand in 2035, up from 1,104 GW in 2008 (EIA 2010a).

12.4 The Vermont Scene: Background

To understand wind energy in Vermont, one must understand the state and its electricity sources. Each state in the United States is different. Vermont may be more different than most.

12.4.1 A Small Rural State on the Canadian Border

Vermont, located in northern New England, has a land area of 9,250 square miles (less than 4% of the size of Texas, the largest state in the lower 48) and a 2000 population of 608,827 (0.2% of the 2000 U.S. population). The rate of population growth in Vermont tracked that of the United States as a whole between 1960 and 2000: Both had a 56% increase. Nevertheless, Vermont remains a small, mostly rural state. Its largest city, Burlington, had a 2000 population of just under 39,000. Its capital, Montpelier, is the smallest in the United States, with fewer than 9,000 residents.

Vermont's population is predominantly non-Hispanic white (95% in 2008, compared with a U.S. average of 66%) and well-educated. (In 2000, 29% of Vermonters aged 25 and over had a college degree, compared with 24% in the US as a whole.) Vermont is not a wealthy state, however: In 2000, its median household income was just over $52,000, virtually the same as for the United States as a whole. Vermont's population is considerably older than that of the United States as a whole: In 2008, only 20.8% were under 18 years old (compared with 24.3% for the US as a whole), while 13.9% were aged 65 or older (compared with 12.8% for the US as a whole). Vermonters tend to stay put. In 2000, 59% had lived in the same house at least 5 years (compared with 54% for the US as a whole), and nearly 70% owned their homes (compared with 66% for the US as a whole).

12.4.2 Electricity Sources

Vermont has 20 electric utilities: three IOUs, 15 municipal utilities, and two rural electric cooperatives. Of these, two of the IOUs – Central Vermont Public Service and Green Mountain Power Corporation – have the lion's share of the state's electricity customers: 44 and 26%, respectively.

Total electricity demand in Vermont is roughly 1,000 MW. Electric utilities in Vermont own few of their own generation resources. Instead, they purchase most of their power. As of 2003, according to the Vermont Department of Public Service (DPS), 35.5% of the state's electricity was supplied by Vermont Yankee nuclear power plant, with 28.2% imported from Hydro-Quebec, a utility owned by the province of Quebec. Most of the remainder came from other purchased power and from in-state hydro. The Vermont utilities' contract with Hydro-Quebec, which began in the early 1980s and was later expanded, was to expire in 2016. Due to increased demand within Quebec, a long-term contract renewal was uncertain, but on August 12, 2010, the contract was renewed until 2038, albeit at 225 MW rather than the former level of roughly 300 MW of assured power. Electricity from the Vermont Yankee nuclear power plant is in jeopardy.

Vermont Yankee, located in Vernon on the Connecticut River a few miles upstream from Massachusetts, may be shut down. A 620-MW reactor currently owned by Entergy, it came on-line in 1972 and its license will expire in 2012. In 2006, Entergy applied to the U.S. Nuclear Regulatory Commission for a 20-year license extension, and its application is under review. In February 2010, however – spurred by concerns voiced by plant opponents, state senators, and others about leaking tritium, the plant's structural integrity, and misstatements by plant officials – the Vermont Senate voted 26 to 4 to retire the plant in 2012. Under Vermont law, both houses of the legislature have to approve the plant's license by issuing a "certificate of public good," and when Entergy acquired Vermont Yankee in 2004, it signed a memorandum of understanding that the certificate would expire with its license in 2012 and a new certificate would be required (Wald 25 February 2010).

12.4.3 Vermont's Version of a Renewable Portfolio Standard

Like a few other states, Vermont has adopted a goal-based rather than a standards-based approach to an RPS. According to legislation passed in 2005, between 2005 and 2012 retail electricity suppliers must meet new demand by adding an equivalent percentage of renewable energy supply. For instance, if demand increases by 5% during this 7-year period, 5% of total supply must come from eligible renewable sources. Eligible sources include, e.g., wind and solar, but specifically exclude all nuclear power. Until a 2010 act (see Sect. 12.7), hydro facilities with a generating capacity of more than 200 MW also were excluded. Power purchased from out-of-state eligible sources may be counted in meeting the goal. In 2012, the state's Public Service Board (PSB) will assess whether the goal has been attained; if not, it will become a mandatory standard.

12.4.4 Wind Energy Projects in Vermont to Date

Vermont's wind resources are along its north–south ridge crests. According to Wind Powering America, a program of the U.S. Department of Energy, wind resources along the spine of the Green Mountains are considered to be especially good; so too are resources along ridges in western Vermont and in the northeastern corner of the state. Figure 12.3 shows Vermont's wind resources at 80 m above ground.

The first utility-scale wind energy project in Vermont went on-line in July 1997. Constructed by enXco, a renewable energy development corporation, and owned by Green Mountain Power, the 6-MW project includes 11 turbines, each with an installed capacity of 550 kW and a total height of roughly 197 ft. The project is located in Searsburg, a small town (2000 pop., 96) in southern Vermont about ten miles north of the Massachusetts border. Located at about 2,800 ft above sea level but not in a visually prominent area, the project apparently has had the support of Searsburg residents but has been fraught with weather-induced difficulties, particularly from lightning and also from high winds that led to the collapse of one turbine in October 2008.

Despite the endorsement of utility-scale wind energy by the state and the Vermont Energy Partnership, a nongovernment organization, Searsburg remains the only utility-scale wind energy project in Vermont. As of mid-2010, a few others were being proposed: Deerfield Wind, a 30- to 45-MW expansion of the Searsburg project; Kingdom Community Wind, a 63-MW project in Lowell, in far northern

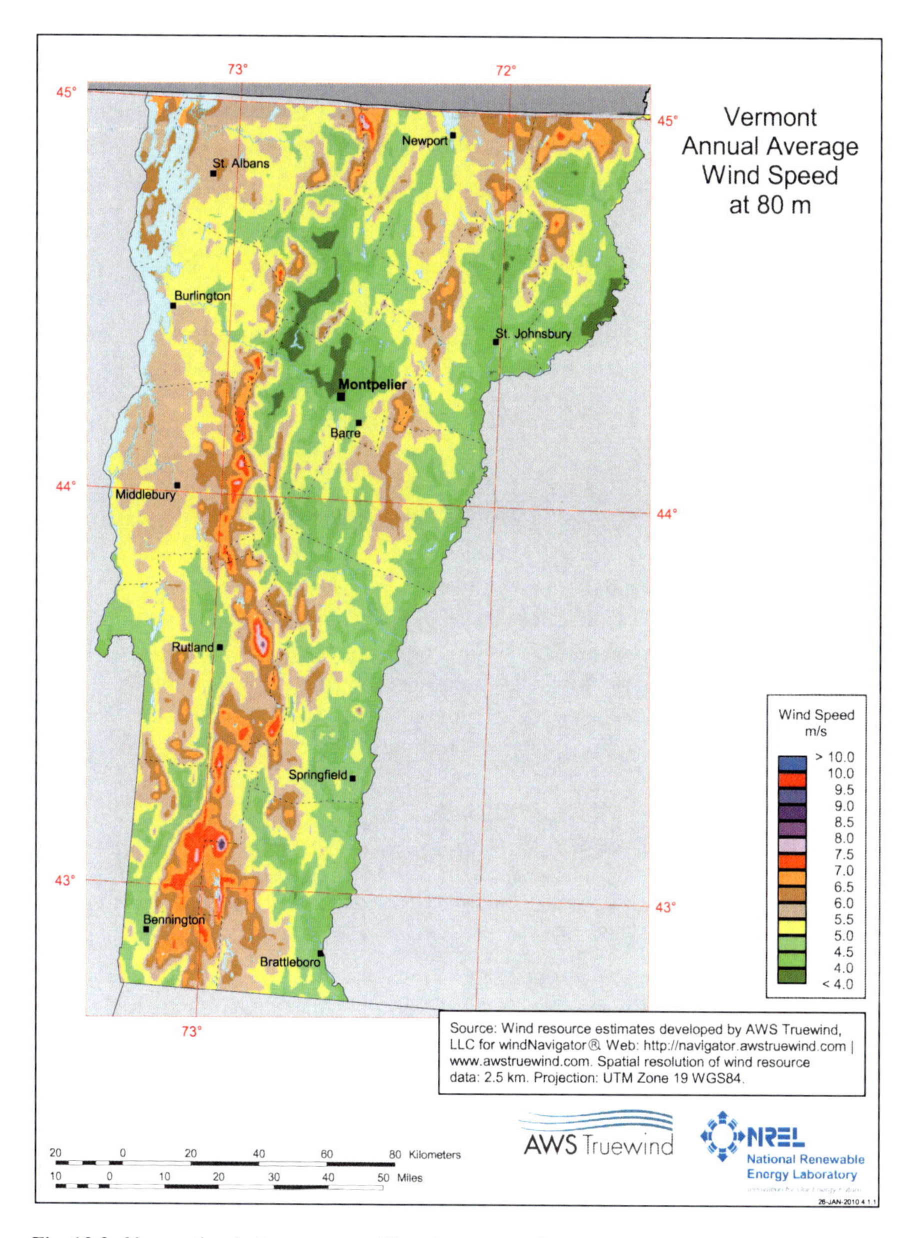

Fig. 12.3 Vermont's wind resources at 80 m above ground

Vermont; Sheffield Wind, a 40-MW project, and East Haven, a 6-MW project, both in northeastern Vermont; and Georgia Wind, a 12-MW project in northwestern Vermont (Page 2010). As of mid-June 2010, only Sheffield Wind and Georgia Wind had received a certificate of public good from the Vermont PSB.

One of the largest projects that had been in the works – the Vermont Community Wind Farm (VCWF) – was tabled in the spring of 2010, shortly before its developer had planned to petition the PSB for a certificate of public good. The remainder of this chapter focuses on that project, because of its scale and the controversy it provoked.

12.5 Vermont Community Wind Farm: Background

To understand the possibilities and limitations of stakeholder involvement in a proposed project, it is essential to understand not only the project but also its regulatory context. This section provides background information on both.

12.5.1 The Developer and Site of the Proposed Project

Vermont Community Wind (VCW) is a wind energy development corporation based in Charlotte, Vermont, with offices in Montpelier and Rutland County. Rutland County, in southwestern Vermont, was the proposed location of the VCWF project.

The VCWF project initially was proposed for ridgelines spanning six small towns in Rutland County (Poultney, Tinmouth, Middletown Springs, Clarendon, Ira, and West Rutland) with a total of 60 turbines. According to Jeffery Wennberg, VCW spokesperson:

> "A series of features makes this site on the Taconic ridgeline of Rutland County one of the top three sites potentially developable in northern New England. It is an extremely high-value site…. There is no other location where you have wind resources of this quality and such immediate access to the grid. There will be no building or upgrading of power lines. This is an exceptionally good location" (Widness 14 January 2010).

Most of the land involved – about 2,900 acres – is purportedly owned by the Yale University Endowment Fund through its for-profit arm, Yankee Forest LLC. (Yale University does not make its investments public.) An additional tract of more than 1,100 acres is owned by NFTI Limited Partnership. Both tracts are managed by Wagner Forest Management, Ltd. of Lyme, New Hampshire. In January 2009, VCW secured a lease with Wagner for 4,000 acres. Most of the turbine sites were to be on Wagner land, but for the project to work land owned by 19 other property owners was needed. Through confidential agreements, landowners arranged to lease their land to VCW in exchange for small annual payments plus a percentage of gross sales (Widness 14 January 2010).

By January 2010, the anticipated number of turbines had been reduced to 45 on sites spanning Poultney, Clarendon, Ira, and West Rutland. Most of the turbines were planned for Ira, where Yankee Forest LLC owns approximately 1,400 acres. All of the sites were under lease (Widness 14 January 2010). Each of the wind turbines was to have an installed capacity of roughly 2 MW and would be more than 400 ft in total height. The electricity generated was to be purchased by Central

Vermont Public Service, whose service area includes Rutland County and most of the rest of central and southern Vermont.

12.5.2 *The State Permitting Process*

Vermont's Section 248 process is key to siting electricity generation and transmission projects. A highly centralized process, Section 248 contrasts with Vermont's Act 250, which governs applications for large-scale and environmentally sensitive developments and allows for extensive regional and local influence over whether these applications are approved.

According to a provision in Act 250, electricity generation and transmission facilities that require a certificate of public good under Section 248 are excluded as "developments" under Act 250 (10 Vermont Statutes Annotated (VSA) Section 6001(3)(D)(ii)). Some people, such as the director of Vermonters for a Clean Environment, argue that Act 250 should apply to wind energy projects (Smith 19 August 2010).

12.5.2.1 Vermont's Section 248 Process

Under Vermont's Section 248, the approval of the PSB is required for in-state electricity generation and transmission projects, including wind turbine projects (30 VSA Section 248). Before site preparation or project construction is begun, the developer must receive a certificate of public good from the PSB. In addition, depending upon the site, additional environmental permits may be needed from other state agencies, especially the Agency of Natural Resources, and from federal agencies. Under Section 248, local zoning permits are not required.

The PSB is a quasi-judicial body that consists of a full-time chair and two part-time members appointed by the governor for staggered 6-year terms. As described in a citizens' guide to the PSB's Section 248 process, the PSB is assisted by staff, including attorneys and others with expertise in financial, engineering, environmental, and policy analysis (PSB n.d.).

The citizens' guide lists the following steps in the Section 248 process (PSB n.d.):

- *Petition*: The entity seeking to construct the proposed project files a petition for a certificate of public good. (According to a PSB rule, a petition for a wind turbine project must provide the maximum vertical and horizontal dimensions of the towers and blades and the maximum decibel level that the turbines will produce at the nearest residence, measured over a 7 PM to 7 AM period.)
- *Prehearing Conference*: The PSB holds a prehearing conference to determine how the case will be managed and to identify issues that will need to be resolved and potential parties to the case.
- *Site Visit*: The PSB usually makes a site visit to get a sense of the project's potential impacts.

- *Public Hearing*: Under Section 248, the PSB is required to hold a public hearing in at least one county in which the project is proposed. The PSB usually opens the public hearing with a brief description of the project by the petitioner; it then takes oral comments that are transcribed by a court recorder. (The PSB also encourages people to submit written comments.)
- *Intervention*: The next step is to identify the formal parties to the case. Those who automatically have standing include the petitioner, the DPS, and the Agency of Natural Resources. Others who may qualify as intervenors include, e.g., landowners, public interest groups, and environmental or business organizations. To become an intervenor, one files a motion to intervene explaining the nature of the interest that may be affected by the outcome of the case. The PSB then issues an order granting or denying the motion. It also may restrict participation to the issues in which the intervenor may be affected. Intervenors may be represented by an attorney or may represent themselves.
- *Prefiled Testimony*: The petitioner files written testimony and exhibits with the original petition. After "discovery" and before the evidentiary hearing, other parties may file written testimony and exhibits.
- *Discovery*: Discovery is an opportunity to ask other parties about their exhibits or testimony; it may be conducted orally or in writing.
- *Evidentiary Hearings*: Prefiled testimony is entered into the evidentiary record; witnesses are called and may be cross-examined.
- *Briefs*: At the close of evidentiary hearings, parties may file briefs with the PSB. Typically, two rounds are filed: an initial brief and a reply brief.
- *Decision*: The PSB issues a decision in the form of a final order. It must be based on the evidentiary record and on findings of fact under the Section 248 criteria as well as conclusions of law. Final orders are subject to motions for reconsideration and may be appealed to the Vermont Supreme Court.

Section 248 includes 10 criteria, which are summarized below:

1. The project will not interfere unduly with the orderly development of the region.
2. The project is required to meet the need for present and future demand for service that could not otherwise be provided more cost-effectively through energy conservation, efficiency, and load management measures.
3. The project will not adversely affect system stability and reliability.
4. The project will result in an economic benefit to the state and its residents.
5. The project will not have an undue adverse effect on esthetics, historic sites, air and water purity, the natural environment, and the public health and safety (regarding this criterion, see Sect. 12.5.2.2).
6. The project is consistent with the principles for resource selection in the company's approved least-cost integrated plan.
7. The project complies with the state electric energy plan approved by the DPS, or there is good cause to permit the proposed action.
8. The project does not have an undue adverse affect on waters that have been designated as outstanding resource waters by the Water Resources Board.

9. With respect to a waste-to-energy facility, the project is consistent with the state solid waste management plan.
10. The project can be served economically by existing or planned transmission facilities.

12.5.2.2 Vermont's Act 250

Within 30 VSA Section 248, Criterion 5 is expanded to incorporate criteria from Vermont's Land Use and Development law, enacted in 1970 and commonly referred to as "Act 250" (VSA Title 10 Chap. 151). Through Act 250, nine district environmental commissions, each with three members and a staff, were created to review applications for large-scale and environmentally sensitive developments, including all construction above 2,500 ft in elevation. Act 250 also created a nine-member state environmental board to hear appeals from the district commissions' permit decisions: but in 2005, this board was eliminated. A natural resources board was created to oversee the Act 250 process, with appeals from district commissions directed to an Environmental Court judge and subsequent appeals to the Vermont Supreme Court.

The district commissions must base their decisions on Act 250's criteria. These focus on the project's expected impacts on air and water quality, water supplies, traffic, educational and municipal services, and historic and natural resources, including scenic beauty and wildlife habitat (Vermont Environmental Board 2000). These are the criteria incorporated into Section 248. Under Act 250 (but not under Section 248), developments also must conform to local and regional land use plans.

Parties to an Act 250 hearing automatically include the municipality and its planning commission, the regional planning commission, and affected state agencies as well as the permit applicant. The district commission also may grant party status to adjoining property owners and to others who qualify under the state environmental board's rules.

12.5.3 Stakeholder Participation: The Local Role in the Permitting Process

Individuals and organizations can participate in the Section 248 process in two ways: as a member of the public and as a formal party to the case. Although not automatically parties to the case, affected town governing bodies (typically, the town's select board) and local and regional planning commissions must receive notice. In addition, construction plans must be provided by the developer to the local municipal and regional planning commissions at least 45 days prior to filing a petition with the PSB.

Local approval of a proposed project is not required for a certificate of public good to be granted, but the PSB must give "due consideration" to "the recommendations of the municipal and regional planning commissions, the recommendations of the municipal legislative bodies, and the land conservation measures contained in the plan of any affected municipality" (30 VSA Section 248 (b)).

12.6 Stakeholder Involvement at the State and Local Levels

Stakeholder involvement is not, or should not be, a one-step process. Especially, as a transition occurs from policies to projects, different phases of stakeholder involvement engaging different people may be needed.

12.6.1 Stakeholder Involvement at the State Level

At the state level, stakeholder involvement regarding Vermont's policies on utility-scale wind energy has been robust. Two efforts stand out: the Wind Siting Consensus-Building Project in 2002 and the Commission on Wind Energy Regulatory Policy in 2004.

12.6.1.1 The Wind Siting Consensus-Building Project

In 2002, the state's DPS launched a project to build consensus on the appropriate siting of utility-scale wind turbines in Vermont, with special attention to aesthetic impacts on Vermont's scenic landscape. The project was made possible by a grant from the U.S. Department of Energy and administered by the DPS, which partnered with Vermont Environmental Research Associates, Renewable Energy Vermont, and the Woodbury Dispute Resolution Center.

The Wind Siting Consensus-Building Project had three parts:

1. Four consensus-building workshops held in spring 2002 at the Woodbury Center.
2. A packet of wind project planning resources for local and regional planners.
3. A state-wide education and outreach initiative.

The workshops engaged approximately 60 stakeholders from around the state, including representatives of state agencies, local and regional planning commissions, environmental organizations, utilities, and wind development corporations and associations (see Table 12.1).

The focus and outcomes of the four workshops were as follows:

Workshop 1: The main aim of the first workshop was to provide information about wind energy and electricity demand in Vermont, to acquaint the participants with each other, and to prioritize the interest areas that had been identified in responses to a preworkshop questionnaire. Following presentations, a good deal of discussion ensued, much of it centering on aesthetics, land use, and land use impacts. According

Table 12.1 Key stakeholders involved in the 2002 wind siting consensus-building workshops

Agency or group	Role
Governmental organizations	
Regional planning commissions	RPCs are staffed with professionals in environmental, land use, transportation, and economic planning; they develop comprehensive plans for their regions and assist municipalities in their planning. (VT has no official county government system.) RPCs must be given advance notice of applications to the PSB for certificates of public good, and their plans must be given "due consideration"
VT Agency of Natural Resources	ANR includes three departments: Environmental Conservation, Fish and Wildlife, and Forests, Parks and Recreation. ANR is automatically a party to a PSB proceeding regarding granting a wind energy project developer a certificate of public good
VT Department of Fish and Wildlife	FWD manages the state's fisheries and wildlife resources and studies and inventories the state's nongame species and natural communities. Some may be affected by wind turbine projects
VT Department of Forests, Parks and Recreation /State Lands	FPR/State Lands is responsible for administration of all of ANR's state lands (state parks, forests, natural areas, etc.); administration responsibilities include leases and special use permits
VT Department of Public Service	DPS is charged with representing the public interest in matters concerning energy, telecommunications, water, and wastewater. One of its responsibilities is to provide long-range planning through the Vermont Electric Plan and Comprehensive Energy Plan. It is separate from the Public Service Board
VT Public Service Board	The three-member PSB (with it staff) serves as the decision-making authority in utility regulatory cases, including proceedings on applications for certificates of public good from developers of proposed wind energy projects
VT Governor's Office	A staff member from the office of Governor Howard Dean participated in the workshops
U.S. Senators' and Representative's Offices	Staff from the Vermont-based offices of U.S. Senators Patrick Leahy and Jim Jeffords and U.S. Rep. Bernie Sanders participated in the workshops
Wind energy companies	
Catamount Energy	Based in Rutland, VT, this company has wind energy projects in the U.S. and U.K. (In 2008, it became a subsidiary of Duke Energy)
EMDC (aka East Haven Windfarm)	This company is based in northeastern VT; in 2003, it proposed a 6-MW demonstration project in East Haven
Northern Power Systems	Based in Warren, VT, NPS was founded in the 1970s; it has expanded from community-scale to utility-scale turbines. (In 2008, it was acquired by Wind Power Holdings, LLC)
Endless Energy	Established in 1987, Endless Energy is a Maine-based wind energy development company with operations in VT
VT Environmental Research Associates	Established in the 1980s, VERA provides technical support services for wind energy development companies
NRG Systems	Founded in 1982 and based in Hinesburg, VT, NRG Systems manufactures wind energy measurement systems

(continued)

Table 12.1 (continued)

Agency or group	Role
Wind energy organizations	
Renewable Energy Vermont	Based in Montpelier, VT, REV advocates and lobbies for using wind energy and other sources of renewable energy in Vermont. Its members include businesses and individuals
Utilities	
Washington Electric Co-op	WEC, based in East Montpelier, VT, serves about 10,000 customers (mainly residential) in north-central Vermont
VT Public Power Supply Association	VPPSA is a private authority enabled by a state statute to buy and sell wholesale power in Vermont and wholesale and retail power outside the state. It sells to municipal and cooperative utilities
Environmental organizations	
VT Institute of Natural Science	Founded in 1972, VINS is a nonprofit organization that does environmental education and research and avian wildlife rehabilitation
Green Mountain Club	Founded in 1910, the GMC protects and maintains the Long Trail (a 273-mile north-south hiking trail with 175 miles of side trails) and seeks to protect other hiking trails and mountains in VT
Catamount Trail Association	Begun in 1984, the Catamount Trail is a 300-mile cross-country skiing trail that runs north through VT on private and public land
Appalachian Trail Conservancy	The ATC has its roots in a 1925 conference in Washington, DC, to establish a 2,000-mile north-south trail through the Appalachian Mountains. The AT coincides with the Long Trail through southern VT until Rutland, where the AT bends east to NH
VT Land Trust	Founded in 1977, VLT seeks to protect Vermont's farms, forests, and wilderness through, e.g., acquiring development rights and conservation easements
VT Natural Resources Council	Founded in 1963, VNRC is a nonprofit organization that seeks to protect and restore the state's natural resources by influencing state and local policies
Natural Resources Council of Maine	Founded in 1959, NRCM's purpose is similar to VNRC's
Audubon VT	Founded in 1901, Audubon Vermont seeks to promote environmental awareness and education and to conserve birds, other wildlife, and essential habitat
Northern Forest Alliance	NFA is a coalition of conservation, recreation, and forestry organization that seeks to protect the forests of Maine, NH, and VT and promote their sustainable use
Forest Watch	Founded in 1994 as Green Mountain Watch, this nonprofit organization seeks to restore wilderness, protect imperiled wildlife, improve public land management, and promote ecological forestry in New England
The Nature Conservancy	Launched in 1960, the VT chapter of TNC seeks to protect ecologically important land and water

(continued)

Table 12.1 (continued)

Agency or group	Role
Other organizations	
VT Farm Bureau	Launched in 1915, VtFB is a state association of county-based farm bureaus whose members are farmers
VT Ski Areas Association	Based in Montpelier, VT, VSSA is an association of downhill and cross-country ski areas
Contractors and consultants	
Woodbury Dispute Resolution Center	Based in Montpelier, the Woodbury Dispute Resolution Center provides facilitation and mediation services and workshops
Landscape architects	Two were involved: Jean Vissering, who developed a draft guide on visual resource considerations in siting wind turbines, and Terry Boyle, whose work has included evaluating the aesthetic impacts of wind turbines
Multiple Resource Management Inc	This company offers wildlife consulting, environmental appraisal, and GPS resource mapping services

Source: information based on Vermont Department of Public Service 2002, appendix 2, list of participants

to a summary of the first workshop, it became apparent that participants were willing to not only identify, discuss, and defend their own positions and interests but also seek to understand the positions of others before forming opinions. The summary noted that the amount of information to be digested and the number of issues to be discussed would make it difficult for the participants to reach consensus (Woodbury Dispute Resolution Center, Appendix I in Vermont Department of Public Service 2002).

Workshop 2: This workshop focused on the potential visual impacts of large wind turbine projects. Jean Vissering, a Montpelier-based landscape architect, presented her working draft of "Visual Resource Considerations in Siting and Designing Wind Facilities Larger than 500 kW," which triggered extensive discussion. A summary of Workshop 2 commented that a noteworthy outcome was the participants' confirmation that aesthetics might be one of the more contentious issues. The report summary also noted that the level of importance of this issue varied, and that reaching a consensus on this issue would be difficult because of the difficulty of establishing a widely acceptable aesthetic norm (Woodbury Dispute Resolution Center 2002).

Workshop 3: The third workshop centered on the potential impacts of wind energy projects on birds and other wildlife. Following a presentation by the Vermont Department of Fish and Wildlife about the species most likely to be affected, extensive discussion ensued. The summary of this workshop noted that as in the other workshops, participants generally indicated a need for more information and discussion before considering reaching consensus. The report also noted that participants might disagree on the level of importance of various issues to their affinity groups (Woodbury Dispute Resolution Center 2002).

Workshop 4: The fourth workshop returned to a revised version of the draft guide presented by Jean Vissering in the second workshop (see Vissering, n.d., for a summary of aesthetic considerations in siting and designing wind turbine projects); the workshop also included discussion of possible contents of the planning resources packet as well as identification of any areas of consensus or widespread agreement from the prior three workshops. The final outcome of this workshop was a statement by participants (Woodbury Dispute Resolution Center 2002, p. 8):

> "Appropriately sited wind energy should be an important part of Vermont's energy future. We are committed to educating our own organizations on issues associated with its development and making appropriate public policy recommendations. The Department of Public Service should continue to provide leadership in education and policy development with the public, the stakeholders, planning organizations and permitting agencies. We believe that this approach is particularly appropriate in the context of Vermont's existing and ongoing commitment in achieving high level of efficiency in energy use."

As the report notes, many of the representatives of various stakeholder groups were unwilling to go beyond this cautious statement. They were "understandably reluctant to move forward with consensus on the issues identified in these workshops until they could report back to their individual organizations, share and discuss the information learned from these meetings and decide what their appropriate next steps might be on an organization level" (Woodbury Dispute Resolution Center 2002, p. 8). The report notes in closing, however, that many participants wanted to see dialogue continue in task-specific working groups as well as larger gatherings.

12.6.1.2 The Commission on Wind Energy Regulatory Policy

In July 2004, the governor issued an executive order creating the Commission on Wind Energy Regulatory Policy. While not composed of stakeholders as such, the commission, although small, had a wide breadth of representation. It was composed of seven members drawn from around the state: the CEO of a bank, a former chair of the Vermont Environmental Board, the president of Norwich University, the executive director of a chamber of commerce, the executive director of a regional planning commission, a representative of a regional development corporation, and an attorney. It was tasked with providing guidance on whether Section 248 establishes a review process appropriate for utility-scale wind energy projects. In its December 2004 report, the commission said that Section 248 was appropriate for siting wind energy projects but suggested several modifications (Vermont Commission on Wind Energy Regulatory Policy 2004). Key recommendations included the following:

- The PSB should host at least two public meetings in the project region: an information session early in the process and a public input session later in the process, after technical hearings but before reaching a decision.
- The PSB should require the applicant to provide (1) advance public notice in all towns wholly or partially within a radius of ten miles of each proposed turbine (roughly, the turbine's "view shed"), (2) initial and ongoing mailings of, e.g., key events to all municipal and regional planning commissions as well as the town

clerks within the ten-mile radii; and initial and ongoing mailings to all stakeholders who sign up on a mailing list.

- The PSB should increase the advance notice to municipal and regional planning commissions from 45 to a minimum of 60 days.
- The PSB should develop requirements for what constitutes "plans for construction" for wind turbine projects and should ensure that the applicant provides municipal and regional planning commissions with user-friendly information that is adequate to understand the various elements of the proposed project (e.g., project conceptual plans; general construction requirements; plans for related new infrastructure such as roads, sub-stations, and transmission lines; and identification of view shed impacts).
- The PSB should establish measures requiring developers to collaborate with local stakeholders prior to initiating the Section 248 process – e.g., by submitting construction plans to and attempting to meet with all municipal and regional planning commissions within the ten-mile radii.
- An ombudsman within the DPS should be appointed to serve as a point of contact for concerned parties in the Section 248 review process.

12.6.2 Stakeholder Involvement in the Vermont Community Wind Farm Project

From the time of the project's inception in January 2009 until mid-2010, key stakeholder groups in the VCWF project included the project's developer (VCW) and individuals and organizations in Rutland County, especially in the town of Ira. Prospectively, if the project were to continue, other organizations would become stakeholders in this project: e.g., the PSB, which would decide whether to grant a certificate of public good; the Agency of Natural Resources, which would be a party to the PSB's proceeding and would make decisions on environmental permits needed for the project; and Central Vermont Public Service, which would purchase electricity from the project. But as of mid-2010, the most active stakeholders were the project's developer and its host area.

12.6.2.1 The Host Area

Rutland County, location of the proposed VCWF, is one of Vermont's fourteen counties and contains just over 10% of the state's population. On average, the county's population is somewhat older and poorer than that of the state as a whole. While Rutland City, with a 2000 population of over 17,000, is large by Vermont's standards, most of the county is made up of small towns. In all, the county has 28 municipalities. (Like other New England states, counties in Vermont are divided into incorporated municipalities; there are no unincorporated areas).

Ira, where most of the VCFW turbines were to be located as the project evolved, had a 2000 population of 455 – the fourth smallest town in Rutland County. One hundred years before, its population was 350. Ira's demographic and economic make-up is similar to that of Rutland County, with a 2000 median household income of $46,875 and with 10.4% of its 2000 population below poverty level. The town is named for Ira Allen, brother of Ethan Allen and one of the founders of Vermont.

12.6.2.2 VCWF: Benefits and Community Interactions

According to the VCW website, the project would result in a number of benefits for landowners, local residents, the state, and the environment (Vermont Community Wind n.d.). As of June 2010, their website stated the following:

> The Vermont Community Wind Farm is estimated to locally produce approximately 240,000 (MWh) Megawatt hours of energy per year. Since an average household in Vermont uses 7110 kWh of energy per year, our wind farm will produce more than enough electricity for all of the 25,683 homes in Rutland County. The Wind Farm will generate electricity which will be fed directly into the local Vermont electrical grid in the service area of the Central Vermont Public Service Corporation. Capturing this bountiful and unlimited resource will provide enormous economic and environmental benefits to the local community for many years.
>
> **Landowner Benefits**
> VCW will create strong relationships with its land owners that are mutually respectful and prosperous. We will:
>
> ***Improve your Land***
> VCW is committed to leaving your land better than we found it. We will improve your logging roads and will strive to minimize the impact we may have on your land while building and maintaining the Wind Farm.
>
> ***Land Usage remains the same***
> What stays the same is your land usage (logging, farming, recreation). What changes? Your revenue! And, by partnering with VCW, you will be contributing to clean renewable energy and a cleaner environment.
>
> ***Increase your income***
> By leasing your land to VCW and designating it as a turbine site with road and power line easements, you will earn a share of the revenue generated by the Wind Farm.
>
> **Community Benefits**
> When you say "YES" to wind power you are saying "YES" to our children. You are joining a growing number of communities who are benefiting from wind power and who are choosing to:
>
> ***Restore environmental health now and for future generations***
> Wind is a limitless natural resource. Using wind energy as an alternative to nuclear and fossil, fuels will significantly improve security and air and water quality.
>
> ***Keep more energy dollars in our state and community***
> A wind farm can generate millions of dollars in new revenue for your town. A percentage of the revenue generated by VCW Farm will go in to the State Education Fund benefiting schools throughout Vermont. (In Vermont, K-12 public schools are funded by the State Education Fund, about half of which is based on revenues from an education property tax.)

> ***Reduce state and property taxes***
> Wind Farms generating millions of dollars in new revenue will help reduce or eliminate local municipal property taxes and fund community projects through "host community payments."
>
> ***Create local jobs***
> VCW will hire local community members for the construction, operation, and maintenance of its wind farm.
>
> ***Help America become energy independent***
> Relying less on foreign energy sources and more on America's unlimited supply of renewable wind power reduces, our nation's dependence on fossil fuel imports while conserving natural resources.
>
> ***Generate power without pollution***
> Clean, renewable wind energy produces electricity without polluting our air and water and by so doing removes health and environmental risks associated with burning fossil fuels.
>
> ***Preserve local land***
> Farmers and loggers receive additional income while continuing the traditional use of the land.

The VCW website also states that

> At Vermont ***Community*** Wind we value the word "Community." We know that solid relationships between developers, land owners, local boards, local town's people, local utility companies, and State and Federal Agencies must be formed in order for a Wind Farm project to succeed. We also know that this same community of people and the environment will prosper and thrive from Wind Farm projects that distribute clean, renewable wind energy – made and used locally....
>
> We consider our Wind Farm projects a success, WHEN:
>
> The relationships formed are collaborative, built on trust and respect, the local production of renewable clean power is made and used in the host community, and the community benefits financially and environmentally

12.6.2.3 The Reaction of the Town of Ira

VCW "talked the talk" of collaborative relationships built on trust and respect, but did it "walk the walk?" Not really, according to the Ira select board and planning commission. An open letter from Christine Tyminski, chairwoman of the select board, and Timothy Martin, chairman of the planning commission, was published in the *Rutland Herald* on March 17, 2010, (Tyminski and Martin 2010) The authors directed the letter to VCW, neighboring towns, and the public, saying that the letter was intended to set the record straight and make the town of Ira's position clear.

According to Tyminski and Martin's letter, residents first learned of VCW's plans in the spring of 2009 when they looked at VCW's website; at that time, VCW had not contacted anyone in Ira about the project nor had it given the select board notice of its intentions. In May 2009, Ira's select board received a petition signed by about 120 town residents asking the board to take all actions possible to oppose the project. At that time, Ira's planning commission had been working on revisions to the town plan. (Vermont state law requires that town plans be updated and readopted every 5 years.) In light of concerns about the VCWF project, the planning commission and select board called two public meetings in June 2009 to hear from town residents in connection with the town plan revision. According to Tyminski and Martin, the meetings were well attended and "the Planning Commission was given a very clear mandate from the floor during the second meeting that the plan's existing provisions restricting development in highland conservation areas (which encompass the ridgelines) be revised to make clear that commercial wind farm development in those areas was not acceptable."

Tyminski and Martin went on to note in their open letter that shortly after the June meetings, VCW hosted a bus tour to Lempster, New Hampshire, to enable residents to see firsthand what a wind farm with 2-MW wind turbines looked like. VCW also sent letters to residents describing the project and revenues the town could expect under a host community agreement (between $247,500 and $632,500 per year, depending on the number and size of the turbines – more than Ira's municipal budget, which is less than $200,000 per year (Kumka 21 February 2010)). In July 2009, VCW made a presentation at the town hall. The meeting was well attended and people were able to pose questions and voice their concerns to VCW's director of state and community affairs, Jeff Wennberg, and its president, Per White-Hansen. White-Hansen is a Danish-born engineer who now lives in Vermont; Wennberg is a Rutland County native and former mayor of Rutland City as well as former commissioner of the state's Department of Environmental Conservation.

During July, August, and September of 2009, Ira's planning commission worked on the town plan revisions in response to the June 2009 meetings on the town plan. In November 2009, the commission held a meeting in the town hall (Fig. 12.4) to enable residents to ask questions and voice opinions about the draft revised plan, and for the planning commission to vote on forwarding the plan to the select board recommending adoption. This was done, with no opposition. In December, the select board held a meeting at Ira's town hall to hear residents' opinions on the plan and to vote on its adoption. The select board adopted the plan with no opposition voiced.

In February 2010, according to Timinski and Martin's open letter, VCW mailed letters to residents that explained aspects of the project and again noted the revenues the town could expect under a host town agreement. In its letter, VCW urged residents to vote "yes" to an upcoming referendum question: "Do the voters of Ira support the opportunity to develop local renewable energy, including wind, in order to create jobs, increase tax revenue, and provide clean energy?" VCW then held an open house at Ira's town hall that lasted 5 hours and was attended by an estimated 200 people.

In Vermont, as elsewhere in New England, municipalities use a direct, "town meeting" form of government in which voters can directly address and vote on local issues. Across Vermont, annual town meetings are held on the first Tuesday in March. On March 2, 2010, at Ira's town meeting, a motion was made to vote on the question that had been posed in VCW's letter. Another motion was made to amend the question to have it read: "Shall the Town of Ira support the opportunity to develop local renewable energy, including wind, in order to create jobs, increase tax revenue, and provide clean energy, *in a manner consistent with the town plan*?" (as quoted by Timinski and Martin 17 March 2010, emphasis added). Prior to the

Fig. 12.4 Ira town hall

vote, the meeting's moderator made it clear that a "yes" vote would support the town plan and its provisions protecting ridgelines from commercial wind farm development. The vote, done by paper ballot, was 89 in favor and 20 opposed.

Figure 12.5 shows a sign that was posted in the front yard of an Ira residence.

Timinski and Martin closed their letter with this request to VCW:

> "At this point, our community has spoken in the most fundamental ways any community can, with a town plan containing specific language, and with a referendum containing specific language in support of the plan. The community's position has been made clear, and this position should now be respected by VCWF. The community has indulged VCWF in such a way that no one should question whether VCWF has had ample opportunities to make its case to Ira residents. Through its mailings, DVD, town hall presentation last July and its open house last month, VCWF has one way or another gotten its message through to every household. Now that the community has spoken clearly through a petition, the town plan, and the referendum vote that it does not approve of commercial wind farm development on its ridgelines, we are calling upon VCWF to respect the wishes of the community, and not pursue this project or any scaled-back version that conflicts with our plan and wishes."

At the February 2010 open house on the proposed project, VCW representative Wennberg was quoted by a reporter as saying, "(Public support) is enormously important from our perspective. The PSB takes it into consideration … clearly the pattern is to get local support" (Kumka 21 February 2010). Nevertheless, VCW continued to pursue the project until late April.

Fig. 12.5 Sign posted in the front yard of an Ira residence

12.7 The Mid-2010 Status of the Vermont Community Wind Farm Project

In addition to the Ira town vote adopting a Highlands Conservation Plan, the Vermont Agency of Natural Resources had raised concerns about the VCWF project's possible impacts on wildlife habitat. Interpreting these concerns as potentially blocking the needed permits, VCW then tabled the project. Wennberg was quoted as saying

> "It [the habitat issue] arose, very suddenly, late in the game. We said all along we did not want to proceed without [Agency of Natural Resources] support" (Dritschilo 27 April 2010).
>
> "There's just so much money that is required to pursue a project like this. There needs to be some reasonable level of certainty that if certain requirements are met, that the project will be allowed to proceed and at this point we don't even know what those requirements are in many cases" (Keck 27 April 2010).

But as of mid-2010, the status of the VCWF project remained unresolved. Although the project was put on hold, VCW retained the leases associated with it. And in June 2010, a legislative remedy to some of VCW's concerns was enacted.

Act 159 – "an act relating to renewable energy" – was passed and signed into law on June 4, 2010. A key provision consolidated appeals of decisions of the Agency of Natural Resources with the PSB's Section 248 proceedings, giving the PSB the authority to hold de novo hearings on issues that have been appealed regarding renewable energy plants. While the PSB must apply the same substantive standards as the

Agency, the consolidation is likely to benefit those petitioning to develop renewable energy, if only by making the permitting process more streamlined. (In contrast, H. 677 – a bill introduced in February 2010 specifying minimum setbacks and maximum decibel levels for wind turbines of 0.5 MW or more – died in committee.)

In May 2010, Central Vermont Public Service and Green Mountain Power contracted to purchase wind-generated electricity from a new 33-turbine project in the mountains of northern New Hampshire. In a joint statement, the presidents of CVPS and GMP commented that "these contracts are in keeping with our historic commitment to green energy, and they extend our portfolios of renewables" (Rathke 19 May 2010). Out-of-state renewable energy sources may take some of the pressure off meeting Vermont's RPS; so will a provision of Act 159 that, effective July 1, 2012, redefines "renewable energy" to include large-scale hydro.

12.8 Possible Paths Forward for Stakeholder Involvement in Wind Energy Projects

The VCWF case suggests the possibilities and limitations of not only this particular project but also others like this one.

There was inadequate follow-through to state-level stakeholder initiatives in 2002 and 2004. The 2002 Siting Consensus Building Workshops concluded with a call for continuing dialogue in task-specific working groups as well as larger gatherings. While some state and regional dialogue has taken place since 2002, it has not been at the scale or consistency envisioned in 2002 (Vissering, personal communication, 27 September 2010). Of the recommendations made by the 2004 Commission on Wind Energy Regulatory Policy, only half were formally adopted through rule changes in 2006. Not adopted were the Commission's recommendations for the PSB to hold a minimum of two public meetings in the project region, to extend the advance notice to municipal and regional planning commissions from 45 to 60 days, to appoint an ombudsperson to serve as a point of contact for concerned parties, to address the unique impacts and needs associated with wind generation projects in the Section 248 process, or to address the need for decommissioning plans and funds.

It also appears that VCW handled local stakeholder involvement poorly. Their promise of host community benefits was not enough: In March 2010, the Ira townspeople, despite a lack of local wealth, voted overwhelming in favor of conserving their mountain landscape. Figure 12.6 shows a ridgeline in Ira where some of the turbines would have been located.

Better follow-through to the state-level initiatives of 2002 and 2004 would have conveyed a clearer message to wind project developers about the need for early consultation with a variety of local stakeholders and about methods for doing so. More astute local involvement by VCW would have helped to cultivate the collaborative relationships built on trust and respect that VCW claimed to want. But would better local involvement have resulted in acceptance of the project being proposed?

Fig. 12.6 View to the west of Ira's town hall

For stakeholder involvement to be truly collaborative, those seeking acceptance of projects – especially *proposed* projects that may fundamentally change the local landscape – need to be willing to consider making significant changes. This may mean scaling back wind energy projects dramatically. This approach is advocated by at least one Vermont group, Energize Vermont, which argues for small-scale, locally distributed wind energy projects. But small-scale projects are not likely to be economically feasible as commercial enterprises, especially when large investments in related infrastructure and in the permitting process are required. This poses a difficult dilemma for both project developers and other proponents of utility-scale wind energy. The answer does not lie in dismissive pejoratives such as calling project opponents "NIMBYs." (See, e.g., the Vermont Energy Partnership's statement that "[wind energy] developers must go through a lengthy approval process with the Public Service Board while battling fierce Not in My Backyard or 'NIMBY' activists" (Vermont Energy Partnership 2006, p. 6).) For people with deeply held values concerning their lives and landscapes, monetary benefits are not likely to be persuasive. They are being asked to sacrifice something they hold dear. A necessary condition (but not always a sufficient condition) for this sacrifice is their certainty that the sacrifice will make a significant difference for society as a whole and that no one else, or few others, can make a similar sacrifice. Can that argument be made about wind energy projects?

Acknowledgments My thanks to Jean Vissering of Jean Vissering Landscape Architecture, Montpelier, Vermont, for her helpful information and advice and to Joanna Burger, the editor of this book, for her excellent editorial suggestions. Thanks also to my husband John Hardwig and to many others – too many to mention here – for thought-provoking discussions over many years about the challenges of stakeholder involvement. The interpretations reported here and the conclusions reached are the sole responsibility of the author.

References

American Wind Energy Association (n.d.) Wind power and energy storage. http://www.awea.org/pubs/factsheets/Energy_Storage_Factsheet.pdf. Accessed 23 October 2010

Bartlett G (2011) Joint Fact-finding and Stakeholder Consensus Building at the Altamont Wind Resource Area in California. In: J Burger (ed.) Stakeholders and scientists. Springer: New York

Database of State Incentives for Renewables and Efficiency (2010) Federal incentives/policies for renewable & efficiency. http://www.dsireusa.org/incentives/incentive.cfm?Incentive_Code=US13F&re=1&ee=1. Accessed 23 October 2010

Dritschilo G (2010) Ira wind farm put 'on hold.' Rutland Herald, 27 April 2010 http://www.wind-watch.org/news/2010/04/27/ira-wind-farm-put-on-hold. Accessed 23 October 2010

European Wind Energy Association (2009) Wind energy – the facts. Earthscan, London

Keck N (2010) Plan for Ira wind farm is tabled. Vermont Public Radio News, 27 April 2010 http://www.vpr.net/news_detail/87886. Accessed 23 October 2010

Kumka C (2010) Wind farm developer: 'We can't touch Yankee's rates.' Rutland Herald, 21 February 2010 http://www.windaction.org/news/25734. Accessed 23 October 2010

National Research Council (2007) Environmental impacts of wind-energy projects. National Academies Press, Washington

Nissenbaum M (2010) Wind turbines, health, ridgelines, and valleys. http://www.wind-watch.org/documents/wind-turbines-health-ridgelines-and-valleys/. Accessed 23 October 2010

Page G (2010) Renewable energy projects in Vermont: a status report. Vermont Energy Partnership. http://www.vtep.org/documents/VTEP%20ISSUE%20BRIEF%20-%20Renewable%20Energy%20Sources%20in%20Vt%20-%20A%20Status%20Report%20May%202010.pdf. Accessed 23 October 2010

Randolph J, Masters G (2008) Energy for sustainability. Island Press, Washington

Rathke L (2010) 2 Vermont utilities OK buying wind power from NH firm. The Boston Globe, 19 May 2010. http://www.mnn.com/earth-matters/energy/stories/vermont-utilities-ok-buying-wind-power-from-new-hampshire-firm. Accessed 23 October 2010

Smith A (2010) New energy needs new process. Valley Reporter, 19 August 2010. http://energizevermont.org/2010/08/valley-reporter-new-energy-needs-new-process/. Accessed 23 October 2010

Tyminski C, Martin T (2010) Ira has spoken with a clear voice. Rutland Herald, 17 March 2010. http://rutlandherald.com/article/20100317/OPINION03/3170310/1039/OPINION03. Accessed 27 October 2010

U.S. Department of Energy (2008) 20% wind energy by 2030. https://www1.eere.energy.gov/windandhydro/pdfs/41869.pdf. Accessed 23 October 2010

U.S. Energy Information Administration (2010a) Electric power industry 2008: year in review. http://www.eia.doe.gov/cneaf/electricity/epa/epa_sum.html. Accessed 23 October 2010

U.S. Energy Information Administration (2010b) Annual energy outlook 2010. http://www.eia.doe.gov/oiaf/aeo/. Accessed 23 October 2010

Ueda Y, Shibata M (2004) Development of next generation 2 MW class large wind turbines. Mitsubishi Heavy Industries Ltd. Technical Review 41 (5): 1–4. http://www.mhi.co.jp/en/products/pdf/vol41_no5.pdf. Accessed 23 October 2010

Vermont Commission on Wind Energy Regulatory Policy (2004) Findings and recommendations. http://publicservice.vermont.gov/energy/ee_files/wind/WindCommissionFinalReport-12-15-04.pdf.Accessed 23 October 2010

Vermont Community Wind (n.d.). http://www.vtcomwind.com/. Accessed 23 October 2010

Vermont Department of Public Service (2002) Wind energy planning resources for utility-scale systems in Vermont: a product of the wind siting consensus building project. Appendix I: Report by the Woodbury Dispute Resolution Center on the wind siting consensus building workshops, July 2002; and Appendix II, list of participants in the wind siting consensus building workshops.

Vermont Energy Partnership (2006) Wind power in Vermont: a primer. http://www.vtep.org/WindPower08_09_06.pdf. Accessed 23 October 2010

Vermont Environmental Board (2000) Act 250: A guide to Vermont's land use law. http://www.nrb.state.vt.us/lup/publications/Act250.pdf. Accessed 23 October 2010

Vermont Public Service Board (n.d.) Citizens' guide to the Vermont Public Service Board's Section 248 process. http://www.state.vt.us/psb/document/Citizens_Guide_to_248.pdf. Accessed 23 October 2010

Vissering J (n.d.) Wind energy and Vermont's scenic landscape. http://publicservice.vermont.gov/energy/ee_files/wind/vissering_report.pdf. Accessed 23 October 2010

Wald M (2010) Vermont senate votes to close Yankee power plant. New York Times, 25 February 2010 http://www.nytimes.com/2010/02/25/us/25nuke.html. Accessed 23 October 2010

Widness S (2010) Rutland County explores wind farm proposal. Rutland Business Journal, 14 January 2010. https://www.vermonttoday.com/apps/pbcs.dll/article?AID=/20100114/RBJ/100119976/-1/RBJ02. Accessed 23 October 2010

Wind Powering America (n.d.) U.S. installed wind capacity and wind project locations. http://www.windpoweringamerica.gov/wind_installed_capacity.asp. Accessed 23 October 2010

Chapter 13
Hydropower, Salmon and the Penobscot River (Maine, USA): Pursuing Improved Environmental and Energy Outcomes Through Participatory Decision-Making and Basin-Scale Decision Context

Jeffrey J. Opperman, Colin Apse, Fred Ayer, John Banks, Laura Rose Day, Joshua Royte, and John Seebach

Contents

Abstract The Penobscot River is the largest river within Maine and historically one of the most important rivers in New England for Atlantic salmon and other migratory fish. For more than a century, the economically and culturally important fish populations have been depressed dramatically due to hydropower dams on the mainstem river that prevented access to spawning habitat. In 2004, a broad coalition of stakeholders – including a hydropower company, the Penobscot Indian Nation, state and Federal agencies, and several conservation organizations – signed the Lower Penobscot River Comprehensive Settlement Accord. The Accord features two primary projected outcomes: a dramatic, ecologically significant increase in

J.J. Opperman (✉)
Global Freshwater Program, The Nature Conservancy, Chagrin Falls, OH 44022, USA
e-mail: jopperman@tnc.org

J. Burger (ed.), *Stakeholders and Scientists: Achieving Implementable Solutions to Energy and Environmental Issues*, DOI 10.1007/978-1-4419-8813-3_13,

the proportion of the basin accessible to migratory fish combined with maintenance of, or potentially an increase in, energy generation. Increased access to migratory fish habitat will be accomplished through removal of two dams and construction of a naturalistic fish bypass around a third, while the energy generation lost due to dam removal will be recouped through structural and operational changes to remaining dams. Here we emphasize two essential conditions that made possible an agreement on the Penobscot that will benefit both energy generation and environmental and social interests. The first condition was the degree and type of stakeholder participation within the Penobscot's decision-making context and the second is the spatial scale of the decision making – the entire system of dams on the lower river. The Penobscot Accord reflects the evolving role of stakeholders in hydropower decision making. Emulating the spatial scale of the Accord, which allowed the stakeholders to select from a broader range of alternatives to benefit both energy and the environment, will require further evolution of stakeholder involvement.

13.1 Introduction

Hydropower dams have generated some of the great conflicts in the history of environmental conservation. The 1965 battle over the Storm King pumped-storage project on the Hudson River is widely credited with inspiring the era of legal protections for the environment, while in the 1970s the Tellico Dam's impacts to the extremely rare snail darter (*Percina tanasi*) triggered one of the first tests of the Endangered Species Act (ESA) (Shabecoff 1993). Today, the rapid expansion of hydropower dams worldwide threatens to dramatically alter thousands of rivers, requiring society to balance the demands for low-carbon energy with the protection of the values that free-flowing rivers provide to ecosystems and river-dependent rural communities (Richter et al. 2010).

While hydropower dams are still controversial in the United States (U.S.), the reoperation of existing dams has more recently provided examples of innovative local solutions that balance energy generation with environmental protection. These outcomes and emerging collaborative approaches to environmental problems could not have occurred without a dramatic evolution in the way various entities have been given access to the decision-making processes of Federal hydropower regulation.

13.1.1 Objectives for This Chapter

In this chapter, we describe one of these innovative outcomes, on Maine's Penobscot River, review the role that broader participation played in this outcome, and suggest how lessons from this example can inform hydropower decision making within the U.S. and globally – and environmental planning and management more broadly.

We emphasize two essential conditions that made an agreement possible on the Penobscot that benefitted both energy and the environment. The first is the degree and type of stakeholder participation within the Penobscot's decision making context and the second is the spatial scale of the decision making. The Penobscot illustrates the evolving role of stakeholders within hydropower regulatory processes. Penobscot stakeholders took advantage of the products of this evolution and worked collaboratively to find creative solutions within the flexibility provided by current regulations (note that here we use the term "stakeholder" broadly to include more than just interest groups but also the regulated entity and representatives of sovereign states, such as Tribes and government agencies).

Hydropower decision making has generally been applied at the scale of a single dam. In contrast, the Penobscot process encompassed an entire system of dams, and its outcomes will influence a major portion of a large river basin, including the key areas for the movements of migratory fish. By adopting this larger spatial scale, stakeholders were able to consider a broader set of possible solutions and thus to select an alternative capable of addressing the needs of both energy production and environmental protection. Collectively, these two essential conditions – expansive stakeholder participation and large spatial scale – offer insights about how hydropower decision making, and energy-environmental processes more broadly, can achieve more sustainable outcomes.

13.2 Hydropower: Energy, Impacts, Regulation and Stakeholders

13.2.1 Hydropower's Energy Benefits and Environmental Impacts

Hydropower, which generates energy from the force of water moving through turbines, provides approximately 20% of electricity generated worldwide and 7% of all electricity within the U.S. (USGS 2010). The amount of energy generated by a hydropower facility is a function of flow and hydraulic head (elevation difference between the upstream water surface elevation and the turbines). Conventional hydropower relies on dams to both increase the hydraulic head and to store and regulate water flow.

While certain types of hydropower reservoirs potentially produce considerable quantities of methane and carbon dioxide (Gunkel 2009), most hydropower operations within the U.S. produce low amounts of greenhouse gases. Because it relies on the hydrological cycle for its "fuel," hydropower is considered renewable and, indeed, is by far the largest source of renewable electricity in the world. However, hydropower dams can produce considerable negative impacts on river ecosystems and the people who depend on the services provided by functioning rivers (Richter et al. 2010). Dams can degrade water quality, alter flow patterns (e.g., timing and magnitude), obstruct or impede the up and downstream movement of aquatic organisms, and inundate important riverine and riparian habitats (Bunn and Arthington 2002; Ligon et al. 1995). Approximately 50,000 large dams (>15 m tall) and orders of magnitude more small dams exist today. The continued proliferation of dams worldwide is a primary contributor to the decline of aquatic ecosystems and freshwater species, which are endangered at higher rates than those from terrestrial and marine ecosystems (Richter et al. 1997).

These losses in ecosystem functions and services can have significant social costs, such as the loss of important fisheries (Dudgeon 2000). The cumulative impacts from the high density of small dams in New England caused dramatic declines in Atlantic salmon (*Salmo salar)* and other migratory fish, while some of the biggest dams in the world in the Columbia River basin have caused populations of Pacific salmon species (*Oncorhynchus spp.*) to plummet (Montgomery 2003). Relatively free of dams, Alaska's thriving salmon fishing industry, worth hundreds of millions of dollars annually, demonstrates the potential economic value of functioning rivers. Though often not easily measured in dollars, healthy rivers also provide immense value to rural communities such as the fisheries of the Mekong River (Mekong River Commission 2005) and flood-recession agriculture in Africa (Adams and Hughes 1986). Richter et al. (2010) reported that, worldwide, approximately 400 million people have been affected by upstream dams and their impacts on flow, sediment, and barriers to fish migration.

These impacts have led to considerable controversy over hydropower dams (Echeverria et al. 1989) and concern over the current proliferation of hydropower dams globally. Worldwide, thousands of new large dams are planned or already under construction, with the majority being built to provide hydropower (Bosshard 2010).

13.2.2 Regulatory Context for Hydropower and Role of Stakeholders

The role of stakeholders in U.S. hydropower decision making is closely intertwined with hydropower's regulatory context. This section will briefly describe this overarching regulatory structure and how it has evolved through time, emphasizing how changes in the regulatory structure have influenced the degree to which various stakeholders can access and influence hydropower decision-making processes.

Within the U.S., approximately half of all hydropower capacity is owned by the Federal Government, operated by the U.S. Army Corps of Engineers, Bureau of Reclamation, and Tennessee Valley Authority. The other half is operated by state or local governments, municipal power providers, and private companies. Nearly all nonfederal hydropower projects are regulated by the Federal Energy Regulatory Commission (FERC or 'the Commission'). The Federal Power Act (FPA; 1920 and expanded in 1935) gave FERC (known then as the Federal Power Commission [FPC]) the authority to issue licenses to nonfederal hydropower projects. These licenses are generally issued for a time period of 30–50 years; projects must undergo a relicensing process prior to license expiration in order to continue operating. Today, approximately 1,800 projects are licensed by FERC (Gillilan and Brown 1997). The licensing and relicensing (collectively, "licensing") of hydropower projects are the primary means through which various stakeholders – including Tribes, agencies, and NGOs – participate to influence hydropower decisions.

Several original sections of the FPA include language that requires FERC to consider a broader set of values beyond energy while making decisions about hydropower. Although these provisions would appear to provide an opening for participation of stakeholders that represent nonenergy values, in practice, these sections had relatively little influence on the degree of participation within licensing processes. Hydropower development was generally not constrained by these broader considerations.

For example, Section 10(a)1 required that hydropower projects be "best adapted to a comprehensive plan for improving or developing a waterway or waterways for the use or benefit" of various values and activities, such as "fish and wildlife (including related spawning grounds and habitat), and for other beneficial purposes including irrigation, flood control, water supply and recreational and other purposes..." Rather than creating these comprehensive plans for waterways (i.e., river basin) themselves, FERC has asserted that the record compiled during a licensing proceeding constitutes a comprehensive plan (Echeverria et al. 1989). Section 10(a)(3) of the FPA required that FERC consult with Tribes and state and federal fish and wildlife agencies regarding licenses such that these agencies can submit recommendations for license conditions to protect the resources under their management. FERC was not required to adopt the recommended license conditions and, in practice, these various provisions of Section 10(a) did not provide for comprehensive or effective participation by external stakeholders (i.e., beyond FERC and the licensee) within licensing decisions.

The lack of effective participation by external stakeholders, such as resource agencies and NGOs, reflected the lack of legal standing and participation by these groups generally within environmental decision-making processes prior to the passage of what are now considered the foundations of modern environmental law, such as the National Environmental Policy Act (NEPA), the Clean Water Act (CWA) and the ESA. A conflict over a hydropower dam helped usher in the era during which much of this modern environmental legislation was passed. A dispute over a proposed license for a pumped-storage plant on the Hudson River at Storm King Mountain resulted in a citizen's group suing the FPC. The Second Circuit Court of Appeals ruled in favor of the citizens' group and ordered FPC to consider alternatives to the project, including a "no action" alternative. This successful lawsuit is credited with both launching the era of environmental litigation and serving as a template for NEPA (Shabecoff 1993).

The expanding foundation of environmental regulation increased various stakeholders' ability to contest licensing decisions. However, FERC generally resisted application of these laws to their proceedings (Pollak 2007). Further, beyond potential legal avenues of protest, the licensing processes themselves did not feature broader participation by stakeholders. Broader participation within licensing processes was facilitated by Congress passing the Electric Consumers Protection Act of 1986 (ECPA). These amendments to the FPA instructed FERC that "in addition to the power and development purposes for which licenses are issued, [FERC] shall give *equal consideration* to the purposes of enhancement of fish and wildlife (including related spawning grounds and habitat), the protection of recreational opportunities, and the preservation of other aspects of environmental quality" (emphasis added). Through ECPA, Section 10(a)2 of the FPA was amended to instruct FERC to consider comprehensive basin plans developed by state and Federal entities and to ensure that licensed hydropower projects were consistent with those basin plans. Section 10(j) strengthened the language for resource agency consultation, stating that licenses "*shall include* conditions" for protecting fish and wildlife (emphasis added; Echeverria et al. 1989).

Both before and after ECPA, environmental organizations sued to overcome FERC's resistance to applying new environmental statues within licensing procedures (Pollak 2007). Over time, the intertwined paths of legislative changes, such as ECPA, and court decisions affirming FERC's obligation to consider environmental laws expanded the role of various stakeholders within licensing processes (Kosnik 2010).

During licensing processes, various state and Federal resource and regulatory agencies can exercise "conditioning authority" under which they issue conditions for FERC licenses (Pollak 2007) and several agencies have mandatory conditioning authority. For example, FPA Section 4(e) decrees that licenses for projects within a Federal reservation – including National Forest or Tribal lands – must include conditions issued by the agency that manages the reservation to ensure that the project does not interfere with the purposes of the reservation. Section 18 of the FPA gives Federal fisheries agencies (U.S. Fish and Wildlife Service (FWS) and the National Marine Fisheries Service (NMFS)) the authority to prescribe fish passage facilities as a license condition. State agencies that implement the Federal CWA can also exert conditioning authority to ensure a project is consistent with water-quality standards (Pollak 2007). Where applicable, FWS and NMFS can also compel mandatory conditions through the ESA.

Following ECPA, river recreation and conservation organizations became more aware that relicensing provided an opportunity to advocate for environmental restoration and licensing frequently led to legal action. As litigation became more common, the various participants (FERC, licensees, agencies, Tribes, and NGOs) began to explore ways to avoid the cost, delays, and uncertainty inherent in legal battles and to collectively find ways to reach agreement outside of the courtroom. A primary mechanism for achieving such outcomes is a settlement agreement – a negotiated agreement and legally binding document that describe how the project will operate, including specific environmental conditions to be followed (HRC 2005).

To better facilitate these negotiated agreements and overcome some process limitations, FERC established the Alternate Licensing Procedure (ALP) in 1997. Although the ALP was intended to foster negotiated agreements, relicensing processes were still often delayed due to jurisdictional disputes between FERC and other agencies (HRC 2005). Realizing that greater improvements were necessary, two groups were formed to advise FERC on further improvements to the licensing process. The first group (the National Review Group) included both the Hydropower Reform Coalition (HRC), an umbrella association of organizations that advocate for protection or restoration of environmental and recreation resources on rivers, and the National Hydropower Association (NHA), which represents the hydropower industry. The second group was the Interagency Task Force, comprised of various federal agencies (HRC 2005). Informed by these groups, FERC developed the Integrated Licensing Process (ILP) in 2003. Beginning in 2005, the ILP became the default process for licensing (HRC 2005). Among other changes, the ILP calls for greater synchronization between the license application process and the environmental review under NEPA and requires stakeholder participation earlier in the licensing process (Kosnik 2010).

Through the ILP, FERC now encourages licensees to negotiate settlement agreements with the primary participants in relicensing (e.g., agencies, Tribes, and NGOs). Functionally, settlement agreements send a signal to FERC that diverse parties agree to a specific set of conditions and FERC can use settlement agreements as the basis upon which to rule on a licensing application. Many settlement agreements have been received favorably by the hydropower industry, agencies, Tribes, and NGOs. For example, the new license for the Pelton Round Butte Project (FERC Docket # 2030-073), a hydropower project jointly owned by Portland General Electric (PGE) and the Confederated Tribes of the Warm Springs Reservation (CTWS), was based on a settlement agreement signed by 22 entities, including PGE, CTWS, American Rivers, NMFS, FWS, the U.S. Forest Service, the Bureau of Indian Affairs, Trout Unlimited, Oregon Trout, four Oregon agencies, and several municipalities and counties. Under the settlement agreement, PGE and CTWS will invest $120 million in fish restoration activities over the 50-year license period. Restoration activities include improvements to fish passage, water temperature, flow, and riparian habitat (PGE 2006).

13.3 Geography, Resources, Culture, and History of Penobscot River

In this section, we examine a specific application of decision making for hydropower in Maine's Penobscot River basin. The Penobscot is the largest river basin within Maine (2.2 million hectares) and second largest in the Northeastern U.S. (Fig. 13.1). The Penobscot is the primary source of freshwater for the Gulf of Maine, one of the largest estuaries in Maine and the U.S. East coast. The basin has been continuously inhabited for at least 9,000 years since the Wabanaki people – four Tribes that include the present-day Penobscot – settled the Penobscot River valley (Sanger et al. 1992).

The Penobscot supported large populations of diadromous fish – species that migrate between salt and freshwater to complete their life cycle – including alewife, American shad, blueback herring, Atlantic salmon, sea lamprey, American eel, rainbow

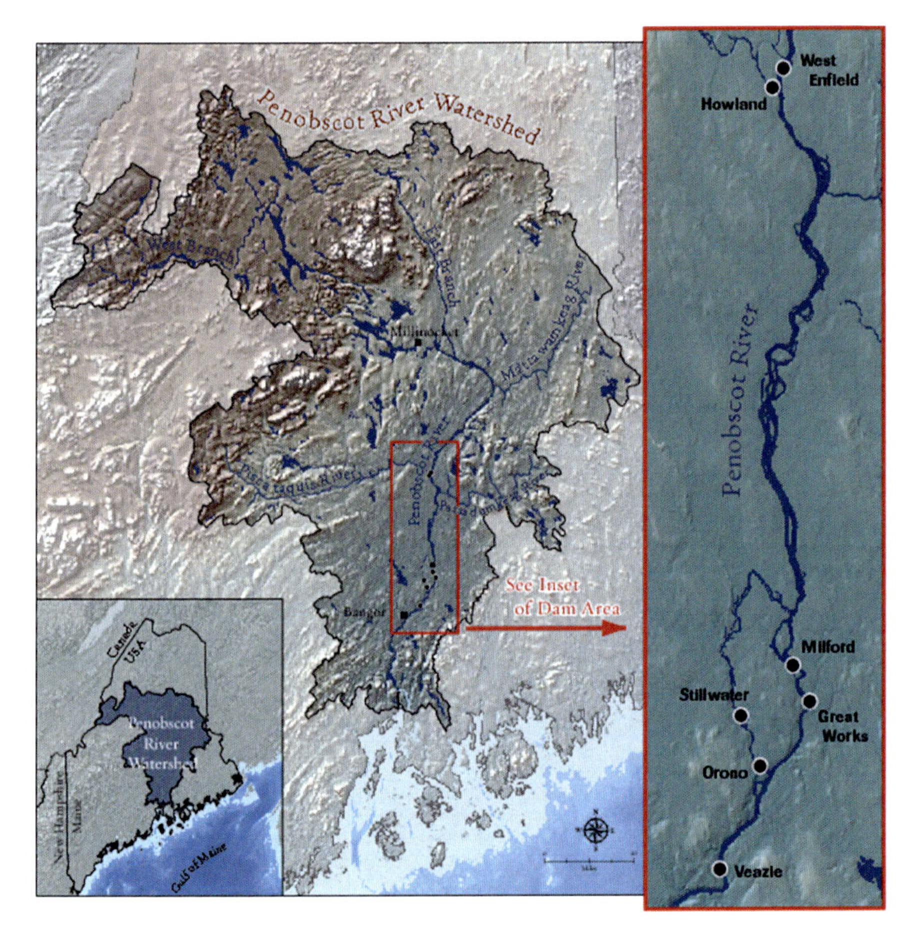

Fig. 13.1 Map of Penobscot River basin

smelt, Atlantic tomcod, striped bass, sea-run brook trout, and Atlantic and shortnose sturgeon. These diadromous fish (hereafter "migratory fish") were an important food source for the Wabanaki and essential part of their culture (Fig. 13.2). European settlers developed commercial fisheries to harvest the abundant populations of migratory fish. American shad supported the most economically valuable fishery (annual runs of over two million fish) followed by Atlantic salmon (runs prior to 1830 exceeded 100,000 fish) (Foster and Atkins 1869). The species' ranges of migration (i.e., upstream limit to migration) varied, with the two sturgeon, smelt, tomcod, and sea-run brook trout all likely restricted by the natural fall line, where Milford Dam is currently located. Other species migrated significantly further upstream to the many headwater streams that lacked impassable waterfalls (Maine Department of Marine Resources 2009; Saunders et al. 2006).

In the early nineteenth century, industrial development began to impact the migratory fish populations and the people and wildlife dependent upon them. Beginning in

Fig. 13.2 A member of the Penobscot Indian Nation demonstrating the use of a fish spear (photo used with permission of Penobscot Indian Nation and the American Philosophical Society)

the 1820s, dams were constructed on the Penobscot mainstem just above head of tide and in the current locations of the Veazie, Great Works, and Milford dams (Fig. 13.1; Table 13.1). These dams stopped upstream movements of migratory fish, eliminating access to the majority of spawning and rearing habitat (Foster and Atkins 1869). Impacts to fish migration were apparent in the first year after the closure of Veazie dam as "a great many shad and alewives lingered about the dam and died there, until the air was loaded with the stench" (Foster and Atkins 1869).

A proliferation of small dams further fragmented the basin as 250 saw mills, each with a milldam, were active by the middle of the nineteenth century. Although laws officially required fish passage, fishways were not built or were generally ineffective. Reflecting broader trends across New England, the Penobscot's migratory fish declined rapidly in the nineteenth century (Montgomery 2003). By 1868, the Penobscot's population of American shad had declined to approximately 5,000 (compared to over two million at the beginning of the century) (Foster and Atkins 1869), and "by 1908, salmon were so rare in portions of the Penobscot River that newspapers reported when one was caught" (Montgomery 2003). The dramatic decline of migratory fish populations had significant cultural and economic impacts, particularly on the Penobscot Indian Nation.

A variety of sources of pollution – including tanneries, pulp and paper factories and municipal wastewater – caused degradation of the Penobscot's water quality (Federal Energy Regulatory Commission (FERC) 1997). Following adoption of the CWA, toxic inputs to the river ecosystem (e.g., dioxin, mercury, DDT and chlorine compounds) declined dramatically. Although water quality has improved, the mainstem dams on the Penobscot remain and current stocks of migratory fish on the Penobscot are at or near historic lows and commercial fisheries have long since shut down (Maine Department of Marine Resources 2009; Saunders et al. 2006). In 2004, the National Research Council of the National Academies of Science identified the Penobscot as the best opportunity to recover wild Atlantic salmon, concluding that dam removal was a necessary component for this recovery (NRC 2004). Also citing dams as the primary threat, NMFS listed the Penobscot River population of Atlantic salmon under the ESA in 2009.

13.4 Regulatory and Stakeholder Context for Penobscot River Restoration Project

The regulatory and stakeholder context for the Penobscot River Restoration Project was a product of both national changes in hydropower licensing (described above), past experiences with multiple licensing processes on the Penobscot, and the evolving status of, and relationships between, stakeholders in the basin. Here we review the specific events and relationships in the basin that created the context in which stakeholders interacted and that set the stage for the eventual negotiations, agreement, and restoration project.

First, in the late twentieth century, dam relicensing and proposals to license new dams on the Penobscot resulted in protracted and divisive processes, demonstrating to all stakeholders that future licensing of individual projects would likely trigger expensive and lengthy conflicts with uncertain outcomes. For example, in the 1980s, intense conflicts arose over proposals to repower the degraded Bangor Dam and to build a new 38 MW hydropower dam, Basin Mills, at the confluence of the Penobscot mainstem and the Stillwater (a major side channel of the Penobscot). FERC ultimately denied both licenses. During this same period, FERC issued new licenses for the Veazie and Milford dams. All of these processes highlighted the important and controversial role that fish passage would play within future licensing processes.

A second factor facilitating a basin-scale approach to licensing was the consolidation of all the hydropower projects on the lower Penobscot under a single owner. In 1999, PPL Corporation purchased all of Bangor-Hydro's hydropower projects in the Penobscot basin, including, the Veazie, Milford, Howland, West Enfield, Medway, Orono, and Stillwater dams (Fig. 13.1 and Table 13.1). PPL's purchase of Great Works dam from Ft. James in 2000 resulted in PPL Corporation's sole ownership of the lower Penobscot dams.

Table 13.1 Dams on the lower mainstem of the Penobscot River

Penobscot River dams included in the settlement accord

Dam	Dam height (m)	Dam length (m)	Electricity generation capacity (Mw)	Distance to ocean (km)	Proposed action for fish	Proposed action for power
Veazie	6	327	8.4	55	Removal	Offline
Great works	6	435	7.7	69	Removal	Offline
Milford	6	427	6.4	73	Fish lift	Increase
West enfield	7	296	13.0	114	Eel passage	Increase
Medway	11	541	19.2	175	Future	Increase
Orono	8	343	2.3	63	Future	New power
Stillwater	6	539	1.9	67	Future	Increase
Howland	6	201	1.9	112	Bypass	Offline

A third important factor that facilitated a comprehensive approach was the legal status of the Penobscot Indian Nation ('the Tribe'). Section 4(e) of the FPA requires FERC to consult with the U.S. Department of the Interior ('Interior') when a licensing decision may impact Tribal trust resources, such as fisheries. Using this conditioning authority, the Tribe, through Interior, advocated for restoration of fish runs and effective fish passage during licensing processes. Although the Tribe is federally recognized by the Bureau of Indian Affairs (reaffirmed by the Maine Indian Land Claims Act of 1980), during the Milford licensing process, the state of Maine and hydropower interests contested the Tribe's status as a Federal "reservation," as defined by the FPA. This dispute loomed as a major unresolved issue between stakeholders that would complicate any subsequent licensing processes.

When consolidation of ownership created an opportunity to address hydropower and fish passage in a comprehensive manner, the stakeholders involved in the dispute over FPA-reservation status agreed to set it aside in this matter, with each government (e.g., Penobscot Indian Nation and state of Maine) maintaining their respective legal positions. This removed a potential legal obstacle to comprehensive licensing and fish restoration.

13.5 Penobscot River Restoration Project

13.5.1 Negotiating the Agreement

As described above, Penobscot stakeholders engaged in numerous debates in the late twentieth century focused on single dams or single issues and did not reach resolution on how best to balance hydropower with the cultural, fisheries and recreational values of the river. This record of conflict between stakeholders, along with the consolidation of ownership and "tabling" of the dispute over the Tribe's reservation status under the FPA, facilitated the possibility of a comprehensive approach to hydropower licensing and fish passage restoration. In 1999, PPL and the Tribe initiated discussions focused on potential alternatives to site-by-site licensing processes that could more effectively balance hydropower production and fisheries restoration.

These discussions were subsequently joined by a broader set of stakeholders (Table 13.2), and the negotiations ultimately included five nonprofit conservation organizations, the Penobscot Indian Nation, the state of Maine (State Planning Office, Department of Marine Resources, Department of Inland Fisheries and Wildlife), the Department of the Interior (Bureau of Indian Affairs, U.S. FWS, National Park Service), and PPL Corporation. The parties structured the discussions to encompass a broad range of values, informed by the diverse representation among the stakeholders as well as a review of the public input from past licensing proceedings, such as those for Basin Mills and Howland. As an example of how the composition of the stakeholders' group reflected broad values for the river, the state of Maine was represented by the State Planning Office, responsible for programs

Table 13.2 Key participants involved in the Penobscot River Restoration Project (both negotiation and implementation of the agreement)

Participant	Role and interests
PPL Corporation#	Energy generation company that owned the mainstem dams on the Penobscot; licensee seeking new licenses from FERC
Black Bear Hydro Partners, LLC	A hydropower company that purchased PPL's Penobscot dams not being sold to the Trust after the agreement was finalized; assumed PPL's interests in and obligations under the agreement
Penobscot Indian Nation#*	Original human inhabitants of Penobscot Valley with culture and economy strongly intertwined with the river and its fish populations. A Federally recognized tribe, focused on restoring migratory fish populations and protection of cultural resources
State Planning Office, State of Maine#	Provided staffing assistance to the settlement process and coordinated other state agencies
Department of Marine Resources, State of Maine#	State agency charged with protection and restoration of anadromous fish habitat
Department of Inland Fisheries and Wildlife, State of Maine#	State agency charged with protection and restoration of freshwater fish habitat; particularly focused on how dam removal would influence freshwater fish populations
U.S. Fish and Wildlife Service (DOI) #	Federal agency that promoted restoration of fish populations; enforces Endangered Species Act, under FPA has mandatory conditioning authority for fish passage
Bureau of Indian Affairs (DOI) #	Under Section 4(e) of FPA, can impose license conditions to ensure a project is consistent with the purposes of establishing a Federal Reservation. On behalf of the Penobscot Indian Nation, promoted restoration of fish populations
National Park Service (DOI) #	Federal agency with responsibilities for hydropower impacts on interests such as recreation and archaeology
American Rivers#*	Nongovernmental conservation organization focused on restoration and protection of rivers and their environmental and recreational values
Atlantic Salmon Federation#*	Nongovernmental conservation organization focused on restoring and protecting populations of Atlantic salmon and their habitat
Maine Audubon#*	Nongovernmental conservation organization focused on conservation of Maine's environment and natural resources
Natural Resources Council of Maine#*	Nongovernmental conservation organization focused on conservation of Maine's environment and natural resources
Trout Unlimited#*	Nongovernmental conservation organization focused on restoring and protecting habitat of trout and salmon
The Nature Conservancy*	Nongovernmental conservation organization that joined the Trust after the agreement was signed for the purposes of contributing to fund raising and adding scientific capacity
National Oceanic and Atmospheric Administration (NOAA) Restoration Center	Federal agency charged with the restoration of sea-run fisheries habitat; major funding and implementation role
Penobscot River Restoration Trust	Nongovernmental organization composed of the Penobscot Indian Nation and six conservation NGOs (membership in the Trust denoted by *). The Trust has the option to purchase dams and is implementing the agreement

Notations indicate signatories to the agreement (#) and members of the Penobscot River Restoration Trust (*)

such as energy and hydropower policy, in addition to the fisheries agencies. In 2001, the stakeholders convened a subset of negotiators to (1) identify a set of common interests and key challenges; (2) develop a structure for negotiation and information exchanges; and (3) commit to working through challenges and reaching a comprehensive settlement agreement. After several initial discussions, this subset of negotiators decided to proceed without a facilitator or mediator. Following decades of vehement and costly disagreement between many of the stakeholders, the negotiators decided to build trust among themselves through frequent in-person interactions.

After negotiating the critical components of the agreement, the stakeholders described the proposed approach in a conceptual agreement, rather than proceeding directly with a formal agreement filed with FERC and other agencies. The conceptual agreement outlined key principles for rebalancing fisheries restoration and hydropower production on the lower river. They released this conceptual agreement to inform the public and to solicit public input prior to legal filing with the regulatory agencies. This process of gathering public input exceeded regulatory requirements and consisted of a series of public meetings in which representatives of the stakeholders presented the conceptual agreement to citizens, organizations and businesses in the region, and solicited further input on the proposed concepts. Meetings with individuals, landowners, town managers, businesses, community groups, conservation organizations, recreational interests, scientists and many others helped inform both the public and the project.

13.5.2 The Agreement

In 2004, the parties filed with FERC the Lower Penobscot River Comprehensive Settlement Accord (Federal Energy Regulatory Commission (FERC) 2004), a multiparty legal agreement designed to reconfigure hydropower production on the lower Penobscot system to restore migratory fish populations while maintaining hydropower production under new licenses at PPL's dams (Table 13.1). Under the agreement, PPL granted a 5-year option to purchase three dams (Veazie, Great Works, and Howland; Fig. 13.1) to the newly created not-for-profit Penobscot River Restoration Trust ('the Trust') for between $24 million and $26 million (membership of the Trust included the Tribe and five conservation NGOs, later to be joined by The Nature Conservancy; Table 13.2). Upon purchase and receipt of required permits, the Trust could then decommission all three dams. The agreement proposed the removal of the two most seaward dams (Veazie and Great Works; Figs. 13.1 and 13.3) and construction of a fish bypass around the third dam, Howland, which is located on the Piscataquis River just 150 m from its confluence with the Penobscot River (Figs. 13.1 and 13.4).

The fish bypass at Howland will be able to accommodate a broad range of flow conditions and its slope (1.5%) is sufficiently low such that relatively poor-swimming species like American shad can use it to reach upstream spawning grounds (Fig. 13.4) (Federal Energy Regulatory Commission (FERC) 2010).

Fig. 13.3 A visualization of dam removal (Veazie Dam) that was used for public outreach and fundraising (images courtesy of MMI engineering)

The stakeholders agreed upon a fish bypass rather than removal for Howland. Engineering feasibility studies indicated that a fish bypass at this location would accomplish safe, timely and effective fish passage and, in combination with removal of the two lower dams, would achieve overall project objectives. Therefore, the bypass was an acceptable approach to both restoring fisheries and addressing the local community's preference, expressed in recent relicensing proceedings, for maintaining the water elevations created by the dam. As part of the Settlement Accord, PPL is required to significantly improve fish passage at four other Penobscot dams (Stillwater, Orono, Medway, and Milford).

Milford Dam, located at the site of the first significant falls and currently the third dam upstream from the ocean, will not be acquired by the Trust and will continue to generate hydropower. After the two lowermost dams are decommissioned, scheduled to begin in 2011, Milford will be the first dam upstream from the ocean and the only one remaining on the lower mainstem of the Penobscot River (Fig. 13.1). Black Bear Hydro Partners LLC, which purchased PPL's Maine hydropower assets

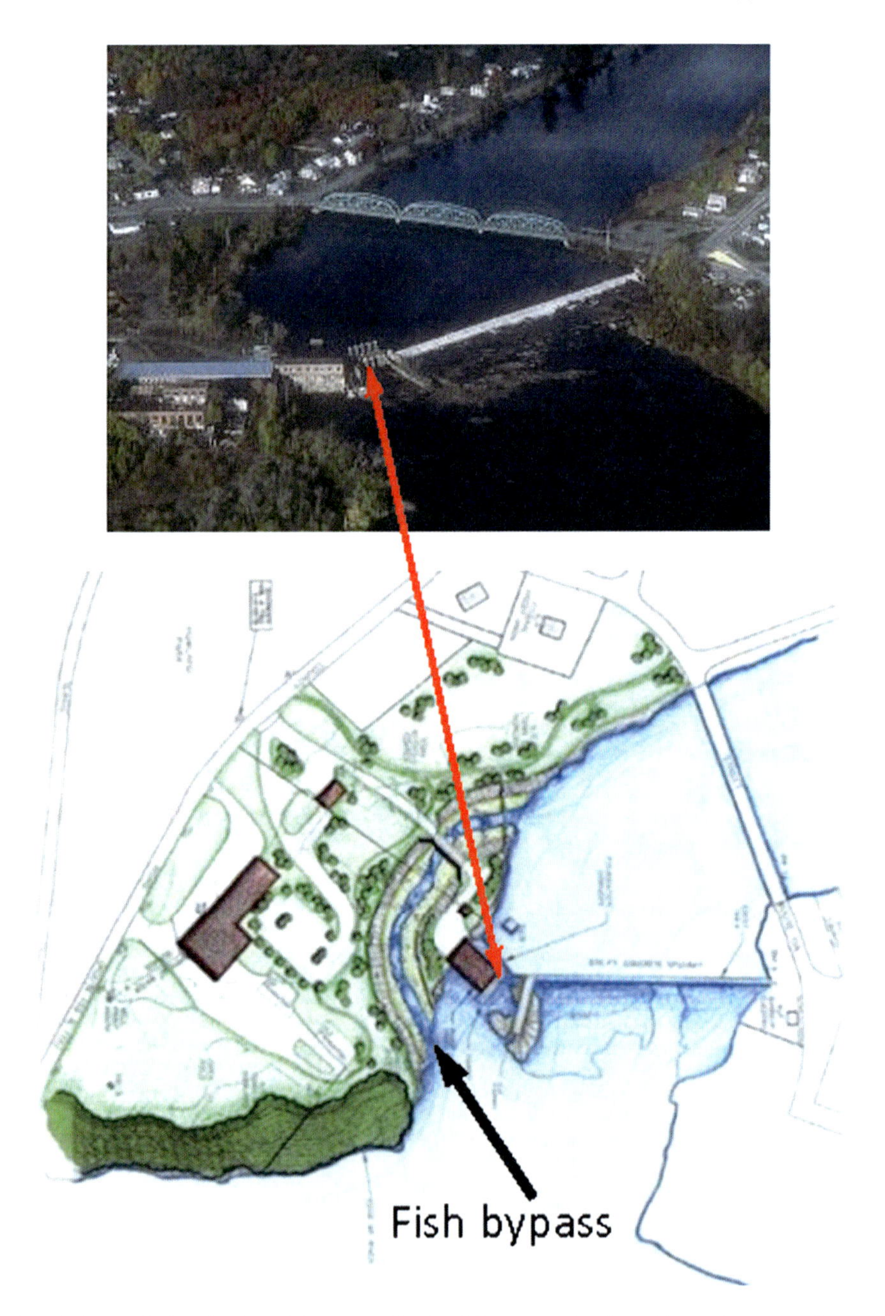

Fig. 13.4 The design for the fish bypass at Howland Dam, seen in photo. The *red arrows* point to the same structure in each image. The *black arrow* points to the downstream entrance to the fish bypass (images courtesy of MMI engineering)

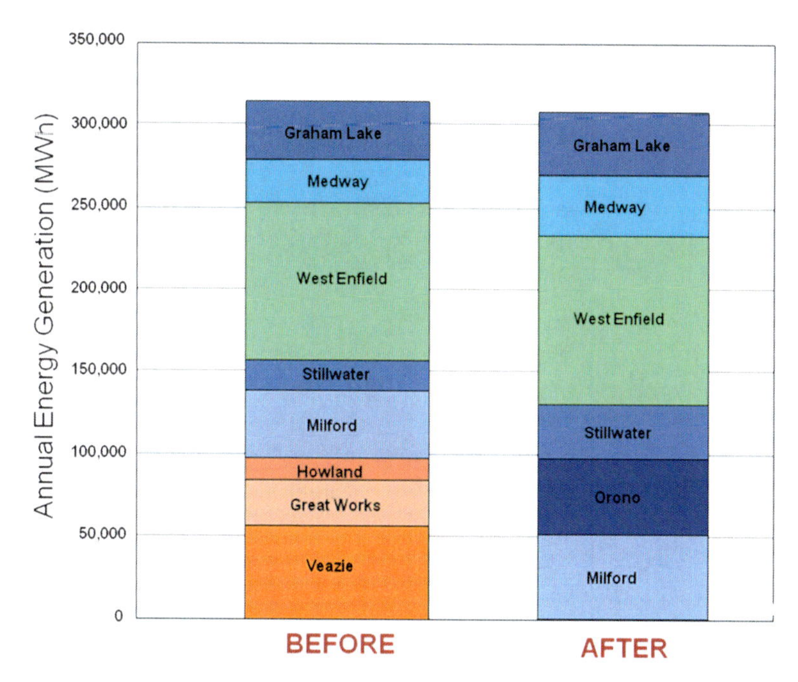

Fig. 13.5 Total annual energy production of Penobscot basin dams before and after project completion. Dams with "warm" colors in the "before" column will be removed and are not repeated in the "after" column. Note that several dams have greater energy production after the project. Although the figure shows a slight net decrease in total energy generation, this reflects conservative estimates. More recent estimates of future energy generation indicate a slight increase in total energy generation (Scott Hall, Black Bear Hydro Partners, LLC, personal communication 2010)

in November 2009, will improve fish passage at the Milford Dam beyond that required in their current license through construction of a state-of-the-art fish lift.

Finally, the Accord provides new licenses for the remaining six dams in the Penobscot basin. Within the Agreement, the stakeholders support increasing hydropower generation at these dams such that, after project completion, the overall system of dams in the basin will maintain energy generation at or above current levels (Fig. 13.5). Project-level maintenance or increase in generation will be achieved through (1) repowering and increasing generation at Orono Dam (on the Stillwater Branch of the Penobscot which parallels the main channel; Fig. 13.1); (2) adding an additional foot of hydraulic head on three impoundments; and (3) increasing hydropower at four other hydropower dams in the system, in some cases transferring turbines from decommissioned dams to achieve the increases (Table 13.1). With these improvements, PPL Corporation projected a slight increase in energy production after project completion (Scott Hall, Black Bear Hydro Partners, LLC, personal communication 2010).

The estimated overall project cost is approximately $55 million and includes purchasing the three dams, removing the two lowermost dams and bypassing Howland, and other implementation costs (not including modifications to increase generation or fish passage costs such as the Milford fish lift). By late 2007, the Trust and its partners had raised $25 million to buy the dams, from a combination of federal and private sources, and in June 2008 the Trust exercised its option to purchase the dams. In July 2010, FERC issued to the Trust the Final Orders for the Decommissioning and Surrender of the Veazie, Great Works and Howland dams and the Maine DEP issued the Final Removal Orders in August 2010. When the Corps of Engineers issues its permit, the Trust will close on the dams and full project implementation will begin.

13.5.3 Biological, Cultural and Economic Implications of the Project

By removing two dams and constructing a naturalistic fishway at Howland, the project will lead to considerable increases in the extent of free-flowing river habitat and overall connectivity in the basin. Opperman et al. (in press) provide a more complete review of the project's forecasted benefits for migratory fish, so here we present only a brief summary. Through the removal of Veazie and Great Works dams alone, the project will restore access to essentially all historic habitat for federally endangered shortnose sturgeon, tomcod, striped bass, rainbow smelt and Atlantic sturgeon (Saunders et al. 2006). With fish passage improvements at Milford and Howland, the six long-distance migrants (alewife, blueback herring, American shad, American eel, sea lamprey, and federally endangered Atlantic salmon) will have access to as much as two-thirds of their historic habitat in this watershed (Fig. 13.6). Federal biologists predict that this increased extent of spawning habitat will boost considerably population sizes of several fish species. For example, Federal fisheries biologists estimate that American shad will increase from a few dozen today to approximately 1.5 million and Atlantic salmon to increase from approximately 2,000 to 12,000.

These forecasted increases in migratory fish populations will likely have considerable positive benefits for the Penobscot Indian Nation. Many Tribal members view the river as a sacred, living entity and believe they are stewards of the basin's inhabitants and resources. The river is central to the Tribe's cultural identity and, recently, the Tribe has begun to revive river-related traditions such as birch-bark canoe-building (Fig. 13.7) and the construction of fishing spears (Fig. 13.2) such that they can continue to be passed on. Penobscot basketry continues to weave together sea-grass from the coast with black ash from the inland river floodplains, an artistic representation of cultural connections between the ocean and inland waters.

These connections have long been severed by the river's multiple dams. For nearly 200 years, the Tribe has been unable to fully exercise their sovereign fishing rights because virtually no migratory fish of value to the Tribe make it beyond the dams downstream of their reservation. Because they view the river and its resources as a gift from the Creator, many members of the Tribe see the Project as an act of reciprocity that contributes to the partial restoration of historic conditions and resources. Consistent with their historic involvement in all aspects of the Penobscot

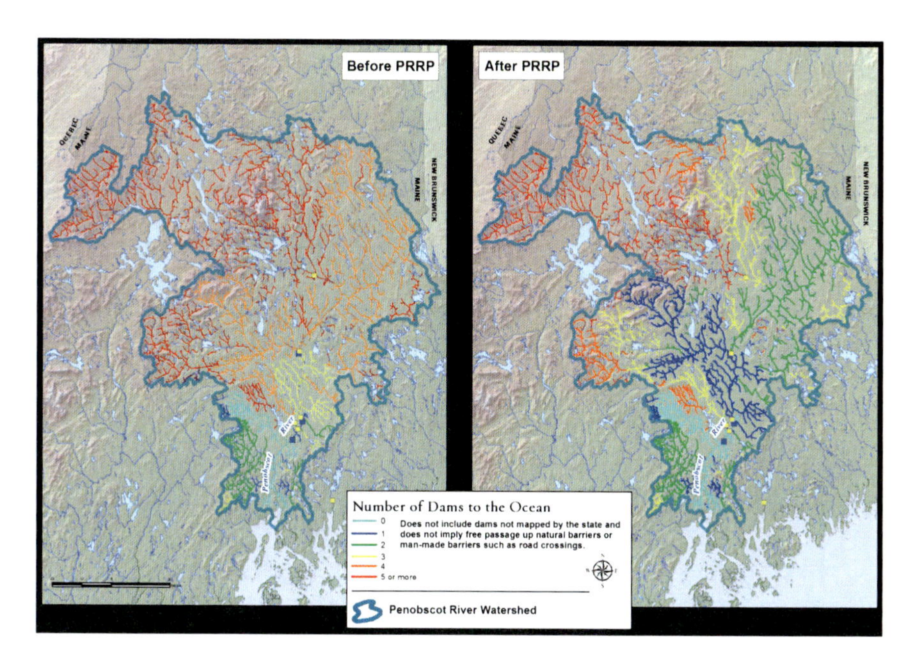

Fig. 13.6 Fish passage scenarios before and after the Penobscot River Restoration Project. Reaches and tributaries are colorcoded to represent how many dams lie between them and the ocean. A key feature of the project is providing state-of-the-art fish passage at the mainstem dams that will remain after project completion. Note the large expansion of *blue* and *green* colored mainstem river and tributaries which denote habitat accessible with passage over one or two dams. These areas previously required fish to pass beyond four dams

River, the Tribe played a leadership role in developing the Penobscot Agreement. Tribal representatives participated in all aspects of the project: they provided fisheries and water quality expertise, directly engaged in communications and community outreach, served on the negotiating team and moved dialogue forward on a government-to-government basis.

In addition to the restoration of economically and culturally significant resources for the Tribe, the Project is expected to have a net positive economic benefit on Penobscot County, a region with a median income in 2000 almost 20% below the national average (FERC 2010). The project is predicted to create almost 200 short-term construction jobs and to add approximately $8 million to the local economy (FERC 2010). Other neighboring dam-removal projects – on the Kennebec and Sebasticook rivers – have demonstrated the potential economic benefits of increasing the opportunity for activities associated with free-flowing rivers including boating, wildlife viewing and recreational fishing (Crane 2009). Further, the Project is predicted to increase considerably the populations of several fish species that are important food sources for cod and other marine groundfish, populations of which are currently greatly depressed in the Gulf of Maine. Thus dam removal may benefit the economically and culturally important commercial and recreational marine fisheries in the Gulf (Ames 2004; FERC 2010).

Fig. 13.7 Historic and current examples of birch bark canoes made by the Penobscot Indian Nation (historic photo used with permission by the Penobscot Indian Nation and the American Philosophical Society; modern photo used with permission by the Penobscot Indian Nation)

13.6 Decision-Making Context for the Penobscot: Stakeholder Participation and Spatial Scale

The Penobscot River Restoration Project features two primary projected outcomes: a dramatic, ecologically significant increase in the proportion of the basin accessible to migratory fish (Fig. 13.6) combined with maintenance or a slight increase in energy generation (Fig. 13.5). Thus, this agreement was able to result in net benefits for *both* energy generation and the environment. Further, the agreement provides greater business certainty for PPL Corporation and now its successor Black Bear Hydro Partners, LLC. The agreement has been recognized as an important precedent by both local and national media (e.g., an editorial in the New York Times) and has been lauded by both environmental organizations and the hydropower industry as a positive example. Here we review the key factors that made this agreement possible and discuss the project's applicability to sustainable hydropower and environmental decision making more broadly.

While many conditions and previous events facilitated the project's outcomes, here we emphasize two key characteristics of the decision-making context: its geographic scope and the extent and type of stakeholder participation. We have discussed the importance of the project's spatial scale at length elsewhere and its applicability to the challenge of improving the sustainability of hydropower both in the U.S. and in parts of the world undergoing new hydropower development (Opperman et al. in press). Here, we will discuss the key role that stakeholders played and the *interaction* between stakeholder involvement and the spatial scale to suggest new approaches to hydropower planning and operations. We posit that more sustainable outcomes for energy, the environment, and cultures can be achieved through decision-making processes that feature the notable elements of the Penobscot stakeholder participation – inclusivity, trust, flexibility and innovation – and apply them to problem solving at broader spatial scales (e.g., beyond that of a single project).

Although basin-scale management is widely recommended and articulated within concepts such as Integrated Water Resources Management and Integrated River Basin Management, the Penobscot Project provides one of the few examples where hydropower decision making has occurred at the scale of a river basin, simultaneously considering all dams on a mainstem river. The scale of the project, which can be described as system- or basin-scale, greatly expanded the set of possible solutions for both energy and environmental health. While the sustainability of individual projects can be improved, providing important environmental benefits (Richter and Thomas 2007), decisions focused on a single site are more likely to encounter zero-sum constraints, whereby substantial gains for either energy production or environmental protection come at the expense of the other. In contrast, the Penobscot's multidam, basin-scale approach allowed for significant environmental gains to come at no cost in terms of lost energy production.

Although we use the term 'basin scale' to describe the Penobscot, we acknowledge that the Project did not encompass every dam or every environmental issue in the basin. However, by addressing all the lower mainstem dams, the Project did address the primary basin-wide constraint to fisheries restoration – the fragmentation of connectivity to the ocean by the mainstem dams. Moreover, it is not realistic to expect that 'basin-scale' approaches will capture *all* projects or issues in a given basin, so here we are using 'basin-scale' to identify a process that sought solutions at the largest scale (of dams, geography and complexity) that remained tractable.

Further, although here we emphasize the Penobscot's positive outcomes for both energy and the environment, we do not suggest that all basin-scale approaches, or hydropower licensing more generally, should always strive to be energy neutral (or energy positive). For example, on the Kennebec River, FERC decided that the removal of Edwards Dam produced sufficient environmental benefits to outweigh the loss of energy. However, because of demands to increase low-carbon energy production, multidam or basin-scale approaches that produce positive benefits for both energy and the environment will have strong appeal. We also do not suggest that payment for dam removal by conservation funding sources will be appropriate in all contexts.

Finally, we acknowledge some of the distinctive characteristics of the Penobscot that are not likely to be replicated elsewhere. First, the river has two primary channels around Indian Island (Fig. 13.1), which facilitated balancing energy production and fish passage. Second, all the dams were owned by a single entity, which is relatively rare in large river basins in the U.S. The project also had a strong regulatory driver (fish passage license conditions) and a clear and measurable environmental performance metric (fish passage and fish populations). These latter two characteristics are not unique to the Penobscot, but will not apply in all river basins. Despite some of these distinctive characteristics, we suggest that the Penobscot provides important lessons in both the benefits of basin-scale approaches and the structure and process necessary for decision making at this scale.

13.6.1 Continued Evolution of Stakeholder Involvement for Hydropower Decision Making

The Penobscot Project featured broadly inclusive, early, and engaged stakeholder involvement and, thus, can be viewed as a product of the ongoing evolution in stakeholder involvement in licensing processes. The Project also reflects FERC's openness to solutions forged by stakeholders and articulated in settlement agreements. The Project's innovative outcomes thus demonstrate the potential benefits of these newer approaches to licensing. However, this evolution in stakeholder involvement alone could not have produced the Project's comprehensive solutions. As described above and in Opperman et al. (in press), the Project's spatial scale was instrumental to its outcomes. Opperman et al. (in press) and many others have proposed that the Penobscot represents an important precedent that should be replicated elsewhere. However, widespread replication will necessarily entail processes within river basins that lack some of the Penobscot's distinctive features (described above) – namely that all the dams were owned by a single entity. We suggest that replicating elsewhere the Penobscot's spatial scale and innovative and comprehensive outcomes will require continued evolution of stakeholder involvement in hydropower decision making.

Below we describe key characteristics of the Penobscot's stakeholder processes. We then discuss potential reforms or incentives that can facilitate application of these characteristics in other settings – including those where there is more than one hydropower stakeholder.

Stakeholder processes were broadly inclusive, with representation linked to the spatial scale of the problem and potential solutions. The Penobscot featured broadly inclusive stakeholder processes but, moreover, the representation of stakeholders encompassed the full spatial scale of the environmental problems (in this case, basin-wide fish passage and the primary dams that contributed to this problem) and the potential energy and environmental solutions.

Stakeholders agreed on a set of objectives that featured a distribution of benefits among the stakeholders. The Penobscot stakeholders were committed to finding a solution that would improve fish passage while maintaining energy generation.

Stakeholders were willing to be flexible and to consider transactions. The hydropower stakeholder did not view the prospect of dam removal as symbolically threatening or injurious but as one potential alternative within a business negotiation. Similarly, the conservation interests were willing to consider transactions that would facilitate environmentally beneficial outcomes. (As described above, we do not suggest that such transactional approaches are appropriate in *all* decision-making settings but should be viewed as a potential option for generating beneficial outcomes under certain circumstances).

Regulatory flexibility with innovation. FERC accepted the solutions negotiated by the stakeholders, reflecting their recent policy encouragement of settlement agreements.

In summary, we suggest that improved outcomes for energy, the environment, and cultures can come about through decision-making processes that share these characteristics. The spatial scale of hydropower decision-making processes should evolve from that which encompasses a single project (or a few projects) to one that more fully encompasses the relevant environmental issues and a broader set of potential solutions. While a river basin will often be the appropriate scale to capture relevant environmental issues, the scale of the solutions may be somewhat broader.

Of these characteristics, the larger geographic scale – and associated increase in stakeholder involvement – is clearly the most challenging to replicate elsewhere. Hydropower decisions (e.g., licensing processes) have rarely been made at the level of river basins because of the inherent complexities that arise at larger spatial scales. These complexities include fragmented ownership – projects are often owned by different entities (or, if owned by the same entity, dams may have different license expiration dates) – and the expanding number of regulatory and environmental issues that may accompany expanding geographic scale and jurisdictions (although it should be noted that the PPL dams on the Penobscot had different license expiration dates and the relicensing of the dams was temporally synchronized to facilitate a comprehensive outcome). Overcoming this understandable avoidance of complexity will likely require either regulatory reform that compels larger scale decision making or regulatory or economic incentives that encourage this approach (Opperman et al. in press).

Expanding stakeholder involvement to larger spatial scales entails two interacting challenges: coordination between two or more FERC-licensed hydropower entities and coordination between nonfederal hydropower and federal hydropower entities, such as the Bureau of Reclamation and the Corps of Engineers. Incentives for coordinated licensing among multiple licensees could be promoted through policy reforms, such as streamlined environmental review processes under NEPA for coordinated licenses. Further, under NEPA, Federal agencies have a duty to fully consider cumulative impacts. To date, FERC has only required mitigation for direct impacts and considers cumulative impacts essentially unmitigable (HRC 2005). If FERC required that greater attention be paid to cumulative impacts and required mitigation for those impacts, multiple licensees within a river basin may have considerable incentive to pursue coordinated, basin-scale review and relicensing.

Section 10(a) of the FPA also provides an intriguing, although heretofore little-used, policy opportunity for coordinated decision making at the scale of river basins. As described in Sect. 13.2.2, Section 10(a)(1) requires that FERC review license applications in the context of a "comprehensive plan" for the basin that considers and balances a full range of resources. Section 10(a)(2), added through ECPA, further requires FERC to consider whether a project is consistent with comprehensive river basin plans developed by other state or Federal agencies. Thus, the FPA already requires that individual licenses be reviewed and adjusted to conform with a variety of other interests and values in the basin – ranging from other power producers to environmental health. Although FERC has consistently concluded that individual licenses in themselves constitute a "comprehensive plan" under Section 10(a)(1), minimizing any meaningful assessment of larger spatial scales, these provisions of the FPA clearly provide an opening for discussion of the role of comprehensive (i.e., multistakeholder) planning for hydropower decision making.

Beyond existing laws and policies, expanding the spatial scale and complexity of stakeholder engagement within hydropower decision making will no doubt require new policy reforms, market incentives and innovative financial mechanisms. For example, markets for carbon credits are one potential policy tool for addressing greenhouse gas emissions, and these credits could be used to provide additional revenue streams for sources of renewable energy generation. Thus far, conventional and existing hydropower has generally not qualified for policy incentives for renewable energy. However, policies regulating carbon markets could stipulate that hydropower produced through coordinated basin-scale decision making, which resulted in net gains for both energy and environmental health, could be eligible for credits.

The Penobscot demonstrated that financial transactions (e.g., the sale of dams for removal) can facilitate positive outcomes. Other basin-scale approaches to managing hydropower will likely identify alternative management scenarios that produce positive outcomes more broadly for energy and the environment but have unequal costs and benefits for various stakeholders (e.g., removal of an inefficient dam to open new habitat coupled with increases in generation at other dams owned by other entities). Transactions or other financial mechanisms will likely be necessary to facilitate agreement in these situations. Finally, although FERC has generally promoted settlement agreements, more recently the Commission has screened settlement agreements for those terms that can be formalized in license conditions and those which are "nonjurisdictional" and thus cannot be placed formally within the license.

While the settlement agreement itself is a legally binding contract, many stakeholders would feel greater certainty if all provisions of a settlement agreement were legally recognized in the license. To encourage further innovative outcomes within hydropower licensing, FERC should be willing to expand what it considers jurisdictional and thus able to be formalized in licenses.

A recent Memorandum of Understanding (MOU) signed by three Federal agencies, and subsequent activities implementing elements of that MOU, potentially signals a new way forward in hydropower decision making (DOE, DOI and DOA 2010). In March 2010, the Departments of Interior, Energy and Army signed an MOU that called on the agencies to implement "integrated basin-scale opportunity assessments." The precedent of the Penobscot strongly influenced the agencies' emphasis on exploring the potential for basin-scale assessments to identify positive outcomes for both energy and the environment (Simon Geerlofs, Department of Energy, personal communication, 2010). The MOU broadly calls for the three agencies to collaborate more closely on hydropower and the basin-scale section states that "the Agencies will collaborate with the environmental community, the owners of Federal and nonfederal hydropower facilities, potentially affected Federal land management agencies, Indian Tribes and other stakeholders to identify river basins where renewable power generation and environmental sustainability could both be increased ..." (DOE, DOI and DOA 2010).

Based on the positive outcomes produced by the Penobscot Agreement, we suggest that hydropower decision making could achieve improved outcomes, for both energy and the environment, were it to emulate the Penobscot's stakeholder involvement and broad spatial context for decision making. To achieve this, we suggest that processes for stakeholder engagement continue to evolve such that hydropower decision making can be made at a geographic scale commensurate with both the primary environmental problems and potential solutions. This will require much greater coordination between federal and nonfederal generators of hydropower along with collaboration with a range of other stakeholders, including Indian Tribes, resource agencies and environmental organizations. The recent MOU provides a mechanism for exploring the feasibility of the approach we outline here and a vehicle for generating potential solutions to achieve this vision.

References

Adams WM, Hughes FMR (1986) The environmental effects of dam construction in tropical Africa: impacts and planning procedures. Geoforum 17:403–410

Ames EP, (2004). Atlantic cod stock structure in the Gulf of Maine. Fisheries 29(1):10–28

Bosshard P (2010) The dam industry, the World Commission on Dams and the Hydropower Sustainability Assessment Forum (HSAF) process. Water Alt 3(2):58–70

Bunn SE, Arthington AH (2002) Basic principles and ecological consequences of altered flow regimes for aquatic biodiversity. Environ Manage 30:492–507

Crane J (2009) Setting the river free: the removal of the Edwards Dam and the restoration of the Kennebec River. Water History 1:131–148

Department of Energy (DOE), Department of the Interior (DOI), and Department of the Army (DOA) (2010) Memorandum of Understanding for Hydropower. March, 2010

Dudgeon D (2000) Large-scale hydrological changes in tropical Asia: prospects for riverine biodiversity. Bioscience 50:793–806

Echeverria JD, Barrow P, Roos-Collins R (1989) Rivers at Risk: a concerned citizen's guide to hydropower. Island Press, Washington

Federal Energy Regulatory Commission (FERC) (1997) Final Environmental Impact Statement Licensing Three Hydroelectric Projects in the Lower Penobscot River Basin, FERC Project Nos. 2403-056, 2312-019 and 2721-020, Washington

Federal Energy Regulatory Commission (FERC) (2004) Submittal of the Lower Penobscot River Basin Comprehensive Settlement Accord with Explanatory Statement. Project Nos. 2403, 2534, 2666, 2710, 2712, 2721, and 10981. Federal Register, Docket No. DI97–10

Federal Energy Regulatory Commission (FERC) (2010) Final Environmental Assessment, Application for Surrender of License, Veazie, Great Works and Howland Projects Nos. 2403-056, 2312-019, and 2721-020

Foster NW, Atkins CG (1869) Report of Commission of Fisheries. Commissioners of Fisheries, Augusta

Gillilan DM, Brown TC (1997) Instream Flow Protection: seeking a balance in Western water use. Island Press, Washington

Gunkel G (2009) Hydropower – a green energy? Tropical reservoirs and greenhouse gas emissions. Clean – Soil, Air, Water 37:726–734

Hydropower Reform Coalition (HRC) (2005) Citizen toolkit for effective participation in hydropower relicensing, Washington

Kosnik L (2010) Balancing environmental protection and energy production in the Federal hydropower licensing process. Land Econ 86:444–466

Ligon FK, Dietrich WE, Trush WJ (1995) Downstream ecological effects of dams. Bioscience 45:183–191

Maine Department of Marine Resources (2009) Operational plan for the restoration of diadromous fishes to the Penobscot River, Augusta

Mekong River Commission (2005) Fisheries Annual Report. Phnom Penh

Montgomery DR (2003) King of Fish: the thousand-year run of salmon. Westview Press, Boulder

National Research Council (NRC) (2004) Atlantic Salmon in Maine. The National Academies Press, Washington

Opperman JJ, Apse C, et al (in press) The Penobscot River (Maine, USA): a basin-scale approach to balancing power generation and ecosystem restoration. Ecology and Society

Pollak D (2007) S.D. Warren and the erosion of federal preeminence in hydropower regulation. Ecol Law Q 34:763–800

Portland General Electric (PGE) (2006) Pelton Butte Fact Sheet

Richter BD, Braun DP et al (1997) Threats to imperiled freshwater fauna. Conserv Biol 11:1081–1093

Richter BD, Thomas GA (2007) Restoring environmental flows by modifying dam operations. Ecol Soc 12(1):article 12

Richter B, Postel S et al (2010) Lost in development's shadow: the downstream human consequences of dams. Water Alt 3(2):14–42

Sanger D, Belcher WR, Kellog DC (1992) Early Holocene occupation at the Blackman Stream Site, central Maine. In: Robinson BS, Petersen BJ, Robinson AK (editors) Early Holocene Occupation in Northern New England. Maine Historic Preservation Commission, Augusta

Saunders R, Hachey MA, Fay CW (2006) Maine's diadromous fish community: past, present, and implications for Atlantic salmon recovery. Fish 31(11):537–547

Shabecoff P (1993) A Fierce Green Fire. Harper Collins, New York

United States Geological Survey (USGS) (2010) Hydroelectric power water use. http://ga.water.usgs.gov/edu/wuhy.html. Accessed 1 Sept 2010

Chapter 14
Using Stakeholder Input to Develop a Comparative Risk Assessment for Wildlife from the Life Cycles of Six Electrical Generation Fuels

Edward J. Zillioux, James R. Newman, Gregory G. Lampman, Mark R. Watson, and Christian M. Newman

Contents

E.J. Zillioux (✉)
Environmental Bioindicators Foundation, Inc., Zillioux Environmental, LLC, School of Public Health and Health Sciences, University of Massachusetts, Amherst, 207 Orange Avenue, Suite G, Fort Pierce, FL 34950, USA
e-mail: zillioux@bioindicators.org

J. Burger (ed.), *Stakeholders and Scientists: Achieving Implementable Solutions to Energy and Environmental Issues*, DOI 10.1007/978-1-4419-8813-3_14,

Abstract An assessment was conducted of the known and documented effects of electricity generation on vertebrate wildlife in the New York/New England (NY/NE) region. A Comparative Ecological Risk Assessment incorporating Life Cycle Assessment ($CERA_{LCA}$) was constructed to make objective comparisons among the six types of electricity generation important to the NY/NE region: coal, oil, natural gas, hydro, nuclear, and wind. Nonrenewable electricity generation sources, such as coal and oil, pose higher risks to wildlife than renewable electricity generation sources, such as hydro and wind. Based on the comparative amounts of SO_2, NO_x, CO_2, and mercury emissions generated from coal, oil, natural gas, and hydro and the associated effects of acidic deposition, climate change, and mercury bioaccumulation, coal as an electricity generation source is by far the largest contributor to risks to wildlife found in the NY/NE region. The focus of this chapter is primarily on the role of stakeholders and how interactions between the authors and these stakeholders influenced and improved the final product. Thus, while the scientific aspects of the study have been much condensed to provide a full accounting of the stakeholder process, we hope that sufficient coverage of the technical aspects has been provided for the reader to fully appreciate the derivation of our conclusions. For those who would like additional information on the original study, we refer them to the March 2009 report available on line at http://www.nyserda.org/publications/Report%2009-02%20Wildlife%20report%20-%20web.pdf.

14.1 Introduction

Demand for electricity in the United States is expected to increase over the next 20 years. The traditional mix of generation sources in the United States includes coal, oil, natural gas, hydropower, and nuclear power, along with a handful of renewable energy technologies. In recent years, there has been increasing debate about how this mix of energy sources, along with energy efficiency, can be adapted to best meet increasing demand and environmental goals. All energy sources have aspects that make them attractive, and each has its own set of issues.

While the economics and jobs associated with fossil fuel extraction, transportation and refining operations make some energy sources important regionally, if not nationally, other factors are gaining sway. Recently, national policy discussions have incorporated aspects other than simple cost, to address considerations of energy sources and their broader implications in national policy. Domestically available sources of energy, e.g., are attractive to those concerned with national energy independence. Concerns about "peak oil" and its implications for the transportation sector have been raised in discussions about the use of oil for electrical generation. Energy sources that produce the lowest emissions are important to states downwind from generators, and energy resources with the potential to lower production of greenhouse gases are rapidly gaining favor.

New York State, in particular, and New England states, in general, have been especially hard-hit with the effects of acid and mercury deposition resulting from emissions from coal-fired power plants in the mid-western United States. As a result of being located downwind from these power plants, coupled with the rising topography of the Catskill and Adirondack Mountains, New York receives substantial loads of mercury and acids through atmospheric deposition. Mercury contamination has resulted in fish consumption advisories in many water bodies in the State and blanket advisories against fish consumption in the Catskill and Adirondack parks. Approximately half of the lakes and streams in the Adirondacks are either chronically or episodically acidified to levels that affect life cycles of aquatic organisms.

In an effort to address changing energy needs and environmental goals, there has been a rapid expansion of renewable energy development, particularly wind power projects in the northeastern United States and throughout the country. As with any electrical generation project, there are many difficult decisions associated with wind projects to ensure that projects are planned and sited appropriately with minimal effects on humans and the environment. One area of particular concern with this renewable energy resource is potential impacts on wildlife, especially birds and bats.

Wildlife risk assessments for wind facilities rarely, if ever, compare the potential risks to wildlife from wind generation vs. the impacts of producing the same electricity from other generation sources. Although researchers have documented a wide variety of environmental impacts from traditional generation sources, including direct and indirect effects from acid deposition, mercury, ozone, extraction of fuel resources, and water use for cooling, comprehensive analyses of environmental impacts from each source, along with comparative analyses with other electric generating sources, have not been conducted.

To better understand the environmental trade-offs and risks associated with various energy sources, the New York State Energy Research and Development Authority (NYSERDA – see original report to NYSERDA available at http://www.nyserda.org/publications/Report%2009-02%20Wildlife%20report%20-%20web.pdf) felt that it would be useful to have a synthesis and summary of existing research pertaining to New York State and the northeast region that would encompass both life cycle environmental impacts of energy resources and electrical energy production. Only by comparing the full life cycle impacts, from resource extraction to generation facility decommissioning, can a logical comparison of the environmental impacts associated with the various energy resources be made.

The purpose of this effort was to introduce the concept of relative environmental risk of wind generation vs. traditional generation into discussions regarding the siting of wind facilities. The resulting report and related materials were developed to translate research results into a format targeting a broad audience, including state regulators, local governments, and communities considering wind power, environmental groups, scientists/researchers, and those conducting risk assessments.

To undertake this effort in a measured and unbiased way, stakeholders representing a diverse range of interests needed to be brought together to define the project scope and to guide the development of a request for proposals (RFP). This included regulators, the energy industry (e.g., PPM Atlantic Renewable Energy Inc.), environmental groups (e.g., Natural Resources Defense Council), and environmental consulting firms (e.g., West EcoSystems Technology, Inc.). These stakeholders also provided feedback at various stages in the project development, and a subset participated in the proposal review and selection process.

The NYSERDA is a public benefit corporation created in 1975. NYSERDA's earliest efforts focused on research and development with the goal of reducing the State's petroleum consumption. Today, NYSERDA's aim is to help New York meet its energy goals: reducing energy consumption, promoting the use of renewable energy sources, and protecting the environment. The following section describes NYSERDA's support of stakeholders and how their Collaborative Process was applied to compare wildlife risks of a variety of electricity generation sources from a life cycle perspective.

14.2 New York State Energy Research and Development Authority's Stakeholder Process

NYSERDA places a premium on objective analysis and collaboration, as well as reaching out to solicit multiple perspectives and share information. NYSERDA's programs and services provide a vehicle for New York State to work collaboratively with businesses, academia, industry, the federal government, the environmental community, public interest groups, and energy market participants. Through these collaborations, NYSERDA seeks to develop a diversified energy supply portfolio, improve market mechanisms, and facilitate the introduction and adoption of advanced technologies that will help New Yorkers plan for and respond to uncertainties in the energy markets.

A major portion of NYSERDA funding comes from state rate payers through the System Benefits Charge (SBC). Part of this funding went into the creation of New York Energy \$mart^{SM}, which helps the State develop competitive markets for energy efficiency, manage electricity demand, provide outreach and educational services, conduct research and development, run technology demonstration projects, provide energy services to low-income New Yorkers, and provide direct economic and environmental benefits to the State. One program designed to advance this mission is the New York Energy \$mart^{SM} Environmental Monitoring, Evaluation and Protection (EMEP) Program.

The primary mission of the EMEP Program is to support research addressing environmental issues related to the generation of electricity. EMEP provides scientifically credible, objective, and policy-relevant research aimed at two primary goals:

- better understanding the nature of energy-related pollution and its impact on the environment and human health;
- characterizing sources of energy-related pollution and defining cost-effective policies to mitigate impacts and opportunities for emissions reduction.

EMEP places particular emphasis on "environmental accountability" by establishing environmental baselines and evaluating changes in the environment as new environmental improvement programs are launched. In addition, EMEP initiatives include elements focused on introducing its latest scientific findings into the policy arena through:

- frequent meetings and conferences with analysts, policy makers and scientists;
- translation of scientific studies into forms useful for a broad audience;
- provision of environmental data and scientific findings in a timely manner.

A Program Advisory Group, comprising representatives from New York State and federal agencies, utility organizations, other public interest organizations, including The Nature Conservancy, the Electric Power Research Institute (EPRI), and others, guides the EMEP program. A Science Advisory Committee, composed of researchers from university-based, federal, and nonprofit organizations including Harvard School of Public Health, USDA Forest Service, Cary Institute of Ecosystem Studies, the Heinz Center for Science, and others, assists EMEP in the development of its multiyear research plan and provides periodic review of critically important research. The EMEP program reaches out to these groups throughout the development and progression of all research projects. Research projects supported by EMEP are peer reviewed, and the principal investigators are required to present project updates to both program and science advisors – leading to high scientific quality of EMEP-funded projects.

The research needs for New York State are greater than the funding available under EMEP. Therefore, program success requires coordination, collaboration, and leveraging with other state and federal agencies and cofunding of research projects. Synthesis and communication of research results are key goals of the EMEP program (i.e., the true test of success is the use of findings by policy makers to improve both environmental quality and human health and welfare). To achieve this goal, research findings are synthesized and translated into understandable formats, forums are provided for scientists and policy makers to discuss issues, and funding organizations seeking opportunities for collaboration. The EMEP program includes an aggressive communication and outreach policy to support science–policy integration and ensure that results from NYSERDA's EMEP research efforts are used.

New York State, along with many states across the nation promoting renewable power generation, is faced with a variety of complex issues related to the siting of wind turbines. Among these issues is the potential impact of wind turbines on birds and bats. There currently is no consensus on the appropriate methods for assessing this impact as part of the siting process. Often lost in the discussion is the fact that if wind power projects are not developed, the energy they could have produced becomes a missed opportunity ultimately resulting in an increased reliance on fossil fuel technologies as the demand for additional generation increases. By disproportionally discouraging wind generation in response to bird and bat impacts, the environmental impacts of fossil fuel extraction, transportation, and combustion have been inadvertently ignored.

To begin addressing these issues, EMEP selected RESOLVE, a nonprofit public policy resolution organization specializing in wind power issues, to develop a series of stakeholder groups and workshops. Table 14.1 provides examples of stakeholders involved in this process as well as those involved in numerous other stakeholder interactions throughout this study.

The first step was the selection of a steering committee to advise two workgroups. One workgroup focused on the issue of bird and bat impacts to develop a prioritized research agenda for NYSERDA and New York State. A subset of this workgroup provided input and recommendations to the New York State Department of Environmental Conservation (NYSDEC) for the development of wind facility siting guidelines. A second workgroup, the "Energy Alternatives" group, developed a framework for a comparative analysis of the environmental impacts of various forms of energy production.

The Energy Alternatives workgroup was comprised of representatives from northeastern state agencies, federal agencies, the power industry, consultants to the power industry, nongovernmental organizations, and academia. This group developed a strategy for a comparative analysis of the impacts of wind power on birds, bats, and other wildlife and the environmental impacts of conventional (fossil fuel) power production. The group identified and worked with experts, both in New York State and throughout the United States, to outline what scientific research and data would be necessary to conduct an analysis of this kind. The ultimate goal of this group was to provide guidance to an RFP process, so that a qualified contractor could be selected to conduct the comparative analysis.

The Energy Alternatives workgroup developed an outline of what the solicitation should include as project requirements. Proposals would be required to synthesize research findings on regional environmental impacts associated with electricity generation, with a focus on risks to wildlife. Based on feedback from the workgroup, the project was to focus on published research and information gained directly from research currently underway, rather than on conducting new field studies. The inclusion of case studies of impacted species in New York State and the Northeast was to be encouraged. Discussions of life cycle or "cradle-to-grave" impacts of various fuels and processes used for generation were also encouraged. The Energy Alternatives workgroup also asked that an overview of human health impacts from generation sources be included where appropriate, drawing on existing studies and review papers.

Examples of impacts to wildlife from traditional forms of generation that the Energy Alternatives workgroup felt should be included were impacts from:

- emissions and deposition, such as acid rain effects on watersheds, habitats, and wildlife, and mercury bioaccumulation in both aquatic and terrestrial wildlife;
- extraction of resources, such as strip mining and well heads;
- climate change and the associated impacts to wildlife habitats, breeding grounds and migratory behavior;
- physical structures, such as smokestacks;
- water cooling, such as fish entrainment and thermal pollution;
- disruption of fish migration, and flooding related to hydropower facilities.

Table 14.1 Examples of stakeholders involved in comparing the reported effects of risks to wildlife from major generation types in the New York, New England region

Agency or group	Role and interests
Audubon New York	Audubon New York promotes the protection and proper management of birds, wildlife, and their habitats. Interested in understanding the relative risk of generation sources to birds and other wildlife over broad landscape scales
New York State Department of Environmental Conservation	NYS DEC is responsible for protecting New York's environment and management of its resources. Interested in the relative trade-offs associated with their regulatory decisions
New York State Department of Public Service	NYS DPS works to ensure safe, secure and reliable access to energy and other utilities. Interested in minimizing cost increases for rate payers and providing funds to improve the energy efficiency of New York State
New York State Energy Research and Development Authority	NYSERDA's three primary goals are reducing energy consumption, promoting the use of renewable energy sources, and protecting the environment. Interested in objective scientific information to better inform energy policy
U.S. Fish and Wildlife Service	The USFWS works to preserve, protect and enhance fish, wildlife and plants and their habitats for their continuing benefit of the American people. Interested in a scientific understanding of the environmental trade-offs associated with the various energy sources
U.S. Environmental Protection Agency	The USEPA's mission is to protect human health and the environment. Interested in how the stages of energy production and associated activities impact the environment and by extension human health
National Renewable Energy Laboratory	NREL is a division of the U.S. Department of Energy responsible for renewable energy and energy efficiency research, development and deployment. Interested in a scientific evaluation of the environmental impacts of all energy sources to develop a balanced view of renewable energy in relation to fossil fuels
General Public and Local Elected Officials	Energy production and use impact the public in a variety of ways. Often concerns about the environmental impacts of wind energy projects are developed without consideration of the energy sources and the commensurate environmental impacts being displaced. An easy to understand and disseminate comparison of environmental impacts by energy sources would better inform these concerns
Power Industry	Committed to providing reliable power to the public and a return on investment for their share-holders. Interested in furthering the interests of their particular fuel source and demonstrating its advantages over other fuel sources

As recommended by the workgroup, the project also should attempt to identify future analyses and data needed to further advance knowledge and understanding of the relative impacts of different types of energy generation, including both fossil fuel and wind. Finally, the products of the project were written in a style which translates scientific findings into a fashion that is interesting, understandable, and appealing to a broad audience, including policy analysts, policy makers, scientists, and the general public.

Following these recommendations, a solicitation was developed and released. A seven-member Technical Evaluation Panel (TEP) was developed to review the proposals. The TEP was comprised of NYSERDA staff along with representatives of state agencies, nongovernmental organizations, and researchers. The TEP recommended the nonprofit Environmental Bioindicators Foundation Inc., along with Pandion Systems Inc., for the project.

14.3 Study Background

Electricity generation causes adverse effects on both humans and the environment, including effects on wildlife and its habitat. In recent years, concerns about global climate change caused by fossil fuel combustion have focused attention on these effects and the need to move toward a mix of electricity generation sources that will reduce adverse effects. The type of effects and relative level of risk vary among the different electricity generation sources. This study compared reported effects to vertebrate wildlife from electricity generation by coal, oil, natural gas, nuclear, hydro, and onshore wind. The scope of the study did not include how mitigation, implementation of new technologies, or future regulations might change these effects, nor did it address human health effects. The study provides a baseline for discussion about cumulative effects.

The focus is on electricity generating sources that are important to New York and the New England states (collectively referred to as the New York/New England [NY/NE] region) and their effects on birds, mammals, fish, reptiles, and amphibians. The NY/NE region relies on six electricity generation sources for the electricity it needs. With the exception of sources in Maine and Vermont, less than 20% of electricity generation in this region is renewable (hydro, wind, solar, etc.). To address this apparent over-dependence on nonrenewable sources (coal, oil, natural gas, and nuclear), many states have adopted renewable energy plans (Cartiedge 2010).

One of the challenges facing the NY/NE region, as well as the rest of the country, is that all sources of electricity generation, including renewable energies, have adverse effects on wildlife to some degree. The effects of electricity generation on people and wildlife have been studied intensively since the 1970s; nevertheless, most studies have focused on fossil fuel combustion sources (coal, oil, and natural gas). Until now, no one has attempted an "apples to apples" comparison of effects on wildlife from the different types of electricity generation. Nor has there been a study to compare all six electricity generation source types using a cradle-to-grave approach on a relative-risk basis.

This chapter is designed to inform scientists, decision makers, and the general public. References to the published literature are primarily confined to the tables, with only limited references cited in the text.

14.4 Comparative Ecological Risk Assessment Life Cycle Analysis ($CERA_{LCA}$)

A literature review was conducted to provide the basis for a Comparative Ecological Risk Assessment (CERA) study of the known and documented effects of electricity generation on vertebrate wildlife. The focus was on peer-reviewed literature and scientifically accepted published reports or documents regarding effects on wildlife from electricity generation. No original analyses of source contributions or effects were made. The results of the literature review were used in the CERA to make an objective comparison of the six types of electricity generation important to the NY/NE region (coal, oil, natural gas, nuclear, hydro, and onshore wind). The Assessment was completed by conducting a Life Cycle Analysis (LCA; Barnhouse et al. 1998; SAIC 2006) within the Ecological Risk Assessment framework (ERA; USEPA 1998; Henderson et al. 2007).

To objectively and thoroughly compare adverse effects to vertebrate wildlife caused by the six electricity generation source types, the total life cycle of electricity generation must be examined. The LCA identified the stages involved in most forms of electricity generation: resource extraction, fuel transportation, construction of facility, power generation, transmission and delivery, and decommissioning of facility. These stages are defined in Table 14.2. Wildlife effects from exposure to stressors encountered at each life cycle stage were identified and compiled from the literature review for each electricity generation source.

Information from the literature review and the LCA was incorporated into an ERA framework (as illustrated in Fig. 14.1) in order to construct a CERA that identified the stressors and receptors (wildlife and/or wildlife habitat) for each life cycle stage of each electricity generation source type. Next, the level of exposure and types of wildlife effects were characterized for each major stressor within each life cycle stage of each electricity generation source. This information was used to characterize the relative level of risk (or likelihood) of an adverse effect occurring.

14.5 Ranking Risk

One of the biggest challenges in this study was to develop a method for comparing risks that would be best understood given the range in backgrounds of the various stakeholders. To do this, the potential risks for each life cycle stage were characterized for each electricity generation source, and cumulative effects for each electricity generation source were established by assigning a relative wildlife risk level to each wildlife effect (Table 14.3). The wildlife risk level system was developed to qualitatively

Table 14.2 Life cycle stages of electricity generation

Life cycle stage	Definition
Resource extraction	Getting the raw materials to make electricity and all the associated supporting activities (e.g., waste disposal, road construction). For example, for coal and uranium this includes surface and underground mining. For oil and natural gas this includes onshore and offshore drilling and extraction.
Fuel transportation	Transporting the raw materials from the mine or well to the electricity generating facility by rail, truck, barge, ship, or pipeline. This includes construction of pipelines.
Construction of facility	Building the electrical generation facility and associated supporting activities. For coal, oil, natural gas, and nuclear facilities, construction includes power blocks, stacks, cooling ponds or towers, lay-down areas and waste areas, and transmission and distribution lines. For hydro facilities, construction includes the dam, power house, impoundment area, and associated transmission lines and roads. For wind facilities, construction includes turbines, transmission and distribution lines, and roads.
Power generation	All aspects of operating an electricity generating facility. For coal, oil, and natural gas this includes the combustion of fuels. For nuclear this includes heat energy production by fission. For wind this includes the action of the wind turbine blades. For hydro this includes reservoir management.
Transmission and delivery	Getting electricity from the generation facility to where it will be used. This includes transmission lines, distribution lines, and substations.
Decommissioning of facility	The demolition and removal of the electricity generating facility. All electricity generation facilities have a lifespan and must eventually be taken offline and removed. This report does not consider repowering.

rank the relative magnitude of potential harm that could be caused by a stressor and the spatial and temporal occurrence of these effects (exposure) for each life cycle stage of each electricity generation source. Continuous (e.g., emissions), periodic (e.g., bird collisions), and episodic (e.g., major oil spill) levels of exposure were considered in assigning life cycle stressors to potential risk levels. *The levels of wildlife risk are evaluated on a relative scale within each electricity generation source and are not meant to infer absolute risks.* The final risk ranking for a single life cycle stage of a single electricity generation type is given as the highest relative risk level among all assigned risk levels within that life cycle stage.

The naming of risk ranking categories presents a special concern. The importance of avoiding subjective and unintended interpretations of assigned risk levels cannot be overemphasized. The naming of relative risk categories, therefore, should use terminology acceptable to all stakeholders and not subject to media or political hyperbole. Although such terminology should ideally be value neutral, the various alternatives all carry some level of social bias. Verbal descriptions are likely to be taken literally; alphabetic scoring is subject to grading bias; numeric scoring may imply a precision that does not exist. This report provides a snapshot in time, with a

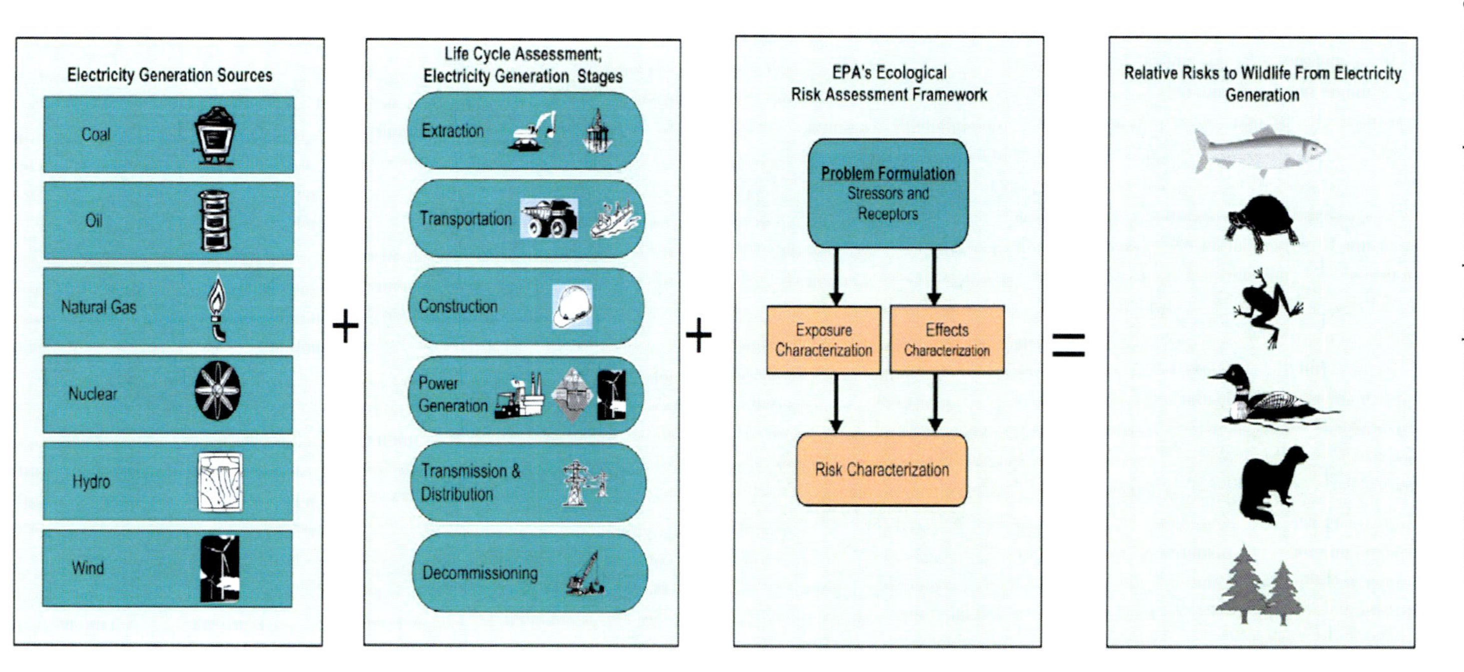

Fig. 14.1 Comparative ecological risk assessment life-cycle analysis ($CERA_{LCA}$)

Table 14.3 Relative wildlife risk levels of potential harm from electricity generation

Relative risk level of potential harm	Potential effects
Highest potential	Large scale, population-level mortality and/or habitat destruction
	Population(s) decline and/or biodiversity is reduced
	A threat to species survival regionally
	Biologically significant mortality or reduction in endangered or threatened species
Higher potential	Limited, but locally to regionally important mortality and/or habitat destruction, with limited population-level effects
	Any biodiversity declines would be local to regional only
	No threat to species survival, but demonstrated effects to physiology and/or behavior of exposed individuals
	Incidental mortality and/or incidental habitat destruction of endangered or threatened species
Moderate potential	Limited and local mortality and/or habitat destruction, with no population-level effects
	Biodiversity declines are unlikely
	Endangered or threatened species may be exposed, but mortality unlikely
Lower potential	Limited to no mortality or habitat destruction affecting populations, but empirical data suggest potential adverse effects on individuals, although not documented in wild populations
	No biodiversity declines
	Exposure of endangered or threatened species unlikely, with minimal adverse effects
Lowest potential	Mortality, if any, limited to individuals; no empirical data to suggest an adverse effect
	No biodiversity declines
	Very limited or no exposure of endangered or threatened species

primary focus on past experience. While future technological advances are not considered in this study, it is recognized that industry responses to existing and anticipated regulations are currently affecting the way electricity is generated and, therefore, the risks associated with that generation. For these reasons, the relative terms covering five separate levels of concern identified in Table 14.3 were selected to describe potential risks that are themselves in the process of continuing change driven by regulatory, technological, and competitive forces.

Highest and Higher Potential risk levels are associated with effects to wildlife individuals and populations, while Moderate, Lower, and Lowest Potential risk levels are associated with only wildlife individuals, without evidence of, or reason to expect, an adverse effect at a population level. This does not mean that wildlife effects to individuals are not important, but if an individual effect does not result in a measurable impact on the population, then it is not considered ecologically significant. However, effects to individual animals can be ecologically significant in at least two situations. First, endangered and threatened species often cannot afford to lose even small numbers of individuals without imperiling the whole population or even the whole species. Second, individuals can become ecologically significant when they are shown to indicate a greater population-level effect.

14.6 Results: Relative Risks to Wildlife from Each Electricity Source

14.6.1 Overall Wildlife Effects and Risks

All life cycles of electricity generation affect wildlife and, therefore, pose risks to wildlife individuals and populations. The degree and extent of the risks depend on the energy generation source, although some effects are common across life cycle stages of many electricity generation sources. Table 14.4 summarizes the highest wildlife risk level for each electricity generation source during each life cycle stage as identified in this study. All risks levels were evaluated on a relative basis across generation types. Coal, oil, natural gas, and nuclear (nonrenewable) have wildlife risks during each of the six life cycle stages, while hydro and wind (renewable) have wildlife risks in only four of the life cycle stages. This difference is because nonrenewable electricity generation sources have to be extracted from the ground and transported to the facility where electricity will be generated. Renewable electricity generation sources do not require resource extraction and may be harnessed at the location where the electricity is generated. The discussions that follow focus primarily on Highest-, Higher-, and Moderate-Potential risk levels for each electricity generation source.

14.6.2 Risks from Coal

Electricity generation from coal has wildlife effects at every stage of its life cycle. Resource extraction and power generation have the greatest number of effects and pose the greatest risk to wildlife. Geographically, the wildlife risks from coal are extensive.

In the resource extraction stage, the wildlife effects and risks from coal are unique because of the way coal is extracted by above-ground and below-ground mining. Above-ground mining includes strip mining, open pit mining, and mountain-top mining and valley fill (USEPA 2000). Above-ground mining poses Highest Potential risks to wildlife populations because of the resulting large-scale habitat destruction. For example, mountain-top mining removes the top of a mountain to uncover the coal seams near the surface. The spoils from the removal are dumped in nearby valleys. The wildlife effects are substantial and impact all types of wildlife and habitats including those in the area of the mining and in valleys where the spoils are dumped. For example, 65,000 acres in West Virginia were permitted in 2002 for

Table 14.4 Highest level of potential wildlife risk within each life cycle stage of each electricity generation source (the overall highest level of potential risk for each generation source is shaded)

	Relative wildlife risk level for potential harm					
Source	Resource extraction	Fuel transportation	Construction of facility	Power generation	Transmission and delivery	Decommissioning of facility
Coal	**Highest potential**	Lower potential	Lower potential	**Highest potential**	Moderate potential	Lower potential
Oil	Higher potential	**Highest potential**	Lower potential	Higher potential	Moderate potential	Lower potential
Natural gas	**Higher potential**	Moderate potential	Lowest potential	Moderate potential	Moderate potential	Lowest potential
Nuclear	**Highest potential**	Lowest potential	Lowest potential	Moderate potential	Moderate potential	Lowest potential
Hydro	None	None	**Highest potential**	Moderate potential	Moderate potential	Higher potential
Wind	None	None	Lowest potential	**Moderate potential**	**Moderate potential**	Lowest potential

mountain-top removal coal mining; this is where much of the coal for the NY/NE region originates. Comparatively, local risks associated with below-ground mining, such as deep shaft mining, are Lower Potential because little habitat is affected compared to above-ground mining.

Both above- and below-ground mining also cause habitat degradation and direct injury and death to wildlife from toxic runoff into waterbodies, which creates Higher Potential risks to wildlife. Mine tailings, mine wastes, and coal processing wastes are highly acidic and often contain trace elements at toxic concentrations. The majority (75%) of acid runoff is associated with underground mining (Mac et al. 1998). This acid runoff from mine tailings (acid mine drainage) can reach streams and injure and kill fish and other aquatic wildlife. It is estimated that about 6,400 streams in the mid-Atlantic and southeastern United States have been affected by toxic mine drainage and runoff, primarily from coal mining in West Virginia and Pennsylvania (Herlihy et al. 1990), which are major coal sources for electricity generation in the NY/NE region.

Underground mine fire is another unique wildlife effect from coal. Although not a common occurrence, these fires release toxic emissions and can last for years or decades. They pose a Lower Potential risk to wildlife and wildlife habitats in the vicinity of the underground mine fire. Local mortality and habitat destruction have been documented in Centralia, PA, where fires have been burning underground since 1962.

Because coal is a fossil fuel, when it burns during the power generation stage it releases multiple emissions (such as SO_2, NO_x, CO_2, Hg, etc.) that cause regional and global wildlife effects. As a result, electricity generation from coal is a significant contributor to acidic deposition, climate change, and mercury bioaccumulation, which are Highest or Higher Potential risks to wildlife. These effects also are common to other generation types. Other wildlife effects associated with power generation from coal include bird/bat collisions with power plant facilities and effects from power plant cooling (once-through cooling) and chemical discharges to surface waters. These pose Moderate Potential risks to wildlife. Effects associated with transmission and delivery includes injury and mortality from collision and electrocution associated with power lines, which pose Moderate Potential risks.

14.6.3 Risks from Oil

Electricity generation from oil has wildlife effects at every stage of its life cycle. Like coal, resource extraction, fuel transportation, and power generation have the most potential wildlife effects and pose the greatest risk to wildlife. Oil is the only electricity generation source that has a Highest Potential risk during fuel transportation. Geographically, the wildlife risks from oil are extensive.

The effects and risks to wildlife during the resource extraction stage for oil are different for onshore and offshore drilling. For onshore drilling, the wildlife risks range from Moderate to Higher Potential. Oil pits containing oil wastes are created in the vicinity of onshore oil wells. Wildlife that contact or ingest the oil from the pits are at risk for death or injury, and this can have regional and local population effects.

Most of the oil used for electricity generation in NY/NE originates from outside the region, but small amounts of oil production occur in western New York (Cattaraugus, Allegany, Chautauqua, Steuben, and Erie counties). Drilling operations in New York are small scale and of short duration (NYSERDA 2005). Thus, the risks from oil extraction in NY/NE region are considered Lower Potential.

Offshore oil extraction can result in injury or death to wildlife and habitat degradation from spillage and discharge of drilling muds, cuttings, and production water (New England Aquarium 1984). Most of these risks are considered Moderate Potential because the effects generally are limited to the vicinity of the drilling range with no population-level effects from mortality and habitat destruction. However, the 20 April 2010 explosion of the British Petroleum Deepwater Horizon drilling rig in the Gulf of Mexico has changed this perspective. Albeit a rare occurrence, the resulting catastrophic oil spill, which has not been fully controlled at the time of this writing, seems likely to rank as the worst oil disaster in American history (see, e.g., New York Times 2010). Although the full extent of damage to habitat and wildlife will not be fully documented for some years, it is expected to surpass that of the 1989 Exxon Valdez oil spill in Alaska (see below). Ranking this risk on a relative basis must take into consideration the frequency of occurrence, the volume of oil spilled and, on a comparative basis, consider the much more common occurrence of oil spills during transportation (see below). The Oil Spill Intelligence Report analysts found that of the 66 spills of at least 10 million gallons, 48 were from tankers while only five were from production oil wells (NOAA OR&R 2010). For this reason, the relative risks from offshore oil drilling are considered of Higher Potential.

Onshore and offshore extraction put wildlife at risk from toxic emissions and from fire from flare stacks, which cause bird mortality and are considered Moderate Potential wildlife risks. The flare stacks and offshore platforms also cause bird collisions. Studies in the Gulf of Mexico show that periodic collisions with oil and gas platforms can occur for migrating birds, primarily neotropical migrants. These are considered Moderate Potential risks with local mortality, but they do not have population-level effects. The exposure is greatest during the migratory seasons and with conditions of low visibility.

Oil is the only electricity generation source that was found to have Highest Potential risks during the fuel transportation stage. Oil is transported to power plants by pipeline, oil tanker, or barge. Injury, death, and habitat contamination are documented effects of fuel spills from barges and tankers. These risks are characterized as Highest Potential with large-scale population-level mortality and habitat destruction (Samuels and Ladino 1984). Although these spills are relatively infrequent, the extent can be widespread, such as in the Exxon Valdez oil spill in Alaska (Burger 1997). The pipelines used to transport oil pose a Higher Potential risk to some wildlife because of habitat fragmentation and destruction. Pipelines can act as barriers to wildlife movement. For example, in Alaska, studies have shown population declines and changes in wildlife behavior, such as in Barren Ground Caribou migration patterns.

Like coal, oil has many wildlife effects during the power generation stage that pose Higher and Moderate Potential wildlife risks. Because oil is a fossil fuel, when it burns it releases multiple emissions that cause regional and global wildlife effects, although to a lesser extent than coal. As a result, power generation from oil contributes to acidic deposition and climate change, which pose Higher Potential risks to wildlife, and to a minor extent mercury bioaccumulation. Because of the relatively low amounts of mercury in oil emissions compared to coal, a Lower Potential risk is assigned. These effects are common to other generation sources. Other wildlife effects associated with power generation from oil include collisions with power plant facilities and effects from power plant cooling (once-through cooling) and chemical discharges to surface waters, which pose Moderate Potential risks to wildlife. Effects associated with transmission and delivery pose Moderate Potential risks. These include injury and mortality from collisions and electrocutions associated with power lines.

14.6.4 Risks from Natural Gas

Electricity generation from natural gas has wildlife effects at every stage of its life cycle. As is the case for coal, fuel extraction and power generation have the most wildlife effects and pose the greatest risk to wildlife. Geographically, the wildlife risks from natural gas are extensive.

As is the case for oil, natural gas has documented population effects during the fuel extraction stage. Gas extraction is similar to oil extraction and is often done simultaneously with oil drilling. The wildlife risks are Higher Potential for oil pits associated with obtaining natural gas from onshore crude oil pumping. Offshore gas extraction can result in injury or death to wildlife and habitat degradation from spillage and discharge of drilling muds, cuttings, and production water; these are Moderate Potential risks. Bird mortality from contact with the toxic emissions and fire from flare stacks can occur and is considered a Moderate Potential risk. Injury and mortality to wildlife (e.g., birds and bats) from collision with offshore gas platforms poses Moderate Potential risks with limited and local mortality and no population-level effects.

As with oil drilling, gas drilling operations in New York are small scale and of short duration (NYSERDA 2005). As a consequence, the risks for fuel extraction in the NY/NE region are considered Lower Potential.

Fuel transportation effects from natural gas are Moderate Potential risks (rather than Highest Potential risks like oil) because gas leaks (e.g., methane) from pipelines do not affect wildlife unless a fire starts. However, methane gas leaks are significant contributors to greenhouse gasses (Litto et al. 2006).

Like coal and oil, power generation from natural gas contributes to risks from acidic deposition and climate change. However, the proportional contribution is less, and thus a risk of Moderate Potential is assigned. These effects also are common to other generation types. Like coal and oil, other wildlife effects associated with the power generation from natural gas include collision with power plant facilities and effects from power plant cooling (depending on the type of cooling, e.g., once-through cooling), which pose Moderate Potential risks to wildlife.

Effects associated with transmission and delivery for natural gas include injury and mortality from collision and electrocution associated with power lines, which pose Moderate Potential risks.

14.6.5 Risks from Nuclear Power

Electricity generation from nuclear power has wildlife effects at every stage of its life cycle. Unlike fossil fuel electricity generation sources, nuclear does not pose any population-level risks to wildlife in the United States. Geographically, the wildlife risks from nuclear are either local or regional, depending on the particular life cycle stage.

Similar to coal, the effects from resource extraction from above-ground surface mining have a Highest Potential risk to wildlife because of the amount of surface habitat that is destroyed. Below-ground mining is considered to have a Lower Potential because of the limited habitat disturbance associated with underground mining compared to above-ground surface mining. Toxic runoff from mining tailings has a Moderate Potential risk for injury and death to wildlife.

During the power generation stage, nuclear power plants, like coal-fired power plants, create very high amounts of heat and require water to cool the generator. If the cooling process involves drawing water from a lake, river, or ocean (such as in once-through cooling), it poses Moderate Potential risks to wildlife. Other wildlife effects associated with power generation from nuclear include collisions with facilities and effects from chemical discharges to surface waters, which pose Moderate Potential risks to wildlife.

Nuclear energy has the potential for accidental or catastrophic release of radioactive materials. In this event, the wildlife risks would be large; however, there have been no such occurrences in the United States and for good reason. The worst example outside the United States was the Chernobyl accident in the former Soviet Union; the associated wildlife effects from this would be characterized as Higher Potential risks. The likelihood of a similar instance in the NY/NE region is virtually nonexistent because the faulty Chernobyl-style reactor design and its lack of containment would not be licensed in the United States. The most serious accident in the history of U.S. nuclear facilities was a partial meltdown of the Three Mile Island-2 reactor core in 1979. This resulted in only very small offsite releases of radioactivity but had a huge effect on regulatory oversight by the Nuclear Regulatory Commission, with an end result of substantially enhanced safety (USNRC 2007; NEI 2007; Rhodes 1993). Therefore, the wildlife effects from a catastrophic nuclear power event were not considered in this study. There is, however, a Lowest Potential risk that injury and mortality may occur during nuclear power generation from accidental release of a small amount of radioactive emissions or effluent discharge.

There may be some bioaccumulation of strontium-90, but this would likely be limited to individuals and not populations.

Like fossil fuels, nuclear energy facilities (e.g., stacks and cooling towers) also can result in collision mortality, posing Moderate Potential risks to wildlife. Effects associated with the transmission and delivery stage include injury and mortality from collisions and electrocutions associated with power lines, which pose Moderate Potential risks.

State wildlife plans did not identify any specific vulnerable habitat and species at risk in the NY/NE region from nuclear power electricity generation.

14.6.6 Risks from Hydro

Electricity generation from hydro has only four stages in its life cycle stage and each has wildlife effects. Like wind, hydro is renewable energy, and the water needed to generate electricity is harnessed at the source. Hydro is the only electricity generation source that has high risks during the construction and decommissioning stages. Geographically, most of the wildlife risks from hydro are local or regional.

The risk to wildlife from construction of a hydro power plant is at the Highest Potential level because of the terrestrial and aquatic habitat clearing and the inundation of these habitats when the reservoir or impoundment is filled with water. The loss of habitat includes not only the inundated terrestrial watershed, but also the stream or river habitats, which pose risks to spawning, foraging, and nesting habitats for fish. This stressor can affect hundreds of acres of terrestrial habitats and tens of miles of stream habitat within the watershed when the reservoir is filled with water. There is also risk of reduction or change in wildlife and fisheries biodiversity. Changes in species composition and populations caused by dams blocking upstream movement of fish can have large-scale reproduction implications for fish (e.g., blocking normal fish movement and migration to spawning habitat). Depending upon the location of the dam, there could be a threat to species survival regionally and biologically significant habitat loss for endangered or threatened species. The consequences of the risk are continuous as long as the dam is in place.

The impounded water in hydro dams is a source of methylmercury formation (Bodaly et al. 1984), the result of flooding of habitats. This flooding mobilizes mercury in the watershed, creating conditions that stimulate bacterial transformation of inorganic mercury to methylmercury, its most toxic form. Natural mercury and atmospherically deposited mercury accumulated over long periods from both natural and anthropogenic sources might be mobilized as a result of disturbance of wetlands systems (Zillioux et al. 1993). Methylmercury formed from bacterial actions in impoundments bioaccumulates in the aquatic and terrestrial food chains and can lead to mortality, injury, and behavioral changes. Mercury emission from coal electricity generation poses a Higher Potential risk. With hydropower, mercury is typically not released in such large quantities into the atmosphere, so the effects are primarily local to the affected Watershed; the risks are considered Moderate Potential.

Greenhouse gases also are emitted from the impounded water of a hydro dam (WCD 2000; Pacca 2007). The greenhouse gases from hydro pose Moderate Potential risks to wildlife from the effects of climate change.

During dam operation, upstream fish are injured and killed during releases of water when they become trapped (entrainment and impingement) in the discharge of water for power. These are considered Moderate Potential risks.

Effects associated with transmission and delivery include injury and mortality from collision and electrocution associated with power lines. These pose Moderate Potential risks.

Hydro is the only electricity generation source that poses Higher Potential risks during the decommissioning of facility stage. Reservoir decommissioning causes mortality to aquatic wildlife and degradation of downstream aquatic habitat from release of sediments during the draining of the reservoir. Risks could be Higher to Moderate Potential for dam demolition (Stokstad 2006). The dismantling also results in the loss of the artificially created upstream lake habitat. Mortality or higher predation rates for fish can occur as drawdown proceeds, leaving fish stranded in shallow pools. The risk is considered Moderate Potential for the fish and other aquatic life that have been using these created habitats.

14.6.7 Risks from Wind

Electricity generation from wind has only four stages in its life cycle and each has wildlife effects. Like hydro, wind is a renewable energy source, and the wind needed to generate electricity is harnessed at the source. Wind is not considered to have population effects, but the risks for some bat species are unknown at this time. Geographically, the wildlife risks from wind are all local or regional.

The most commonly cited effect from wind power generation is injury and mortality to birds and bats from collision with wind turbines. For birds, these risks are considered Moderate Potential, and they are limited to the site. Local mortality to individuals is likely to occur with no population-level effects and a high degree of species recovery (NRC 2007). Biodiversity declines are unlikely for birds. Endangered or threatened bird species in the NY/NE region may be exposed to potential injury or mortality, although they are at no more risk than other species.

For bats, especially tree bats, the risk posed by wind turbines may be Higher Potential, but this is uncertain because of the lack of accurate population information and mortality studies at wind farms. Ongoing research is looking at the effects and risks to birds and bats from wind farms, but at this time there are no documented population-level effects. However, based on the few available studies, there is general consensus from the scientific community that bats are likely to be at the greatest risk.

Effects associated with transmission and delivery include injury and mortality from collision and electrocution associated with power lines. These pose Moderate Potential risks.

14.7 Stakeholder Review of Draft Report

Copies of the draft report were provided to a subset of the members of the original stakeholders group formed at the inception of the project and NYSERDA staff for review and comment. Individuals were selected for their expertise, involvement in activities related to energy and related environmental implications, and ability to review and provide comments in a timely manner. This latter point proved to be a fairly consistent issue for stakeholder groups; often the preferred reviewers were not always available, requiring the Project Manager to continually balance individual reviewer involvement with the need for timeliness. Stakeholder thoughts and comments were collected by the NYSERDA project manager for synthesis and refereeing.

A few of the comments provided by the stakeholders were redundant while others offered unique insights into the issues. Some reviewers focused on technical issues while others seemed to be considering the implications of data presented and how it would be interpreted, and possibly cited, by the reader. Many of the reviewers requested additional information that would have required an expansion in scope. This was not unexpected, as the findings of the project in general tended to elicit questions rather than just simple answers. To the extent that requests were appropriate and possible, comments were provided to the authors for consideration.

The final synthesized set of comments varied widely, but the most complicated issues to contend with focused on the way the information was categorized and presented. For example, there was a good deal of discussion of how to present the relative risk table in a way that fairly and accurately represented the findings of the project. Since this would be the take-home message from the project, it was important to make it easy to understand while accurately reflecting the findings and limitations of the work. The project's limitations were highlighted by many reviewers. While described in various ways, most reviewers emphasized the need to clearly convey that the findings were *relative risks* associated with a particular fuel source and life cycle stage and not quantitatively derived. Additionally, there was concern that the table on its own does not provide other important caveats contained in the supporting text. One such caveat was the retrospective nature of the research. Since the project findings were based only on available, published literature, it does not take into account changes taking place in the various industries that would reduce the associated relative risks (e.g., double-walled oil tankers, emissions-control technologies). Without knowing these details, the table could be misinterpreted as a predictor of potential risk going forward rather than the risks based on past experiences. Similarly, some reviewers pointed out the relatively low risk associated with nuclear power generation. This seemed to run counter to commonly held beliefs. There was concern that the lack of a catastrophic nuclear incident to date had led to an underrepresentation of the relative potential risk and raised concerns for how this might be received by industry and the public.

Finer points raised by individual reviewers generally concerned the specifics of the generation source. For example, one reviewer questioned whether the stoppage of fish passage at hydroelectric dams should be considered a "Power Generation" impact or a "Resource Extraction" impact. Similarly, another stakeholder was concerned that the increased risk of exotic species with reservoir construction, or other prolonged habit alterations, should be generally accounted for under "Power Generation" rather than "Resource Extraction." Another reviewer questioned which life cycle stage is represented in the operation of oil refineries and natural gas cleanup facilities.

In general, the diverse stakeholder reviewers provided a wide array of insights into potential issues with the report as well as the project in general. In a few cases, similar issues were described by the reviewers in different ways, demonstrating the unique perspective of the reviewer. Ultimately, by using such a diverse group of stakeholders to consider the report's goals, uses, implications, and technical issues, the value of the project was greatly improved.

14.8 Stakeholder Reactions: Potential Uses for Project Results

During this study, numerous opportunities were taken to solicit input from stakeholders in the form of criticism and direction for further project development and improvement. As discussed earlier, one of NYSERDA's EMEP Program primary goals is to provide a better understanding of the nature of energy-related pollution and its impact on the environment and human health. In attempting to achieve this goal, the EMEP Program introduces its latest scientific findings into the policy arena through frequent meetings and conferences with analysts, policy makers and scientists. For example, at its annual EMEP meeting on 15 November 2007 in Albany, New York, the authors presented a paper entitled *Electricity Generation Effects on Wildlife Populations: A Synthesis*, which was a progress to date description of the project and its goals and objectives. The audience was comprised of scientists, federal and state policy makers, nongovernmental organizations, and the general public. It was the first public presentation of the project. The central theme of the stakeholder input after the presentation was a better understanding by them that in discussions of impact of electricity generation, comparative risks to wildlife cannot be simply stated. Each source of electricity generation has its own unique effects to wildlife. In addition, any discussion of such effects needs to look at a life cycle approach to understand the cause and effect relationships. Finally, they developed a better understanding that "gee whiz" numbers on the magnitude of mortality for a particular type of electricity generation, e.g., avian mortality at wind farms, provide little information for decision making. Several stakeholders provided research studies on different examples of wildlife impacts that occurred in the NY/NE region, e.g., local harbor oil spills.

The first public presentation of the study for the wind industry was in June 2008 at the annual meeting of the American Wind Energy Association (AWEA) in Houston, Texas. Prior to this study, when opponents of wind brought up wildlife impacts, the wind industry was only able to argue that wind had significantly less wildlife impacts, particularly bird collision, as compared to buildings and windows, cats, cell towers, automobiles, and transmission lines. Unfortunately, while these comparisons are true, they are not related to energy and also only compare collision mortality which is only one of the many risks associated with energy development. In addition, they don't provide comprehensive comparisons of the complete life cycle for energy development.

The authors gave a 15-min overview of the study and presented the preliminary results comparing the risks to wildlife for the life cycle of each electricity generation type studied. The presentation emphasized that all electric energy development have impacts on wildlife and that by comparing risk within and across life cycles, better information can be provided to decision makers and stakeholders to understand those risks.

After the presentation, attendees agreed that this was the type of analysis that was needed. All asked to be sent the final report. Included were representatives from the following: Bat Conservation International, American Bird Conservancy, Boston University Ecology and Conservation Biology Department, US Department of Energy, Nuclear Energy Institute, Babcock and Brown, RES America, St. Croix Environmental Association, Bluewater Wind, Wind for Illinois, and Nuclear Energy Insight. Fowler (2008) from the Houston Chronicle wrote an article on the session and presentation, thus generating additional interest in the study.

Some skepticism concerning the study was also expressed. A sampling of these comments is given below.

> *Given the very limited amount of coal and the extremely small amount of oil that is involved with generating electricity for NY and New England, I wonder how these fuels could have a significantly greater impact upon wildlife and wildlife habitat compared to wind energy - which likely will impact adversely far more wildlife and forest habitat if extensively developed (as is planned). I also wonder how "smokestack emissions" associated with the burning of coal were evaluated for their harmful risk to wildlife - particularly in the context of the current "cap and trade" emissions program and the newly implemented CAIR program, which the EPA has begun to administer for all the fossil-fueled power plants located in the east and midwest of the US.*
> *Consequently, I would like to obtain a copy of your presentation to the AWEA meeting in order to learn more about the scope of and methods used in your analysis, as well as find out what data sources you relied upon.*

This comment came from a staunch opponent of wind power who is well known for his opposition to wind in the Mid-Atlantic States. He identified himself as a "Consulting Conservation Biologist" and worked for both federal and state wildlife agencies prior to consulting.

> *I did not have an opportunity to see this presentation while attending the WindPower conference on Tuesday, but I am even more curious about how impacts to the animals were actually measured and compared among the sources of energy. My understanding of the data and methods would lead me to believe that is an impossible analysis...qualitative and speculative at best since actual fatalities are rarely if ever found at or near other such sources and the habitat impacts on individual deaths and population impacts, while clearly an issue and concern, would be littered with assumptions and little supporting evidence. But, I could perhaps be convinced with a thorough understanding of the methods and data, so would appreciate seeing all of this as well.*

This comment came from a conservation group that is very concerned about wind–wildlife interactions but is also researching those issues and has worked at both the state and federal level to develop siting guidelines.

> *I am looking forward to the actual paper. These comparative papers provide useful information, but the underlying data used to calculate impacts of various energy projects is the most critical aspect. Unfortunately, this information is never provided in a media release of the results.*

This remark was from the director of a prominent national conservation group who has expressed publicly concerns about impacts to wildlife from wind energy development.

The authors gave four more presentations to the wind industry prior to the study being finalized. In October 2008, the authors gave the same presentation to CanWEA (Canadian Wind Energy Association) in Vancouver, Canada. That presentation generated no response or follow-up from any of the attendees. In fact, the authors received no questions from the audience following the presentation.

Also in October 2008, the authors were asked to present to a small group sponsored by the National Wind Coordinating Committee (NWCC) in Washington, DC. It was a forum on the Environmental Benefits and Costs of Wind Energy. In attendance were representatives from the Department of Energy, National Renewable Energy Lab (NREL), Pacific Northwest National Laboratory, EPRI, National Academy of Sciences, Stanford University, USEPA, AWEA, and Clean Energy States Alliance. The authors gave a similar presentation but focused more on the process and discussion of the life cycle analysis.

The authors also gave the presentation at the AWEA Wind Energy Fall Symposium in November 2008 in Palm Desert, California. Again, there was general interest in the results but there was no follow-up from the event.

An additional presentation was given by the authors in January 2009 as a part of a webinar series on Wind and Renewable Energy. The particular webinar in January was on Wind Power Siting and Environmental Issues. The webinar was sponsored by a consortium of organizations ranging from the National Rural Electric Cooperative Associate, American Public Power Association, NREL, Utility Wind Integration Group, AWEA, and NWCC. Although this presentation was remote over the Internet, it generated some follow-up from participants similar to the presentation to AWEA in June 2008. Attendees followed up with requests for the study. They included the Natural Resource Defense Council, Northwest Wildlife Consultant, Merlin Environmental, Illinois State University-Center for Renewable Energy, TRC, Colorado Springs Utilities, Cooper Erving & Savage LLP, and Penn State Erie Behrend College. All wanted a copy of the study and response to the presentation was positive.

In November 2008, the authors gave a presentation at the Society of Environmental Toxicity and Chemistry (SETAC) annual meeting in Tampa Florida in a direct attempt to obtain input from stakeholders on a persistent problem. The presentation was entitled *Ranking Wildlife Risks from Multiple Stressors: A Communications Conundrum*. The audience was a near even mix of scientists from academia, government agencies (mostly USEPA), and industry consultants. The "conundrum" referred to the problem we had wrestled with throughout the study on how to rank risks from dispirit activities while avoiding apples-to-oranges comparisons and selecting risk descriptive terms that avoid unwarranted subjective interpretations while being equally acceptable among all stakeholders. Input from open discussion following the presentation resulted in the use of value-neutral comparative terms not subject to media or political hyperbole. The resulting approach to ranking risks used in this study is described in detail in Sect. 14.6 above.

Following NYSERDA's posting of the final report on their website; AWEA (2009) issued a press release about the study, again resulting in the authors being contacted by numerous stakeholders.

The most recent in this series of presentations was given on 16 January 2010 at Florida Power & Light Co. (FPL) headquarters in Juno Beach, Florida. The invitation was given as a result of the AWEA press release mentioned earlier. The author was the featured speaker in a webinar scheduled weekly with attendees primarily from FPL and NextEra Energy. The audience was composed of industry representatives all supporting the acceptance and advancement of wind energy development. They appreciated that the comparative study, providing relative risks from all life cycle stages of six different electricity generating sources, was an important tool for advocating their position with respect to wind energy initiatives.

14.9 Overall Improvement Derived from Stakeholder Involvement

Stakeholder input from the presentations provided useful insight in writing the report. In particular, it assisted with how to present the results in a way that emphasized how the study could be interpreted while also being explicit about the limitations of results due to the underlying scope of the study and the availability of data.

As expected, the stakeholder review process provided numerous suggestions on the draft report that improved the final report. This is discussed in detail in Sect. 14.8.

As discussed in Sect. 14.8, the risk ranking process was greatly improved as a direct consequence of the discussion with government, academic, and consultant scientists attending the November 2008 SETAC annual meeting presentation.

14.10 Summary and Conclusions

14.10.1 Electricity Generation Source Risks

The following overview conclusions can be drawn concerning the comparative risks among the various electricity generation options available in the NY/NE region.

Based on the comparative amounts of SO_2, NO_x, CO_2, and Hg emissions generated from coal, oil, natural gas, and hydro, and the associated effects of acidic deposition, climate change, and mercury bioaccumulation, coal as an electricity generation source is by far the largest contributor to these risks to wildlife in the NY/NE region.

Overall, nonrenewable electricity generation sources, such as coal and oil, pose potentially higher risks to wildlife than renewable electricity generation sources, such as hydro and wind.

Major risks by source are as follows.

- Coal has risks that range from Lowest to Highest Potential, including unique risks during the resource extraction stage (e.g., Highest Potential risks associated with effects of strip and mountain top mining). The combustion of coal during the power generation stage contributes to acidification, mercury bioaccumulation, and potential climate change effects causing Highest Potential risks to wildlife.
- Oil has Lowest to Highest Potential risks, with unique risks during the resource extraction and fuel transportation stages owing to the potential for oil spills. Oil contributes to acidification risks and potential climate change effects during the power generation stage.
- Natural gas has Lowest to Higher Potential risks for wildlife. A number of the types of effects associated with the power generation life cycle stage are similar to oil generation sources, but the magnitudes of these risks are less, e.g., Moderate Potential risks from habitat change from greenhouse gas emissions compared to Higher Potential risks from oil because of the lower magnitude of the contribution of natural gas emissions.
- Nuclear presents Lowest to Highest Potential risks. Some of these risks are not unique to nuclear, and they also are found with other nonrenewable electricity generation sources, such as bird collisions with stacks and cooling towers associated with coal and oil generation sources.
- Hydro has Lowest to Highest Potential and unique risks during the construction, power generation, and decommissioning stages, such as loss of large areas of terrestrial and aquatic upstream and downstream habitats and blocking fish migration due to reservoir or impoundment construction.
- Wind has Lowest to Moderate Potential risks during operation (i.e., bird and bat collisions with wind turbines). No population-level risks to birds have been noted. Population-level risks to bats are uncertain at this time.

Since there are more conditions, by-products, and actions in the resource extraction and power generation stages that act as stressors to wildlife, higher risks to wildlife generally are associated with these life stages than in other life cycle stages.

Construction, transmission and delivery, and decommissioning stages generally have fewer stressors that affect wildlife. However, the construction, operation, and decommissioning of dams pose relatively Higher Potential risks to ecosystems, fish, and habitats.

The degree and extent of the risks depend on the electricity generation source, although some effects are common across life cycle stages and electricity generation sources. See Table 14.4 for a summary of the highest potential wildlife risks (Highest, Higher, Moderate) levels for each electricity generation source during each life cycle stage.

14.10.2 Variability and Uncertainty

A detailed discussion of variability and uncertainty in the source data was included in the original report of this study to NYSERDA. This includes limitations of data accessibility and interpretation based on project scope and assumptions. Based on this analysis a number of data gaps were also identified. The reader is referred to the original report, http://www.nyserda.org/publications/Report%20 09-02%20Wildlife%20report%20-%20web.pdf, for the full discussion.

14.10.3 Suggestions for Future Studies

The following opportunities for future comparisons of wildlife risk were identified during this study. They are not presented in any order of importance.

1. Discuss and rank recovery potential of affected populations and habitats. Various at-risk wildlife groups have different abilities to handle risks. Some populations have the reproductive potential to offset losses that might occur. Some habitats can readily recover once the stressor is removed (e.g., spill in a stream), while other habitats may have changed so much that recovery is not possible (e.g., mountain top mining habitat loss and climate change effects to sensitive habitats).
2. Consider relative risk from the improvement in air quality (e.g., decrease in acidic deposition and mercury) in the last 20 years related to recovery potential.
3. Compare the existing wildlife risks to future technologies. For example, clean coal technologies should reduce the wildlife impacts from power generation. Discuss to what extent this can occur.
4. Evaluate the wildlife risks associated with other renewable energy technologies, such as offshore wind, biomass, solar, etc.
5. Discuss contributive risk. Not all electricity generation sources in the NY/NE region are equally prevalent. A state-by-state analysis of wildlife risk could be conducted. This would be useful in looking at long-term trends to wildlife risks in the NY/NE region as shifts in the electricity generation portfolios occur.
6. Quantify comparative wildlife risks from different facilities of the same electricity generation size.
7. Discuss policy implications of the wildlife Comparative ecological risk assessment life cycle analysis ($CERA_{LCA}$), including identification of the best use(s) of available data.

Acknowledgments We are indebted to many for the ideas, guidance and plain hard work that brought this project to completion. Without the support and vision of the NYSERDA staff, there would have been no project; without the stakeholders mentioned throughout the chapter there would have been no purpose for the project. Our colleagues, Peter Colverson, Christine Denny, Karen Hill and Susan Marynowski of Pandion Systems, Inc., and William Warren-Hicks of EcoStat, Inc. deserve our lasting gratitude for their insight, direction, and attention to detail throughout the project. A special thanks also goes to the editors of the original report: Diane Welch of NYSERDA, Jayne Charles, Deian Moore, and Timothy Sullivan of E&S Environmental Chemistry, Inc., and Zywia Wojnar of Pace Energy and Climate Center, Pace Law School.

References

AWEA (American Wind Energy Association) (2009) AWEA Calls New NYSERDA Wildlife Study a "Welcome Look at an Important Issue." Press release 12 May 2009. Contacts: Julie Clendenin (202):384-3090 and Heather Caufield (212):255-8478.

Barnhouse L, Fava J, Humphreys K et al. (1998) Life-Cycle Impact Assessment: The State-of-the-Art, 2nd Edition. Report of the SETAC Life-Cycle Assessment (LCA) Impact Assessment Workgroup, SETAC LCA Advisory Group. Society of Environmental Toxicology and Chemistry (SETAC) and SETAC Foundation for Environmental Education, Pensacola.

Bodaly RA, Hecky RE, Fudge RJP (1984) Increases in Fish Mercury Levels in Lakes Flooded by the Churchill River Diversion, Northern Manitoba. Canadian J Fisheries & Aquatic Sci 41:682-691.

Burger J (1997) Oil Spills. Rutgers University Press, New Brunswick.

Cartiedge J (2010) Renewable Electricity (Portfolio) Standards. BrighterEnergy.org. http://www.brighterenergy.org/3972/faq/faq-legislation/renewable-electricity-portfolio-standard. Accessed 20 Jul 2010

Fowler, T. 4 (June 2008) Research finds wind power poses least risk to wildlife. Available online: http://www.chron.com/disp/story.mpl//5819709.html. Accessed 20 Jul 2010

Henderson RF, Datson GP, Duke CS et al. (2007) BOSC Workshop on USEPA Risk Assessment Principles and Practices. Human and Ecological Risk Assessment 14(1):39

Herlihy AT, Kaufmann PR, Mitch ME et al. (1990) Regional Estimates of Acid Mine Drainage Impact on Streams in the Mid-Atlantic and Southeastern United States. Water, Air & Soil Pollution 50(1-2):91-107

Litto, R, Hayes RE, Liu B (2006) Capturing Fugitive Methane Emissions from Natural Gas Compressor Buildings. Abstract. Journal of Environmental Management 84(3):347-361

Mac, MJ, Opler PA, Haeker CEP et al. (1998) Status and Trends of the Nation's Biological Resources, Vols. 1 and 2. U.S. Department of the Interior, U.S. Geological Survey, Reston.

NEI (Nuclear Energy Institute) (2007) Resources and Stats: Nuclear Statistics. Available online: http://www.nei.org/resourcesandstats. Accessed 20 Jul 2010

New York Times (2010) Gulf of Mexico Oil Spill. Available online: http://topics.nytimes.com/top/reference/timestopics/subjects/o/oil_spills/gulf_of_mexico_2010/index.html. Accessed 20 Jul 2010

NOAA OR&R (Office of Response and Restoration) (2010) Oil Spills in History. Available online: http://response.restoration.noaa.gov. Accessed 20 Jul 2010

NRC (National Research Council) (2007) Environmental Impacts of Wind-Energy Projects: Prepublication Copy. The National Academies Press, Washington.

NYSERDA (New York State Energy Research and Development Authority) (2005) State Energy Planning. Available online: http://www.nyserda.org/Energy_Information/energy_state_plan.asp. Accessed 20 Jul 2010

Pacca, S (2007) Impacts from Decommissioning of Hydroelectric Dams: A Life Cycle Perspective. Climatic Change 84:281-294

Rhodes, R (1993) A Matter of Risk. Chapter 5 In: Nuclear Renewal. Penguin Books. USA. Available online: http://www.pbs.org/wgbh/pages/frontline/shows/reaction/readings/chernobyl.html. Accessed 20 Jul 2010

SAIC (Scientific Applications International Corporation) (2006) Life Cycle Assessment: Principles and Practice. National Risk Management Research Laboratory, Office of Research and Development, USEPA, EPA/600/R-06/060. Cincinnati.

Samuels WB, Ladino A (1984) Calculations of Seabird Population Recovery from Potential Oil Spills in the Mid-Atlantic Region of the United States. Ecological Modelling 21:63-84.

Stokstad E (2006) Environmental Restoration: Big Dams Ready for Teardown. Science 314(5799):584

USEPA (U.S. Environmental Protection Agency) (1998) Guidelines for Ecological Risk Assessment. USEPA Risk Assessment Forum, EPA/630-R095/002F. Washington

USEPA (U.S. Environmental Protection Agency) (2000) Mountaintop Mining/Valley Fill Environmental Impact Statement: Preliminary Draft. U.S. Environmental Protection Agency, Region 3, EPA/903/R-00/014, October 2000. Philadelphia

USNRC (US Nuclear Regulatory Commission) (2007) Fact Sheet on the Three Mile Island Accident. Available online: http://www.nrc.gov/reading-rm/doc-collections/fact-sheets/3mile-isle.html. Accessed 20 Jul 2010

WCD (World Commission on Dams) (2000) WCD Press Releases and News Announcements: 27 November 2000 – Does Hydropower Reduce Greenhouse Gas Emissions? Available online: http://www.dams.org/news_events/press357.html. Accessed 20 Jul 2010

Zillioux EJ, Porcella DB, Benoit JM (1993) Mercury Cycling and Effects in Freshwater Wetland Ecosystems. Environmental Toxicology and Chemistry 12:2245-2264

Chapter 15
Institutional Void and Stakeholder Leadership: Implementing Renewable Energy Standards in Minnesota

Adam R. Fremeth and Alfred A. Marcus

Contents

Abstract Many state-level policies in the United States have been adopted in an effort to reduce carbon emissions, reduce exposure to fuel price volatility, and encourage economic development by creating a renewable energy industry. Experience with such instruments, however, has been mixed. In this chapter, we argue that a series of obstacles prevent a single actor to take the lead in designing the rules necessary to fill the institutional void that is created by the introduction of novel command and control energy policies. Using case study evidence from the state of Minnesota, we

A.R. Fremeth (✉)
Richard Ivey School of Business, University of Western Ontario,
1151 Richmond St. N, London, ON, N6A 3K7, Canada
e-mail: afremeth@ivey.ca

J. Burger (ed.), *Stakeholders and Scientists: Achieving Implementable Solutions to Energy and Environmental Issues*, DOI 10.1007/978-1-4419-8813-3_15,

find that the collective action problem we describe in this chapter tends to impede the implementation of renewable portfolio standards despite the new and additional certainty that has been provided by a legislated mandate.

15.1 Introduction

Renewable portfolio standards (RPSs) have become a state-level policy of choice as energy and environmental regulators seek to reduce carbon emissions, reduce exposure to fuel price volatility, and encourage economic development by creating a renewable energy industry. Experience with such instruments, however, has been mixed. Implementation challenges have prevented electric utility firms from undertaking many of the actions necessary to meet the objectives set by these policies. We argue that one of the greatest challenges presented is the hesitancy of any single actor (public or private) to take the lead in designing the rules necessary to fill the institutional void created by this command and control approach to energy policy. Governments have created a legal framework that demands action, yet the institutional leadership to effectively implement the framework is often weak and deficient. The framework is not a surrogate for the rules that would lay out the appropriate actions and prescribe the dealings between utility firms and other vested interests (transmission management, municipal government, independent power producers [IPPs]).

A collective action problem exists in which the individual incentives for the major players (i.e., the technically leading and dominant firm and key regulatory body) to act are insufficient to meet the demands of the policy. Each of them has reputational, political, and authoritative concerns that keep them from taking the lead in mobilizing the stakeholders who must be brought together if policy is to be effectively implemented. Prior collective action coalitions can provide indications for how a solution can be found, but their ephemeral nature and the introduction of new parties can limit their suitability for the focal problem. Using case study evidence from the state of Minnesota, we find that the collective action problem we describe in this chapter impedes the implementation of RPSs despite the new and additional certainty that has been provided by a legislated mandate.

This chapter begins by defining the problem of stakeholder leadership in industry formation. Stakeholder leadership requires overcoming what we call an "institutional void." Relatively new industries require rules and standards of practices for them to flourish. Transaction problems abound among those involved in such an industry, one that is just emerging and is amorphous in form and configuration. Those involved need a way of organizing themselves. Self-organizing is insufficient given the complex tasks that the players in the industry face and the coordination problems they confront. The collective action problem is to work together to overcome this institutional void. Stakeholder leadership is needed to define the rules and norms for the relations among the players in the emerging industry so that they can work together effectively, and new business models can take hold. Legitimate, taken-for-granted patterns of behavior, however, are lacking and must be forged even though it may not be in the interests of any single actor to create these norms and the standards of interaction. The incentive for each player is to avoid being a leader of the stakeholder coalition and to shift the burden onto other players. The collective benefits of acting are great, but the individual costs are perceived to be even greater. Thus, stalemate and delay take place, which results in frustration among the players with their incapacity to act in a timely fashion when time is of the essence. Nonetheless, it is necessary to note that the sense of urgency varies significantly among the relevant parties.

The second section illustrates this problem by providing a brief description of the impediments we have identified in the development of the renewable energy industry in the United States, and the third section tests this through a case study. Focusing on the state of Minnesota and the role of Northern States Power (NSP), an operating utility of Xcel Energy (Xcel), we show that the state's attempt to expedite the renewable sector with a stringent energy policy may fall short unless stakeholder leadership is effectively exercised and the institutional void is overcome. Finally, the chapter concludes with a brief discussion of the public policy implications of this analysis.

15.2 The Problem

The classic case of a collective action problem presented by Olson (1965) identifies how members of a group, such as an industry, are unlikely to cooperate in group action so long as the benefits and costs of a collective good are unevenly shared by the parties. A collective good is one that each member of the group can share in and one member's use of that good subtracts from another member's. Of particular interest is the situation of nonatomistic players in a small group where there can be exploitation of the "great by the small," and it is likely that one or a few members value the collective good to such an extent that they may be willing to bear more than their share of the cost. Within the strategic management literature this propensity to tolerate free-rider behavior has been explained by the differences among firms and the internalization of the related costs (King and Lenox 2000). However, the likelihood that a dominant actor, such as the market share leader, will independently coordinate the collective action problem and disproportionately bear the costs assumes that not only they are in a position to do so, but the institutional arrangements exist to support such action. The institutional arrangements may include, but are not limited to, forums for communication or norms of interaction within a group. When these institutional arrangements are weak or simply do not exist, we are left with what we term an "institutional void" whereby the likelihood decreases for the dominant player to assume collective action.

In the case of an emerging industry, as in the development of the renewable energy sector in the United States, dominant actors may be called upon to take on the important role of stakeholder leaders and organize the coalition of actors needed to effectively implement the policy. Generally, these firms must cradle a technology from infancy to commercialization, coordinate the necessary actions, and disproportionately bear the burdens as they wait patiently for long-term gains. Murtha et al. (2001) show how the birth of the flat panel display (FPD) industry in Japan took the leadership and coordination of major firms (such as NEC, Sharp, and Toshiba). These firms had to take on the task of bringing about a significant shift in industry norms and knowledge creation. These stakeholder leaders could not immediately reap appropriate benefits from the activities in which they engaged. Although they broke the deadlock to action, this process was not simple. In this case, the firms in cooperation with Japanese Ministry of International Trade and Industry (MITI) worked to develop rules for collaborative efforts that were necessary to bring the new industry into existence. They overcame the institutional void that in the end permitted the industry to flourish. It is not uncommon for government agencies, such as MITI, to have an essential role in new industry creation as they generally have a responsibility for the economic development. The Federal Energy Regulatory Commission (FERC) had an important role in the creation of independent power production industry in the United States as it set out the rules and direction that these firms should follow in establishing their facilities (Russo 2001; Sine and David 2003). There are many parallels in the energy arena.

In either the case of the firm-led flat panel industry or government-led independent power industry, the dominant actors helped to define the rules of the game (North 1990) that were essential to the design of the industry's architecture. In other words, these actors not only were willing to internalize the costs to overcome potential collective action problems, but filled the institutional void that would have otherwise made market transactions difficult. Institutional voids such as those faced by industries before they coalesce and have legitimacy make it costly for individual firms to deal with critical product, labor and capital markets because of information problems, imperfect contract enforcement, inability to enforce property rights, and flawed regulatory structures (Khanna and Palepu 2000). Therefore, new industries are more likely to be successful only when the collective action problems are solved and institutional voids are overcome. Mainly, this takes place when a dominant actor is willing to take on the role of organizing the critical stakeholders (competitors, collaborators, and others) and filling the institutional void.

The interrelated actions necessary for bringing together interests to establish an emerging industry is further complicated when there are multiple actors who share necessary resources, capacity, and authority, but none with exceptional clout or capacity to command. In these complex cases, a dominant firm may be unwilling to internalize the costs associated with solving the collective action problem and filling the institutional void. For instance, there may be reputational concerns for having a private firm determine the "rules of the game" when the issue is politically sensitive, such as those relating to the natural environment. Similarly, there could be competitive concerns as a dominant firm may run into obstacles created by other firms that understandably prefer not to have the institutions designed by their competitor. This problem of who sets the rules is a concern not only of private firms that may have the resources or capacity but also of government agencies and politicians that tend to have statutory authority yet few other resources to ensure implementation.

The public actors may be capable of directing action, as in the case of FERC with the development of the IPP industry, where the public actors had legal and regulatory authority on their side, but they may not want to be seen as setting standards that could constrain certain industrial activities or prove politically dangerous. An example would be a technology-discriminating policy that defines exactly what actions or investment a firm would need to undertake to be compliant (see Yao 1987 for a discussion of the automotive emission standards in the United States). Legal and regulatory authority does not necessarily mean that public actors will take on the role of stakeholder leader and as a responder to the institutional void. When neither private industry nor policymakers are willing or able to take the lead in overcoming the void and developing the institutions necessary, the collective action problem is likely to persist and prevent, block, and delay industry development.

If there is a dominant firm, that is likely to benefit from the public policy and emerging industry, this conundrum can leave it in a "catch 22" dilemma where it is in its best interests to promote the creation of the industry, but it feels unable to act on its own and to fill the institutional void. At this point, the problem emerges as to *who makes the rules of the game when no one wants to*. The rules are a public good that belong to everyone. They are essential. They are like highways. Without them the new industry cannot succeed, but the question remains who will bear the cost of organizing the industry so that there are accepted rule of transacting?

As we develop in our case study later, firms that are successful in these situations must rely on or develop an aptitude for stakeholder leadership (Sharma and Henriques 2005). This capability involves coalition building not only in the lobbying and design of public policy but also in its implementation as the rules for interaction in the industry must be developed. The dominant firm, in essence, must act as an institutional entrepreneur (Aldrich and Fiol 1994), a builder of the industry itself and a leader in its legitimization. These skills do not come easily to private or public leaders who tend to be increasingly cautious about their exposure and fearful of media critique or alienating important interests. The careers of private and public leaders tend to be short-lived, while playing the role of institutional entrepreneur and stakeholder leader that fills an institutional void is a long-term proposition. Being the ongoing coordinator of various stakeholders in an effort to develop a consensus on how the industry will develop is not a task that most private or public leaders will want to take on because it is rarely a simple matter of being consistent with their interests. We now turn to our illustration of this problem, first by focusing generally on the U.S. renewable energy industry and then moving to the specific Minnesota case.

15.3 Impediments to the Creation of a Renewable Energy Industry

The creation of a utility-scale renewable energy sector in the United States, beyond small-scale and experimental projects, has been in the works for many years. Collaborative efforts between the Department of Energy's (DOE) National Renewable

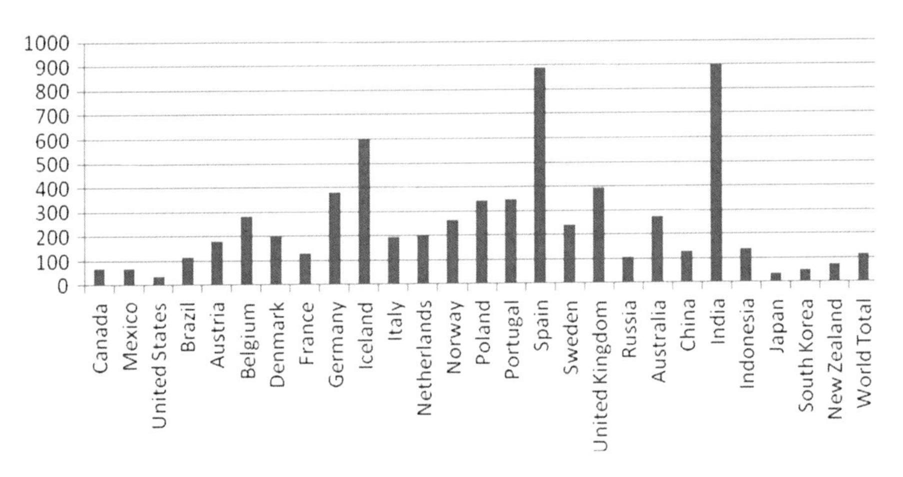

Fig. 15.1 Global renewable energy adoption—growth rate in KwH for 1997–2006

Energy Laboratory, academic institutions, private firms, and electric utilities laid the basic groundwork for the development of this industry. Despite the significant initiative and investment in renewable energy technology, the investor-owned utilities (IOUs) in the United States had made very little progress to include wind, biomass, solar, and other clean generating technologies in their portfolios of fuel mix technologies. This lack of adoption was despite the growth in use of such technologies in both developed and developing countries. Figure 15.1 illustrates how the United States lags far behind many other countries in the adoption of renewable energy.

The lack of initiative in adopting renewable energy sources by IOUs came from several sources. First, the utilities were required to operate within the constraints of their regulated environments, which often placed a priority on providing consumers with a reasonable price for electricity. As a result, decisions to invest more costly energy sources (such as solar or wind) in new and in most cases faced a significant external hurdle as Public Utility Commissions (PUCs) were sensitive to passing costs on to ratepayers (Fremeth and Holburn 2012). Similarly, well-organized interests, such as industrial customers or conusmer advocates, would likely create opposition to increased energy costs. Second, the culture among senior executives at many utilities had been formed around the traditional generating technologies that were based on boilers and steam-powered turbines. An environmental manager at a major electric utility remarked how this much engrained culture originated in the Navy following WWII and had been instituted by the executives and engineers that had transferred their knowledge from the powering of warships to the electric industry (Jim Turnure, Xcel Energy, pers. comm, Sept. 2007). As a result, the majority of new power plants that were built in the United States over the past 50 years have been nuclear plants followed by natural gas turbines that have filled both the demand for baseload and peak capacity, but never replacing coal as the core technology. Finally, in addition to the cost and organizational constraints there was

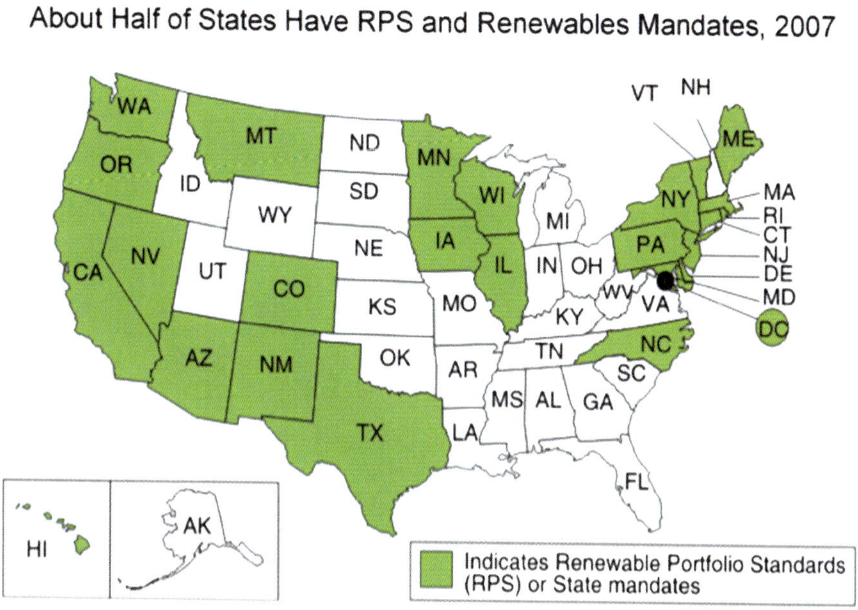

About Half of States Have RPS and Renewables Mandates, 2007

Indicates Renewable Portfolio Standards (RPS) or State mandates

Fig. 15.2 Renewable portfolio standards in United States. Source: Energy Information Administration

considerable uncertainty in the available renewable technology. IOUs are relatively slow-moving organizations that are responsible to many stakeholders, and prior decisions to enter new yet uncertain technologies had backfired in the past (Lyon and Mayo 2005).

As an emerging industry, the renewable energy sector would need to overcome the obstacles put forth by the positions of the IOUs. Although this created a significant hurdle, there had been some initiative put forth in the mid-to-late 1990s as states had begun to implement retail and wholesale deregulation (Delmas et al. 2007), and small- and large-scale renewable energy developers were taking advantage of the qualifying facility status provided under the *Public Utility Regulatory Policies Act* (Sine and David 2003). However, the legitimacy of this industry was limited due to the lack of familiarity and credibility with the technology and many of the new players (i.e., developers, rural business leaders, technology providers). Beginning in the late 1990s, states had begun to mandate objectives, known generally as "RPS" that compelled utilities to include increasing numbers of renewable power in their generation or procurement portfolio. These objectives varied widely throughout the United States (Fig. 15.2) and acted as a "command-and-control" policy that levied penalties for utilities unable to meet its objective. A similar mandate had been discussed nationally in Congress ranging from 10 to 15% of power sold for all utilities but had yet to pass as of the close of the 2008 session. The Energy Information Administration (EIA) published a thorough report on the subject entitled the "Impacts of a 10-Percent

Renewable portfolio Standard" that was requested by Senator Frank Murkowoski, the ranking member on the Senate Energy and Natural Resources committee.

Although criticized by some, these policies were seen as a means to "prime the pump" and enable a wider market for renewable power (Cory and Swezey 2007). In total, these state-level policies were expected to increase the role renewable energy played in the United States by more tenfold. However, the ultimate impact of these policies was still to be determined as many of their goals were expected to bear fruit only 15–20 years in the future. Figure 15.2 illustrates the states that had adopted RPS policies as of 2007.

15.4 Implementing Minnesota's Renewable Energy Standard

This case study will focus on the role that NSP, a division of Xcel Energy (Xcel), has had in the development and implementation of a far-reaching state-level environmental initiative. NSP is the dominant electric utility in Minnesota as it provides electricity to 78% of the state's customers. The firm dates back to the electrification of the Midwest United States when Henry Marison Byllesby, a protégé of Thomas Edison, established an electric power company in Stillwater, Minnesota, in 1909. The firm was renamed Northern States Power Co. in 1916. In 2000 NSP joined with New Century Energies to create the Xcel Energy holding company. Three separate utility companies operate under the Xcel brand, NSP, Public Service Company of Colorado (PSCCo), and Southwestern Public Service Co. (SPS) in Texas. Each operating company is independent of one another and has its own executive staff.

This Minnesota policy is partly the result of the utility leveraging long-standing relationships it has with outside stakeholders. Its ability to work with its stakeholders has helped to advance the cause of environmental sustainability in Minnesota. The collaborative effort in which NSP has been engaged has helped produce new policies, but the current situation will require further collaboration with stakeholders to refine these polices further so that the utility industry can move forward with its sustainability efforts. In short, there now exists something of an institutional void and collective action problem which is hindering further efforts to create a sustainable renewable energy industry—one that we argue is essential if the state's far-reaching renewable energy objectives are to be met.

15.4.1 The Stakeholders

A variety of stakeholders are active in the development of energy policy in Minnesota (see Table 15.1). The state has historically chosen to be inclusive, and the stakeholders tend to have long-standing relationships with one another. It is not uncommon to have an environmental advocacy group spearheading a technical initiative that brings together diverse public and private interests. This likely has historical and

Table 15.1 Key stakeholders involved in energy policy in Minnesota

Agency or group	Role and interests
Northern States Power, subsidiary of Xcel Energy (NSP)	Dominant investor-owned utility in the state and the Midwest United States It is focused on profit-seeking objectives while meeting regulatory compliance and improving environmental performance. A national leader in the sale of wind power. Has a major stake in the development and implementation of energy policy as it can have a great impact on its operations and long-term strategy
Great River Energy (GRE)	Provides wholesale electric service to 28 distribution cooperatives. As a cooperative it is owned and operated by its members and is less focused on profitability as on providing services to its members at a reasonable cost. The organization has begun to increase its focus on renewable energy and environmental performance. Its exposure to new policy is less than that of an investor-owned utility
Minnesota Office of Energy Security (OES)	The state agency in the Department of Commerce that coordinates energy and climate issues. It is responsible for implementing the governor's policy objectives. Responsible for ensuring a rational process for the development of an emergent renewable energy sector in Minnesota
Midwest Independent System Operator (MISO)	The not-for-profit organization that is responsible for managing transmission infrastructure in the Midwest United States It is committed to reliability, the nondiscriminatory operation of the bulk power transmission system, and to working with all stakeholders to create cost-effective and innovative solutions for our changing industry. It implements federal standards for transmission interconnections and upgrades
The Izaak Walton League	Environmentalist group that prides itself on its "grassroots, commonsense approach to solving local, regional and national conservation issues." It has been involved in energy issues in Minnesota for decades and has acted as a broker in the past in resolving public–private concerns. Recognized for having a balanced understanding of tradeoffs between economics development and social concern
Citizens Energy Task Force	Upstart consumer group with NIMBY concerns that support the development of renewable power but is opposed to the construction of new transmission lines. Seeks rural economic development and acts with a high degree of urgency. It has used the local media to express its concerns
Renewable energy developers (FPL Energy, National Wind, Iberdola Renewables, NAVITAS Energy)	Profit-seeking firms that manage the construction and development of wind farms. They were attracted to Minnesota when policy catalyzed the renewable energy sector. They are seeking long-term contracts with investor-owned utilities. Most are located out-of-state or are even international firms

cultural roots to the state's commitment to the natural environment that is an outcome of its use of the outdoors for boating, fishing, and hunting. As a result, the state has been a leader in adopting far-reaching policies that situate the natural environment as a key concern for energy-producing firms.

Despite a historically well-developed line of communication between stakeholders, there remains divergence on some issues. In addition, there has been a growing contingent of new entrants that had not previously participated in the debate. As a result, reaching consensus on how the renewable energy sector should be structured and developing the rules for interaction would confront new challenges. To further complicate matters was the fact that the more salient issues would extend beyond the state's borders and involve stakeholders that were not only located outside the state but in some cases outside the country. As a result, coming to a resolution on the collective action problem would involve interactions at the local, regional, national, and international levels.

15.4.2 Formulating the State Environmental Initiative

In February 2007, the state of Minnesota adopted one of the most proactive and demanding energy policies in the United States. All major electric generating utilities in the state now faced a legislated requirement to generate a substantial portion of their energy from renewable sources. The main element in a series of laws aimed at reducing the state's carbon dioxide emissions was the Minnesota Renewable Energy Standard (RES), which required that NSP generate 30% of its power from renewable sources by 2020. Besides NSP, there were four other utilities that were subject to RES legislation in Minnesota. This policy had been well researched and recommended by the Minnesota MPUC, which was responsible for regulating the state's electric utilities. At the time, this was well above the demands set by the policies of other states and beyond the 10–15% that had been debated as a national target in Washington (Nogee et al. 2007). With legislative backing for the measure in place, all parties compromised and the policy passed with virtually unanimous support. The collaborative effort included political representatives from the legislature and the governor's office, the Minnesota Public Utility Commission (MPUC), the Department of Commerce, the state's electric utilities, major environmental groups, and rural economic leaders. The aligning of interests in the lead up to the adoption of this path-breaking result was crucial in its design and acceptance by stakeholders. The president of NSP publicly embraced the initiative in the local media and outside of the public limelight.

15.4.3 Creating a Collaborative Atmosphere in Minnesota

A number of institutional factors played an important role in the development of the agreement. Under Minnesota law for open meetings and administrative processes, the MPUC made environmental concerns and rural economic development priorities in fulfilling the federal mandate to set rates. The rates that MPUC and other state commissions set are supposed to allow utilities to earn a "fair and reasonable" return on their assets. Federal legal precedent establishes that PUCs must set rates that enable utilities to earn a fair and reasonable return on "used and useful" assets, though methodologies for assessing such criteria are not specified. The MPUC's ongoing commitment under state law has been to allow key environmental stakeholders to be present at important meetings and actively participate in its hearings and epitomizes the commission's desire to ensure sustainable environmental stewardship (Dowrkin et al. 2006). According to James Turnure, environmental manager at NSP, such access for environmentalists to bureaucratic decision making in Minnesota was quite unique. It led to a situation in which NSP negotiated openly with other groups and took their concerns seriously.

It is necessary to note that this arrangement had historical precedents. A key turning point in this history took place in the early 1990s when NSP applied for the right to store spent nuclear fuel at its Prairie Island plant. The MPUC was legally required to be inclusive in its deliberations and encouraged participation and input from the utility, key environmental groups, and other stakeholders.

Allowing continued storage of radioactive waste at Prairie Island was a controversial issue. Relationships between NSP and key stakeholder groups were frayed. A newspaper article in the *Minneapolis Star Tribune* on April 3, 1994, stated that the controversy:

> provided ... good theater and a splendid view of how the political process works, or doesn't work, depending on where one stands. Throw away those dry brochures on "How a Bill Becomes a Law," and witness the real thing: Opposing packs of lobbyists, ... Endless hearings and dueling scientific experts. Celebrity advocates.... Daily demonstrations ... even death threats.

NSP claimed that without legislative approval, it would be forced to close the Prairie Island plant, thereby putting 500 people out of work and causing electric bills to skyrocket. Environmentalists argued that continued operation of the plant was an unacceptable risk that demonstrated NSP's failure to pursue alternative energy options. The antinuclear coalition was large and surprisingly powerful. It included such prominent local players as Robert Hentges of Faegre & Benson, a well-known law firm, and public affairs consultants Pat Forciea and Ann Mulholland. In addition, the antistorage group had help from the Sioux and Ojibwa Tribes, which had growing clout in the state legislature because of Indian gaming and casino operations. Finally, a broad coalition of antinuclear groups, Citizens for Nuclear Responsibility, charged NSP with trying to thwart the will of the people. However, on the other side of the issue, large labor groups in the state, including the AFL-CIO, backed NSP. They were worried about both actual jobs that might be eliminated and potential reduction in job growth that could take place if the utility raised rates.

With the AFL-CIO on NSP's side, the utility won the right to continue to store radioactive waste at Prairie Island. Nonetheless, the concession that Citizens for Nuclear Responsibility won was that a new multimillion-dollar fund would be created, which would give wind, solar, and other renewable sources of energy a significant boost in Minnesota. Under a May 1994 agreement, the governor of Minnesota allowed NSP to store spent nuclear fuel in above-ground dry casks in exchange for the creation of the fund, the purpose of which was to explore the potential for greater renewable energy power in the state and to build or purchase at least 825 megawatts (MW) of wind generation. There were also a mandate to introduce power generated by biomass and requirements for greater demand side management.

In Minnesota, this collaborative arrangement jump-started the development of wind power. From 1994 to 1998 more wind power was put in place in Minnesota than in any other state (EIA 1999). The amount of wind power generated in Minnesota grew from 25 MW in 1994 to close to 900 MW in mid-2007.

The potential for wind power generation in Minnesota and adjacent states was very significant. Utilities were looking for ways to add capacity to their generating systems, as other alternatives, such as coal and nuclear, were blocked for environmental or political reasons. The costs of generating electricity from wind, moreover, were dropping because of technological progress. At the time that a deal was reached to store nuclear waste, Carl Lehmann, manager of public affairs at NSP, saw no problem in finding common ground with environmentalists about the need to develop additional sources of energy that were environmentally sound (Smith 1994). Environmentalists realized that wind was connected to jobs and economic development. Diane Jensen, a spokesperson for the Sustainable Energy for Economic Development coalition that helped to negotiate the agreement with NSP, started to frame the once exclusively environmentalist cause in terms of the "potential" for economic development. She pointed out that wind farms in southwestern Minnesota benefited local economies, produced jobs, and expanded the tax base.

The Prairie Island nuclear storage deal set the stage for further collaboration in Minnesota between environmentalists and the utility. The views of these historic adversaries began to converge, not completely but enough to result in important compromises. This convergence of views manifested in the "Wind Integration Study" that led to the passage of the Minnesota RES. In doing this study, the Energy Reliability Administrator at the MPUC brought together major utilities, wind power and environmental advocates, and technical consultants to determine how much wind power could be included in the state's energy mix without substantially increasing electricity costs. The conclusions of the study, the joint product of these groups, were released in December 2006. This study acted as a catalyst for the passage of RES by state government a mere 2 months later.

Rural economic development was an important part of the deliberations that led to the passage of the RES. The MPUC was obligated to consider it in the deliberations. Burl Haar, the executive secretary of the MPUC, saw the RES as an extension of an earlier Community-Based Energy Development (CBED) program that had encouraged major investor-owned utilities to work with small-scale energy producers in rural areas (Burl Haar, Minnesota Public Utility Commission, pers. comm, Sept. 2007). This experience had implications for the creation of RES, as there were existing relationships between major investor-owned utilities, like NSP, and rural actors where most of the renewable energy would be harvested and the transmission infrastructure would be built.

Despite these connections, the utilities and large co-ops did not share this enthusiasm for using energy policy to spur rural development. Gary Connett, director for Environmental Stewardship at Great River Energy (GRE), claimed that the idea was interesting as a concept but not fully thought through as rural partners generally lack the resources or capabilities to bring energy projects to fruition (Gary Connett, Great River Energy, pers. comm, June 2008).

15.4.4 NSP's Role

These collaborative considerations paved the way for a demanding renewable standard in Minnesota. With the addition of holdings in Colorado (PSCCo) and Texas (Southwestern Public Service Company), NSP's parent company, Xcel Energy, had grown to become one of the largest utilities in the United States After it merged with New Century Energies utility of Colorado in 2000 and changed its name from NSP to Xcel, it integrated various fuel and technology types into its generation mix. It had new leadership at the top, a CEO, who was sympathetic to alternative sources of power generation. David Sparby had taken over as president and CEO at NSP in January 2007 and had risen through the ranks over 25 years at the utility. Sparby was a lawyer that had spent most of his time dealing with regulatory issues at both the state and federal levels. As a result, he was not tied to the old utility culture that was focused on boilers and turbines. He also had a deep understanding of the complexities involved in regulatory approval, the rate-making process, and long-term contracting with IPPs.

Rather than considering a future wherein it would be able to function within the status quo, management at NSP now took seriously the prospect of operating in a "carbon constrained economy, the backdrop of which would be an aging infrastructure and rapidly escalating prices for raw materials" (David Sparby, NSP Minnesota, pers. comm, February 2008). Internal strategic planning documents placed environmental issues on par with earnings targets and employee safety issues. See Fig. 15.3 for NSP's priorities and mission in 2008. This document set the agenda for spring 2008 meeting between Sparby and Xcel's board of directors where he was hoping to be able to set the direction for the utility firm's future.

The RES would have significant effects on NSP's operations. In 2007, it had just 1,035 MW of wind energy capacity on-line or 9% of its total generating capacity (Xcel Energy 2007). All of this capacity was procured by purchase power agreements with small IPPs. In 2007, NSP also had other renewable resources including 111 MW of biomass energy capacity, 277 MW of hydro, 15 MW of landfill gas, and 100 MW of refuse derived fuel. Despite having developed a particular skill at managing a diverse portfolio of energy types, almost tripling the amount of renewable energy used in 12 years was a daunting challenge. It would involve not only finding new sources of power but developing the transmission lines to move the power from outlying and mostly rural regions to larger metropolitan areas. Furthermore, with wind as the preferred means of generation, NSP would need technologies to store

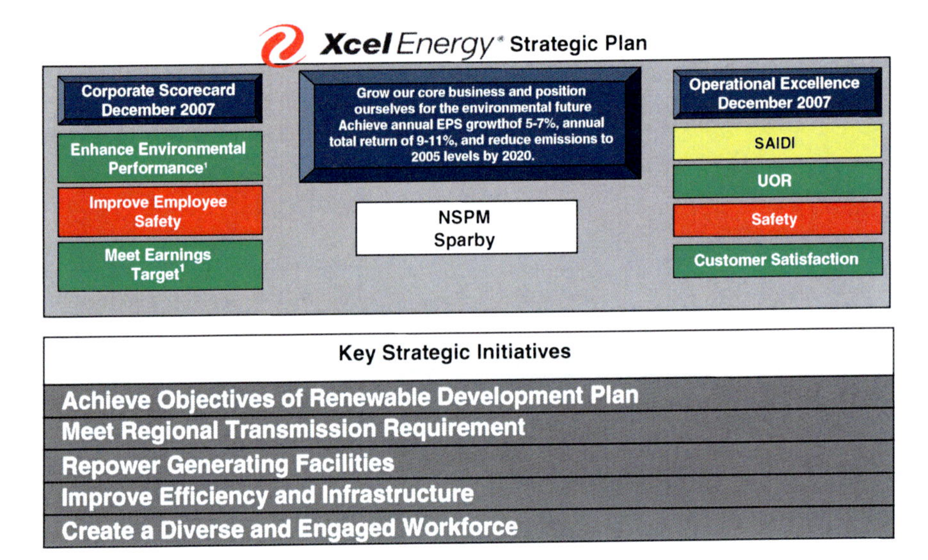

Fig. 15.3 NSP strategic planning documents—winter 2008

the wind, for though wind had great potential, its liability was its episodic and intermittent character. A utility could not count on the wind being there when the utility most needed to generate power. The wind in Minnesota and surrounding states was most plentiful in fall and spring, while the need for it was highest in summer. NSP, therefore, had to explore ways not only to generate wind and transmit it from mostly rural to urban areas but also to store the power. Battery technologies existed, but they were still experimental. Despite some limited use internationally, it was unclear when these battery technologies would be ready for commercial use in the United States (Brooks 2008). Another potential roadblock was that the supply of wind turbines was inadequate. The major suppliers were in Europe, and they had large backlogs of orders that they were trying as best they could to fill.

Furthermore, no company had ever managed so much renewable power and placed this much intermittent wind on its grid before, and it was unclear how well or even if NSP could do it. It had to maintain the integrity of its service at all costs. The flow of power to customers could not be interrupted. Therefore, the predictable development of this resource was imperative to make this work and to do so in a profitable manner.

15.4.5 Implementing the State Environmental Initiative

NSP embraced the RES despite these challenges. It proclaimed that RES set "the foundation for a reasonable cost and environmentally sound energy future (Xcel 2007)." It recognized the value of being a leader in renewable energy and started to tout this fact, albeit cautiously, in its marketing efforts. Its internal strategic planning documents clearly stated that the firm's mission was to "grow" the "core business" and "position" the company "for the environmental future." Environmental leadership was a core business objective with equal standing to the firm's financial objective of 5–7% earnings growth. Sparby claimed that the state had set the firm's strategic initiatives for it and now it just had to find some way to implement them (David Sparby, NSP Minnesota, pers. comm, February 2008).

The immediate issue was how the utility would be able to accomplish its ambitious environmental and financial goals. How could it simultaneously pursue environmental stewardship and maintain its financial integrity, when there were no guarantees that these goals would be mutually reinforcing? NSP, under the RES, was not in a favored position with regard to creating renewable energy power production facilities itself. As it stood, the RES allowed small and mid-sized developers to feed renewable power into its grid, essentially forcing NSP to purchase the power, and thus not realizing the same degree of return if it actually owned the generating assets.

As a regulated utility, the incentive for NSP was to own renewable power-generating assets, when it made financial sense, and to find a balance with the renewable power resources it purchased from others. Senior executives at NSP felt that it was important to participate materially in the ownership of wind generation and that such participation would be necessary to comply with the RES. Paul Bonavia, the president of the Utility Group at Xcel, which managed the operating companies and who Sparby reported to, had remarked that ownership over the generation would prevent costly renegotiation from IPPs in the future and would have an important financial impact as capital markets view long-term contracts with IPPs as imputed debt on the firm's balance sheet (Paul Bonavia, Xcel Energy, pers. comm, February 2008). Furthermore, ownership of the renewable generating assets would allow for the efficient design of a centralized transmission grid, which brought power from rural areas where it was generated to large metro regions where it was used.

Rather than responding in a disorganized and piecemeal basis to the initiatives of a large number of independent renewable generators seeking interconnection to its system, NSP desired a more coordinated, planned process that provided greater predictability over the construction of transmission lines. A rational process would more efficiently bring on board the large amounts of required renewable energy. The RES had not adequately considered these issues. The collaborative structure that had started to come into existence in Minnesota would need further elaboration and refinement. In working through these issues, its robustness would be tested. NSP was fully committed to collaboration. It now would need a mechanism for moving this collaboration beyond goal setting to implementation, but feared what such leadership may bring with it.

15.4.6 The Onslaught of Interests and Investment

Farmers with windy fields, entrepreneurs that recognized a new opportunity, and major energy development corporations with the know-how and experience had all expressed interest in participating in the generation of wind power by entering the

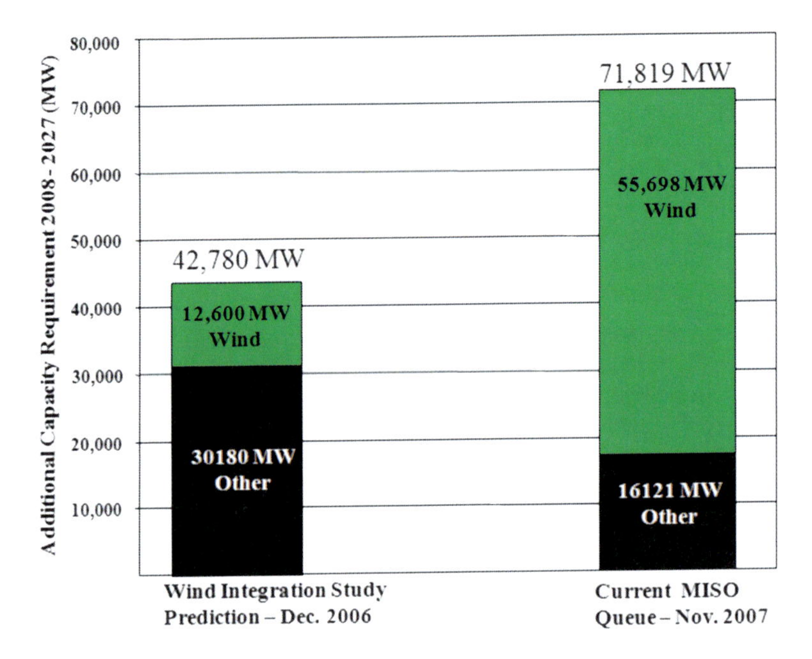

Fig. 15.4 Market response to Minnesota RES

Midwest Independent System Operator's (MISO) queue for interconnection to the transmission network. MISO is the nonprofit organization that is responsible for managing the power grid and transmission of power in 15 states in the Midwest United States and Manitoba in Canada. Approved interconnections represented the first step in being able to develop a renewable energy project. If all these projects were brought on line, the RES goals for renewable energy generation would be more than met. In fact, Minnesota might be in the enviable position where it would have more than its required amount of renewable power and it could export this power to other regions. However, there was a hitch that could derail the entire process and prevent even the IOUs in Minnesota from meeting the RES objectives. Renewable energy projects could not be initiated until approval by MISO, and the great demand for new development had created a backlog that was estimated to take 612 years to clear at the current rate (Cummins 2008). The MISO approval process was mandated and developed by FERC when energy projects were larger (i.e., coal or natural gas plants) and undertaken by more credible and well-funded actors. It is a rules-based, first in–first out queue that is blind to whom the party is requesting the interconnection to the transmission grid. According to Clair Moeller, the vice president of Transmission Assets at MISO, the process is entirely nondiscriminating even "no matter how many governors write a letter in support (Clair Moeller, MISO, pers. comm, July 2008)."

The "Wind Integration Study" calculated that there might be a need for 12,600 MW of wind power; however, the RES had attracted developers wishing to bring 56,000 MW of wind generation on line. Figure 15.4 graphically and clearly

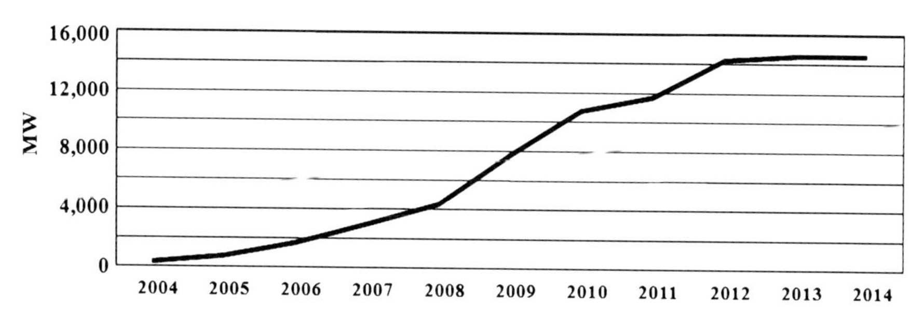

Fig. 15.5 Anticipated growth in wind power development in Buffalo Ridge, Minnesota

highlights this issue. To put this into context, NSP, the largest utility in the region, owned approximately 8,400 MW of total generating capacity of various fuel types in four states in the Midwest. Despite being involved in the wind study, MISO had been left out of the creation of the RES, and Moeller notes that much of the problems that were faced were technical and not political. He claimed that "legislation is easy to pass but making it work is a different issue altogether" and that challenge is what can bring down the entire initiative in Minnesota. Figure 15.5 illustrates what this means to Buffalo Ridge, a wind-rich region in southwestern Minnesota. In fact, the key interconnection in Buffalo Ridge had not been upgraded in 50 years and required a significant upgrade to interconnect the wind power projects.

The RES' overincenting the development of wind projects had bred conditions that could potentially lead to chaos as supply would far outstrip the legislated demand. MISO was not ready to authorize this level of construction nor were the IOUs capable of absorbing the wind generation on its existing transmission lines. If some order was not introduced, there would be bottlenecks that would prevent Minnesota from meeting its ambitious renewable energy goals. The combination of administrative backlog in the queue process and the IOU's inability to absorb and transmit the energy generated would lead to long lead times, higher costs, duplication, litigation, and frustration for all the participants. Such an eventuality would tear asunder the harmony that had come into being with RES's passage. MISO had made some changes to slow down the growth of the queue, such as increasing the cost to enter the queue and aggregating projects together when evaluating their potential. However, these piecemeal solutions would not settle the larger problem that the transmission grid was old and investment of unparalleled scale would be required.

For NSP the problem was exacerbated by the fact that it had not really entered MISO's queue with projects of its own. This is because the main role to which it had been assigned was to be the receptacle of the power generated by others, not the originator of that power itself. Despite an obligation to purchase wind power and develop transmission lines, the utility had little control over wind power generation and would not see a significant return on the investments that were being made in this area. The incentives in the system were not well-aligned for NSP to optimally participate in the process. The company was at the mercy of the market and those who had acted quickly and opportunistically to enter the MISO queue. This result was the natural outgrowth of electric utility deregulation that tried to limit existing utilities' role to the transmission, distribution, and marketing of power as well as to encourage IPPs to compete with one another to provide this power. Despite not being in the queue and seeing limited return from renewable energy investment, NSP would have an important role to play if a solution to this backlog would be found. The utility has experience in siting and constructing new transmission lines and had recently led a consortium of interests to build CAPX 2020 that was a multistage project that laid out 1,600 miles of transmission lines. But gaining approval for new transmission lines can take 5–25 years to get through the bureaucratic process and the RES objectives required significant action in the next decade. Moeller had remarked how the political horizon and the transmission project horizon are mismatched and that real leadership would be required to get over this hurdle (Clair Moeller, MISO, pers. comm, July 2008). He recommended that a "coalition of willing" interests may be needed for the necessary action, but bringing such groups together could be a challenge.

15.4.7 New Interests in Minnesota

A real challenge for the creation of the coalition that Moeller had recommended was that the composition of vested interests had changed dramatically. In a short period the parties that had supported the Prairie Island deal and the RES were now joined by new energy developers and a variety of vocal special interests.

NSP had seen the renewable energy developers as partners in the development of the emerging industry. Sparby was comfortable with the arrangement and believed that there was "enough room for everybody to play"; however, the power purchase agreements that NSP was required to develop with the developers were treated as debt on NSP's books by bond rating agencies. This situation was not ideal for NSP considering its reliance on public financial markets. Specifically, these contracts are viewed as imputed debt on NSP's balance sheet and are based on the size, type, and provisions of a purchase power contract. This raises costs for customers as NSP faces higher debt costs and is required to make the appropriate changes to its capital structure. In 2006 NSP-Minnesota had over $2.5 billion in long-term debt and an S&P credit rating of BBB+. Similarly, these developers had a very different timeline from that of the IOUs as they were on short-term basis and had little incentive to work on improving the state's transmission grid.

The renewable energy developers in Minnesota included many very large firms such as NAVITAS (owned by the Spanish energy conglomerate GAMESA), Florida Power & Light, PPM, EnXco, and Invergys. According to Wanda Davies, director of development at NAVITAS, Minnesota became extremely attractive place for operations once the RES was passed and that it helped convince her company to move from Illinois and Wisconsin. Of particular importance to NAVITAS was the higher average price of power that the RES would bring as it would force utilities to pay a higher price for power and that the MPUC would allow these costs to be passed on to customers (Wanda Davies, NAVITAS, pers. comm, June 2008). Many of these developers have positions in MISO's queue and treat them as "lottery tickets" that if they come up they can be quite lucrative and the more positions you have the better positioned you are. However, these developers have little to no role in improving the MISO process or investing in the transmission grid. What we observe from this is that developers clearly play an important role in the emerging renewable energy market and teaching IOUs how the industry and technology works, but they are less inclined to participate in the collaborative relationship that Sparby once perceived.

15.4.8 Restructuring the Market

In the winter of 2008, senior executives at NSP, drawing on ideas that once had been considered by officials from the state Department of Commerce, crafted a new plan. The plan was to use the collaborative relationships that had been instrumental in the development of RES to restructure the Minnesota energy market to better support the development of renewable energy. NSP's blueprint, entitled the "Central Corridor Concept," was designed to better manage the movement of renewable energy from outlying regions of the state toward the metropolitan areas where it would be consumed. The plan envisioned three energy development corridors that would better link the urban center of Minneapolis–St. Paul to the northwest, southwest, and southeast regions of the state, where most wind and other alternative energy sources were found. Figure 15.6 depicts NSP's plan.

The first step in the plan was to collaborate with key governmental and nongovernmental stakeholders to promote the creation of the "energy development zones" in three regions of the state. These zones would be hotbeds of wind generation, biofuel production, and related research and development. As a result, the plan would foster significant economic development in rural communities, and thus NSP could expect support from political and business interests in these areas.

The second step would be the construction of a transmission network that would be able to deliver the power from these outlying regions to the load center in Minneapolis–St. Paul. This element of the plan was an enhancement of a prior investment in transmission that was named CAPX 2020 that was being jointly conducted with the other utilities in Minnesota. A decision to upgrade the power lines was needed to incorporate the massive amounts of wind power that were to come online in the next 15 years. Use of such lines could be expected to carry 10,000 MW of wind power to urban centers. This was necessary to create predictability in a system that was previously lacking such certainty. Moreover, this part of the plan would be essential for the stringent RES policy to be successful; for without it, NSP, and for that matter the other utilities, would be unable to sell the requisite amount of wind power to consumers. Finally, this part of the plan would piggyback on the idea of state transportation corridors, another politically salient issue. This aspect of the plan would allow NSP to nurture a more rational approach

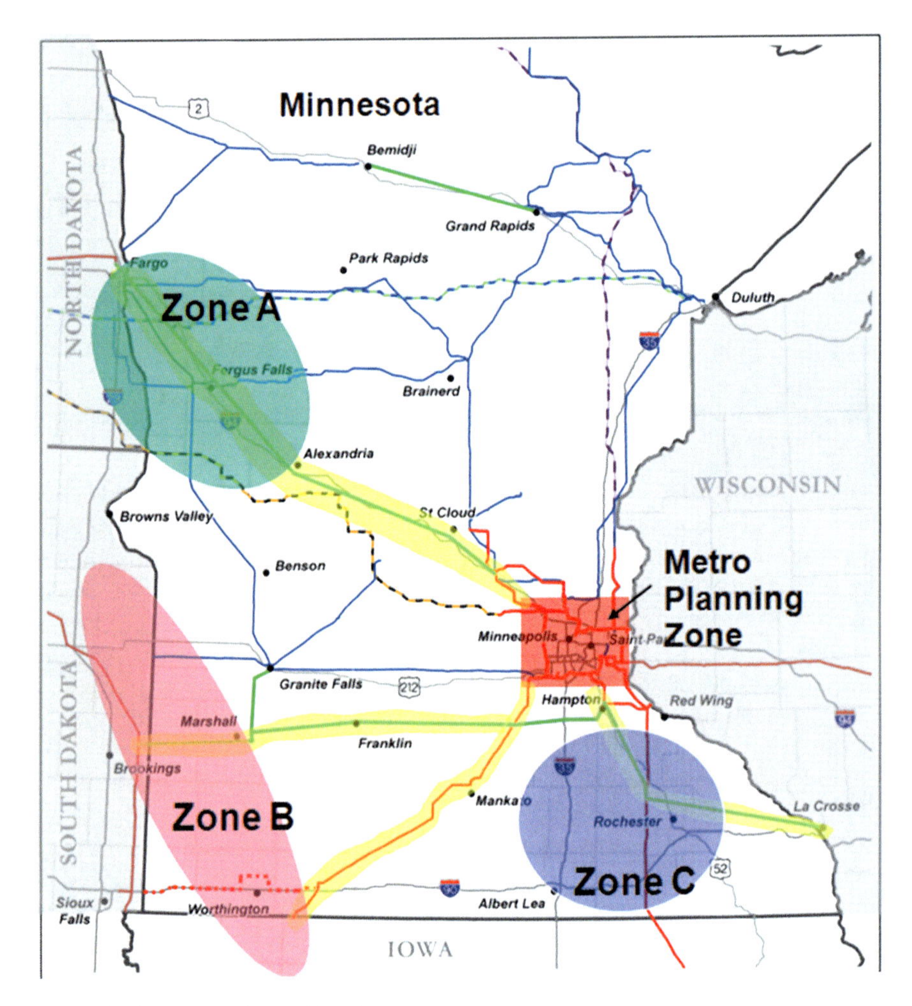

Fig. 15.6 Elements of the corridor plan

to bring on wind projects. For this element of the plan, NSP could hope for collaborative support from other utilities that would be able to also take advantage of the transmission lines and environmental groups that would recognize the necessity of this investment to reduce carbon emissions. All of these groups had been essential in the earlier development of RES and the Prairie Island agreement. Therefore, the MPUC would likely approve this necessary and prudent investment in the firm's rate base.

The final step of the plan would take place in urban centers. NSP would create "urban energy production corridors" that not only enhanced the urban transmission network but also lined the proposed light rail lanes in Minneapolis–St. Paul with solar panels. This urban initiative would be complemented with the installation of advanced metering and smart grid devices at commercial locations that could be used to enhance energy conservation efforts. Energy customers living in the city would be an active participant in a newly restructured energy system focused on environmentally sustainable sources of energy. Linking the plan to the transportation and light rail expansion debate that was taking place in Minnesota, NSP could avoid costly problems in the future by assisting in the development of this public infrastructure program. By incorporating an urban element to the plan, NSP could expect greater support from public-at-large. Greater public support had direct implications on the degree of political support that the firm could expect from the state government.

Political support for the proposed plan was critical, and executives at NSP were more than happy to allow the politicians leading the effort to take the credit. NSP wanted a win for everyone in the implementation of Minnesota's ambitious sustainable energy plan—a win that built on the collaborative spirit that achieved in the formulation of the state's policies. However, it did not perceive it to be in its interests to take extraordinary steps to mobilize the stakeholders needed to make this plan become a reality. Sparby and others at NSP were overly concerned with the appearance of heavy-handed self-serving behavior. The Central Corridor plan presented a new direction for the firm and perhaps the entire Midwest energy market, but it was not NSP's primary responsibility. Further, it had a series of pending or upcoming regulatory decisions on electricity pricing, nuclear waste storage, and generating plant repowering that took priority as they would have an immediate impact on firm performance. As a newly minted president and CEO, Dave Sparby had to place present concerns over those in the future, and his regulatory experience gave him insight into how the MPUC would have the "final word on what the firm will be no matter its mission statement" (David Sparby, NSP Minnesota, pers. comm, February 2008). Therefore, managing these relationships and finding a way to make it appear that NSP is considering the needs of the region as a whole and not just itself was essential.

NSP's hesitancy to take a leadership position in restructuring the market despite its ambitious plan was accompanied by instability among the key government actors that could lead such action. In January 2008, Governor Tim Pawlenty (Rep.) set up the Office of Energy Security (OES) within the Department of Commerce that would oversee larger concerns and coordinate energy and climate issues throughout the administration. Pawlenty was very keen on developing a renewable energy industry in the state and saw the political benefit that it could create in both the urban center and rural communities. The OES' mandate would also foster easier access to energy information and technical assistance that was deemed essential for market development. Initially in charge of this office was Ed Garvey, a former MPUC commissioner who had extensive experience working with the key players in the Minnesota energy market. Garvey saw his role as implementing the laws that were already on the books. He preferred an "ambiguous working environment" and was willing to let the system "hang loose but not collapse" (Ed Garvey, Minnesota Office of Energy Security, pers. comm, May 2008). As well, he was overly concerned with "screwing things up" and took a very high level perspective on the key issues such as transmission investment and the MISO queue. This working style and approach to key investments in the state frustrated many people at NSP who felt a more rational approach was necessary for action. Despite NSP's willingness to allow the OES to take credit for the central corridor program, it would appear that it would not come under Garvey's leadership. As a result, government leadership to streamline necessary investment and develop the rules of interaction would be lacking.

If leadership was not going to come from the key bureaucratic agency or the renewable project developers then an alternative could be to partner with environmentalists to help promote the central corridor project. The Izaak Walton League, a key environmental group in the Midwest, had played an important role in brokering the Prairie Island nuclear waste deal and had actively participated in the Wind Integration study that provided the technical support for the RES. The group had been led by Bill Grant, who took a pragmatic approach to environmental issues and recognized the necessity for new transmission investment if renewable power would be able to come on line. Grant had already well-established linkages to NSP and had an NSP construction hat hanging in his office for facility inspections. However, Grant was nervous over the influx of new parties into the Minnesota energy debate and was especially concerned with development groups such as Florida Power & Light that took a much less cooperative approach to development (Bill Grant, Izaak Walton League, pers. comm, Feb. 2009). He believed that the cooperative approach used in the past would be a good model for the future, but brokering this deal would be much more difficult as diverse groups need to be "cut into the deal."

To further complicate the issue, transmission line siting (such as NSP's Central Corridor idea) had a controversial past in Minnesota among environmentalists as the late former Senator Paul Wellstone had come to prominence in the 1970s in a battle to prevent new transmission lines from being built. As a result, the issue had been divisive in the environmental community and Grant's Izaak Walton League was not representative of other environmental interests. New groups had used the media to gain prominence and to wage their battle against investment in the transmission grid and utility-scale renewable energy development. In a February 2009 interview on Minnesota Public Radio, Jeremy Chipps, a member of the Citizens Energy Task Force, questioned the necessity of large-scale investment in transmission and what it would mean for his picturesque backyard in La Crescent, Minnesota (Stachura 2009). He then went on to compare renewable energy developers and IOUs to door-to-door salesmen that shop around needless products. Chipps states, in reference to the corporate interests, "He's quite burly. He says, '*here I'd like to show you my new vacuum*.' I say wait a minute, we had one here last week, and we told him to get stuffed. '*Oh, but this is much better, it goes much further and until you see it*.' I said look, get lost." The Citizens Energy Task Force is one of several civil society groups in Minnesota that promotes environmental issues and renewable power but vehemently opposes the transmission investments that would be necessary. Further, these groups suspect that the transmission lines would not be used solely for wind-generated power but could open the door to interconnections with new coal burning power plants. Figure 15.7 illustrates a leaflet that the group distributes to represent their position. This group, like others, distrusts private interests and the linkages that exist between firms and government. They had

ARE YOU CONCERNED ABOUT THE ***CAP-X 2020*** *HIGH-VOLTAGE TRANSMISSION LINE BEING PLANNED THROUGH MINNESOTA?*

CITIZENS ENERGY TASK FORCE

GOALS OF CETF:

1) **To stop** the **Cap-X** high voltage power lines unless the utilities can prove **they are needed and cost-effective**;

2) To **change the practice of eminent domain** by utilities so that **property owners are treated fairly** and **offered fair value** for any land taken for power lines or other utility infrastructure;

3) To **emphasize the importance of conservation** to **reduce the demand for energy** and **reduce the need to build** high voltage power lines;

4) To **promote clean, renewable energy** and reduce the likelihood that high voltage power lines will support the construction and operation of coal plants in Minnesota and neighboring states;

5) To **promote community-based renewable generation** throughout Minnesota not only to **reduce environmental impacts** and costs of power plants and power lines, but to **strengthen Minnesota's rural economy**;

6) To **minimize the environmental harm** of constructing power lines, including **climate change impacts**, damage to aesthetics and natural features, effects of electromagnetic fields and adverse impacts to farms and property;

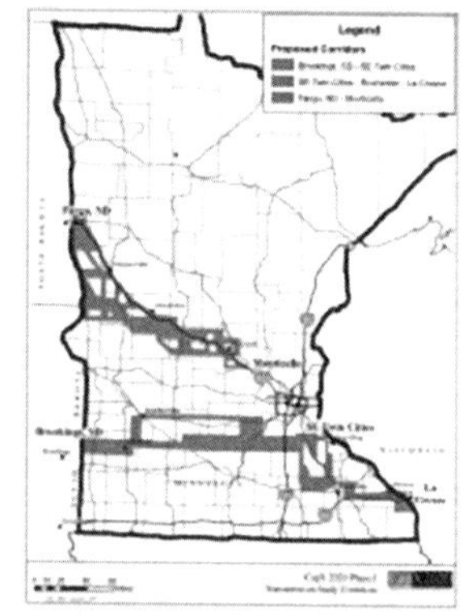

DID YOU KNOW THAT UTILITIES ARE EXEMPT FROM PROTECTIONS THAT ENSURE LANDOWNERS CAN GET A FAIR PRICE IN EMINENT DOMAIN? EVEN STATE AND COUNTY GOVERNMENTS ARE NOT EXEMPT! MINN. STAT. §117.189

The permitting process is happening through this summer. Testimony and public input are starting in just a few weeks. To be effective in the legal proceedings we need to provide expert testimony, evidence and informed citizen input.

WE NEED YOUR SUPPORT NOW.

CETF has hired **Paula Maccabee**, a public interest advocate and independent attorney who is **highly experienced** in Public Utility Commission processes and has a **successful history** dealing with utilities on behalf of landowners. We have **become a legal party** in the process which determines if the Cap-X power lines are needed and, if they are needed, how they will be built.

SUPPORT THE PROMOTION OF CLEAN RENEWABLE ENERGY.

YOU BELIEVE citizens should have an influence on what type of energy is produced for Minnesota's energy needs and how far it travels.

ADD YOUR VOICE to the CapX public hearing in your community. Speak up and write - your opinions matter.

ADD YOUR VOICE to change the unfair eminent domain law, which exempts utilities from laws helping landowners to get a fair price in eminent domain.

WE NEED YOUR HELP to raise money, inform citizens and get the word out.

THIS IS YOUR OPPORTUNITY to have a voice in Minnesota's energy future and to impact whether or not these lines will be built.

PLEASE GIVE NOW.

SEND A DONATION TO:
Citizen's Energy Task Force
P. O. Box 601
Castle Rock, MN. 55010

PLEASE INCLUDE YOUR EMAIL to receive updates about when and where you can speak up, write letters or volunteer.

For more information, contact:
Bev Topp:
eurekatopp@gmailcom
952-469-4859

Atina Diffley:
atinagoe@frontiernet.net
952-469-1855

TELL YOUR NEIGHBORS – SPREAD THE WORD

Fig. 15.7 Leaflet of the citizens energy task force

no role in the coalitions of the past in Minnesota, are not represented by more pragmatic environmental groups, and would likely benefit from greater delay that would result as the rules wait to be developed.

Neither the government leaders, renewable developers, nor various public interest players in civil society seemed to be willing to risk coming forward to help organize and coordinate the actors and overcome the collective action problems that hindered implementation of NSP's ambitious initiative. All groups had a vested interest in seeing renewable energy developed and for the most part felt strongly in purveying a public good in the cleaner natural environment that would result. Nonetheless, the inability for a coalition to come together as in the past posed a serious threat to the entire project. The result was the existence of an institutional void that threatened the achievement of the RES goals and the development of an emerging renewable energy industry in Minnesota.

15.5 Implications and Lessons Learned

This case illustrates the key but often forgotten point that the collaborative efforts among public groups (firms, customers, special interests) do not end at the creation of environmentally sustainable objectives. It is not enough for public participation, through institutional forces and cooperative relationships, to be at the foundation of environmental policies. Although this is essential for getting the process going, to move the process along collaborative ties must continue as events evolve and new developments take place. In sum, a policy's success is contingent on continued collaboration during implementation and not just policy formulation. During implementation the institutional void and collective action issues just grow, and unless someone is willing to step in and take on the mantel of leading the various stakeholders and coordinating their efforts, a stalemate likely occurs, and there will be an inconvenient delay precisely at a time when the project needs to continue to move along.

For the implementation of the RES to succeed and Minnesota to become a dominant player in renewable energy production, NSP needs to apply the lessons it has learned about stakeholder leadership from the past. A number of lessons stand out from Minnesota's experience:

1. Business interests, other utilities, IPPs, MISO, and politicians/bureaucrats *need to recognize that the current need for collaboration is based on the lack of predictability in the current system and that there is a need for a more rational approach.* This recognition must lead them to action that will overcome the institutional void.
2. The conditions for collaboration must be motivated by *convergence around technical facts*. This is hard to do since interests color these interpretations.
3. The issues have to be structured in a way that state officials are given *appropriate legal authority*. This may be derived from actions by the governor or another party that has the power to knit together implementation of RES. Without adequate legal authority all the parties are constrained.
4. Implementation depends on more than ad hoc muddling. A *clear schedule* must be in place to account for the RES' timetable and builds predictability into the process. A schedule often is the start in the path to overcoming the institutional void.

As it stands, NSP's success in restructuring the energy market in Minnesota is still a work in process. What is certain is that the success of the state's environmental sustainability objectives will be contingent on further collaborative efforts to overcome what has become a sticky collective action issue.

These lessons, we believe, are general ones that will have to be learned and applied throughout the United States as a nascent renewable energy industry is forming. Institutional voids will arise. This is inevitable. There will be serious collective action problems, and stakeholder leadership will have to be exerted when it may not be in the interests of any of the players to come forward and assume this role.

Acknowledgments We would like to thank the many individuals and organizations that contributed their time to the development of this chapter. This includes representatives at Xcel Energy, NAVITAS, Great River Energy, MISO, The Izaak Walton League, Minnesota Public Utility Commission, and the Minnesota Office of Energy Security. The chapter has benefited from the comments at the University of Minnesota, Yale University, Washington University, and Concordia University.

References

Aldrich HE, Fiol M (1994) Fools rush in? Conditions affecting enterprenereurial strategies in new organizations. Acad Manage J 19: 645–670
Brooks SL (2008) Xcel's plan to store the wind. Pioneer Press. Feb 29, 2008: C1
Bonavia P, president of the Utility Group at Xcel Energy (2008) Interview by author, 8 Feb. Minneapolis
Cummins HJ (2008) And the wind waits…and waits… Star Tribune. Jan. 27, 2008: C1
Connett G, director for Environmental Stewardship of Great River Energy (2008) Interview by author 17 June, Minneapolis
Cory KS, Swezey BG (2007) Renewable Portfolio Standards in the States: Balancing Goals and Rule. Electr J 20: 21–32
Davies W, director of Development at NAVITAS (2008) Interview by author, 30 June, St. Paul
Delmas M, Russo MV, Montes-Sancho MJ (2007) Deregulation and environmental differentiation in the electric utility industry. Strateg Manage J 28: 189–209
Dowrkin M, Farnsworth D, Rich J, Klotz JS (2006) Revisiting the Environmental Duties of Public Utility Commissions. Vt. J Environ Law 7: 1–69
Energy Information Administration (EIA) (1999) Renewable Energy 1998: Issues and Trends. (DOE/EIA-0628(98) Washington. U.S. Department of Energy
Energy Information Administration, Renewable Energy Annual 2005 and Database of Status Incentives for Renewables and Efficiency. http://www.dsireusa.org/. Accessed March 13, 2008
Fremeth AR, Holburn GLF (2012) Information Asymmetries and Regulatory Decision Costs: An Analysis of Electric Utility Rate Changes 1980-2000. J Law Econ Organ (forthcoming)
Garvey E, director of Minnesota Office of Energy Security (2008) Interview by author, 6 May, St. Paul
Grant B, associate executive director of the Izaak Walton League of America – Midwest Office (2009) Interview by author, 13 Feb. St. Paul
Haar B, executive director of the Minnesota Public Utility Commission (2007) Interview by author, 25 Sept. St. Paul
Khanna T, Palepu K (2000) Is Group Affiliation Profitable in Emerging Markets? An Analysis of Diversified Indian Business Groups. J Finance 55: 867–891
King A, Lenox M (2000) Industry self-regulation without sanctions: The chemical industry's responsible care program. Acad Manage J 43: 698–716
Lyon TP, Mayo JW (2005) Regulatory opportunism and investment behavior: Evidence from the US electric utility industry. RAND J Econ 36: 628–644
Moeller C, vice president of MISO (2008) Interview by author, 21 July, St. Paul
Murtha TP, Lenway SA, Hart JA (2001) Managing New Industry Creation. Stanford UP: Palo Alto

Nogee A, Deyette J, Clemmer S (2007) The Projected Impacts of a National Renewable Portfolio Standard. Electr J 20: 33–47
North DC (1990) Institutions, Institutional Change, and Economic Performance. Cambridge University Publishers, Cambridge
Olson M (1965) The Logic of collective action. Harvard University Press, Cambridge
Russo MV (2001) Institutions, exchange relations, and the emergence of new fields: Regulatory policies and independent power production in America, 1978-1992. Adm Sci Q 46: 57–86
Sharma S, Henriques, I (2005) Stakeholder influences on sustainability practices in the Canadian forest products industry. Strateg Manage J 26: 159–180
Sine W, David, R (2003) Environmental jolts, institutional change, and the creation of entrepreneurial opportunity in the US electric power industry. Res Policy 32: 185–207
Smith D. Prairie Island: A nuclear fight full of fear (1994) Star Tribune. Apr 3, 1994: A1
Stachura S (2009) High-voltage power lines could criss-cross Minnesota. Minnesota Public Radio. http://minnesota.publicradio.org/display/web/2009/02/20/powerline_proposal/. Accessed 25 February 2009
Sparby D, president and CEO of Northern States Power (2008) Interview by author, 6 Feb. Minneapolis
Turnure J, environmental manager at NSP (2007) Interview by author, 7 Sept. Minneapolis
Xcel Energy (2007) Renewable Energy Plan. Minneapolis
Yao DA (1987) Strategic responses to automobile emissions control: A game-theoretic analysis. J Environ Econ Manage 15: 419–438

Chapter 16
Communicating Between the Public and Experts: Predictable Differences and Opportunities to Narrow Them

Michael R. Greenberg and Lauren C. Babcock-Dunning

Contents

Abstract Communications between experts and the public are often fraught with misunderstandings and approached with trepidation by both groups. This chapter aims to improve these communications by providing readers with a better understanding of who the "public" and "experts" are, the unavoidable differences between experts and the public that can lead to misunderstandings and friction, and suggestions for bridging the public–expert gap.

16.1 Introduction

Two or three times a year, the first author has the opportunity to address a public audience about risk-related issues, such as cancer and the environment, remediation of brownfield sites, nuclear power and nuclear waste management, and other environmental health subjects. These talks address a broad range of audiences, from small

M.R. Greenberg (✉)
Edward J. Bloustein School of Planning and Public Policy, Consortium for Risk Evaluation with Stakeholder Participation (CRESP), Rutgers University, 33 Livingston Avenue, New Brunswick, NJ 08901, USA
e-mail: mrg@rutgers.edu

J. Burger (ed.), *Stakeholders and Scientists: Achieving Implementable Solutions to Energy and Environmental Issues*, DOI 10.1007/978-1-4419-8813-3_16,

Table 16.1 Stakeholders in environmental health policy issues

Elected officials and their staff, including agencies
Not-for-profit organizations that may be based locally, regionally or nationally with an important role in the decision-making and information dissemination processes
For-profit organizations with an important role in the decision-making and information dissemination processes
The media, both traditional and new (bloggers, YouTube, Twitter, etc.)
Special populations with important political leverage positions
The public
Experts from universities and consultants

groups of retired college professors who share an interest in environmental health to groups of 500 people worrying about a facility proposed for their community. Public-expert events are an opportunity to address public concerns. But direct communications between the public and scientists, engineers, and representatives of business are rare. Much of what the public learns about environmental hazards is mediated by the mass media (see following chapter). Even when the public and experts have an opportunity to interact, the public's participation is often constrained by legal requirements and/or limited by gatekeepers who manage meetings.

With this caveat noted, the purposes of this paper are first to describe the "public" and "experts," emphasizing their diversity. Second, we describe the unavoidable differences between experts and the public that lead to misunderstandings and friction. Third, we offer some suggestions for bridging the public–expert gap.

16.2 The Stakeholders

Every environmental health issue has stakeholders' groups, including elected officials and their staff, appointed bodies charged with representing the public, not-for-profit and for-profits with agendas, the media, unaligned experts, and the public (Table 16.1). Each of these has a niche in environmental health policies. In this chapter we focus on interactions between the general public and experts. We will not focus directly on elected officials and their staff or on the media, and will use energy sources and related waste management as examples throughout the chapter to illustrate some of the points.

16.3 The Publics

There is no single public; rather there are multiple publics with different demographic attributes, values, knowledge, and perceptions. With the caveat that the following descriptions are generalizations, not descriptions of individuals, we briefly describe four public groups that have been found in survey research about energy sources and waste management (Greenberg 2009a–c).

One group is educated and affluent white males, middle aged and older, who disproportionately favor increasing the use of nuclear power, approve of locating

Fig. 16.1 Sun tracking heliostats at solar two, Daggett, California (Department of Energy (DOE) 1996)

new nuclear power plants and waste management facilities in their area, and are strongly opposed to relying more on oil and coal. This group is more knowledgeable about energy-related issues than their counterparts and relies on books, magazines, web searches, and personal contacts for information more than other groups. The white male group disproportionately trusts energy facility owner–operators and government agencies that are responsible for managing these facilities. And the white male group as a whole worries less about environmental hazards than all other groups. They tend to view the social environment through the lenses of their individual needs (rather than society as a whole) and believe in hierarchical decision making (rather than widespread sharing of authority). As a whole, the white male group has controlled the U.S. economy and political system, and hence has a deep stake in being aware of, and promoting, policies that they believe will benefit the existing economic and political system.

In strong contrast is a second group composed of educated and relatively young white women who oppose nuclear and fossil fuel energy sources, and just about every other energy source that requires massive, centralized facilities (Fig. 16.1). They strongly prefer decentralized wind, solar, and other renewable energy sources. They tend to be less trusting of site managers and hence, unsurprisingly, do not like the idea of having large facilities in their communities, including transmission towers and lines as well as generating and waste management facilities. Disproportionately, the white female group is knowledgeable about energy issues; however, there are some common misconceptions this group subscribes to. For example, members of the white female group substantially overestimate the current use of solar and wind energy and

Fig. 16.2 California Bureau of Land Use Management public meeting (Bureau of Land Use Management, California 2008)

underestimate the use of coal (Greenberg and Truelove 2010). They also have some consistent misunderstandings regarding the disposal of energy waste; for instance, many believe that spent commercial nuclear fuel is shipped to Yucca Mountain for permanent storage rather than stored on-site. The white female group tends to be active in environmental organizations. Finally, in contrast to their white male counterparts, this group is more likely to favor the needs of society as a whole over their own.

The third group consists of older (60+ years old) and poorer people, who are often African or Latino American. This group is less willing to support a reduced reliance on fossil fuels, and is slightly less enthralled with solar, wind, and other renewable sources than its counterparts. This relatively economically disadvantaged group is motivated to lower energy prices. Members of this group embrace a history of a half century of dependence on fossil fuels that brought affluence and a higher standard of living. Hence, they do not readily agree to abruptly reduce fossil fuel use.

The fourth public group is "stealth." They are invisible in surveys because they are small in numbers. However, the less visible group comes to the forefront in the policy arena because of their substantial influence. Their influence stems from the capital they possess in the social, financial, or political spheres. Some examples of "stealth" group members may include owners of large parcels of land, individuals with considerable wealth and/or control of media that they use to influence others (including decision makers), and those who possess other attributes which make them power brokers in a variety of issues related to environmental health. In some cases, "stealth" members' power has been legalized, and they have an official role in decision making. For example, in Hanford, Washington, American Indian Tribes residing near the U.S. Department of Energy's (DOE) former nuclear weapons facilities have the power to influence DOE's policies (Burger et al. 2006) (Fig. 16.2).

These four groups represent a sample of the many publics that experts and policymakers will encounter while addressing environmental health issues. The more we seek to understand members of the public by talking with them and by conducting survey research, the more likely we will be to be able to broach the divide between our perspectives.

16.4 The Experts

"Expert" is a label reserved for someone who brings uncommon wisdom and knowledge to the table. Experts typically align with four groups. One is for-profit companies. For-profit experts are often mistrusted by the public because they are invariably viewed as supporting the financial interests of the companies that pay their salaries. Nevertheless, the scientists, engineers, lawyers, and other experts who receive financial support from for-profit organizations often produce accurate and useful information.

Experts who work on behalf of environmental and public health organizations align with their organization's objectives. These scientists, engineers, biologists, chemists, and others often clash with their for-profit counterparts. While many members of the public trust the expertise that not-for-profits bring, others find that some not-for-profit experts are more interested in their organization's position than the facts of the issue.

Experts also work for government. While they are supposed to represent their public constituencies, their work may be influenced by the elected officials and senior staff to whom they report. With state and local budgets under great pressure (Conservation West 2009; Nagy 2002), the capacity of government agencies to sustain research and high-level expertise has declined and in the near future we believe will continue to do so. Public reliance on state and local government for sound scientific information clearly has been compromised.

The fourth group of experts is "nonaligned." Typically, this group comes from universities and is expected to be independent of any political or financial incentive. However, it is not uncommon to find professors and other independent experts who have been aligned with for-profit or not-for-profit organizations. Assuming that truly independent experts exist, the distinction between levels of expertise is then important. Anyone who is a member of the National Academy of Sciences (NAS) is by definition an expert. Academy members and others chosen by the NAS to serve on committees, as well as fellows of major professional societies, should represent the cream of the expert crop. However, these experts may have relatively little knowledge about a particular issue that affects an area. The local high school chemistry teacher may know more about that problem in that setting than the fellow of the international society.

16.5 Different Realities and Predictable Differences

Three differences pose formidable barriers between experts and the public: (1) who they rely on for their information, (2) what kinds of information they trust, and (3) how much they trust statistics and probability. Experts trust members of prestigious scientific organizations and their refereed publications. The public has no such inclination. Their trusted messengers are people and institutions they know and respect, including their friends, relatives, neighbors, spiritual leaders, local physicians (NCI 2002), and others who may not have much expertise in the issue at hand. The first author, for example, has been at meetings regarding the destruction of chemical warfare agents at which local publics were hostile and did not trust the expertise of some famous scientists because they were "outsiders" and because the public played no role in choosing them. One group's scientific guru is another's outsider telling them what is important and not important.

Second, experts have been trained in the scientific method and to use deliberative reasoning when developing and interpreting information. An expert, even one who starts with a position to defend or to make, should expect credible science to be based on studies with clear and specific research questions, grounded in theory, on acceptable data gathering and analysis procedures, and on studies that corroborate findings.

The public does not restrict their search for credible information to that derived from methodical research processes, nor do they attempt to restrict their reactions to disinterested observations. Anecdotal information and analogies that experts struggle to explain are commonly used by the public. For example, the first author has been asked dozens of times to explain why the events at Chernobyl could not happen in the United States. There are multiple reasons why that sequence of events could not occur in United States, but he is never able to convince everyone that they could not. So much time is spent focusing on the public's worst scenario that often little time remains to discuss realistic problems and how their likelihood and consequences can be reduced (Fig. 16.6).

Third, the role of statistics, especially probability, separates experts and the public. Many publics accept the assertion that statistics are used to lie and obfuscate, whereas scientists rely on them. In fact, many members of the public do not actually want to hear about statistics when discussing risk. For example, in a survey regarding chemical hazards, Kraus et al. and her colleagues (Kraus et al. 1992) found that nearly 85% of the public respondents endorsed the statement "When some chemical is discovered in food, I don't want to hear statistics, I just want to know if it's dangerous or not." In contrast, when a scientist sees that some fact is certain with 95 or 99% confidence, s/he typically is more than happy to consider that fact credible. But publics may not trust it, and/or may be concerned with the low likelihoods outside the confidence limits. Whereas scientists look for consistency and universality, the public focuses on specific people and places. They are not interested in impersonal numbers with very low probabilities; they want to know the risk to them and their loved ones.

The first author vividly recalls a failed attempt by a state government official who came before a population to explain that the risk associated with a nearby hazardous waste site was negligible. The government official's expression of this reassurance was a statement of the numerical risk, which indeed was very low. However, the public did not understand his mathematical exposition nor did they understand the basis of these numbers. By the time the explanation was provided, the public was extremely agitated and had lost confidence in the official. It took some time for that official and his organization to recover from that mistake.

We see no obvious way to overcome these three cultural and experiential differences. This is not to say that some members of the public are not themselves experts with the same kind of training and experiences as the experts. However, in matters that are personally threatening, these same experts sometimes abandon their attempts to process information with psychological distance, relying instead on emotion, which limits their capacity to assess the validity of what another expert tells them.

In addition to these three cultural and experience-based differences, a set of interactions that occur between experts and the public often increases the distance between them and sometimes leads to confrontation. Experts arrive with sets of risk-related factors to present to the public. The public arrives with a different reality to consider. Sandman (1989) and Lowrance (1976) described about two dozen factors that Sandman called "outrage" factors that cause the public to evaluate many risks differently than experts. For example, when experts estimated the impact of a waste management facility to process medical waste in Chester, Pennsylvania, they focused on the risk associated with the facility, which they found to be minimal. The experts did not take into account the reality that the neighbors did not want the facility in the first place, as it was one of many waste management facilities proposed or already built in the area. Additionally, there were several outrage factors at play which augmented community members' perception of the risk posed by the facility and in turn their opposition to it, namely, they felt that the risk was unfair, out of their control, and unfamiliar. To the local public the facility was an unfairly imposed risk, given their existing burden of waste facilities, and a violation of federal environmental justice laws. Additionally, they felt they had no control over the operation of this facility and the risk it posed. Furthermore, despite the area's familiarity with waste processing facilities, this was an unusual kind of facility, and because of that it presented unknown risks and caused more dread than better-known risks would have. The outraged community sued in federal court under the Civil Rights Act of 1964, and after much debate, the facility was not built (Greenberg and Schneider 1996).

More generally, when experts and publics are asked to compare risks, their perceptions differ with regard to many risks. Perhaps the most well known comparison of expert and public risk perceptions was conducted by Paul Slovic and his colleagues (Slovic et al. 1985). In this study, 15 national experts on risk assessment and 40 members of the League of Women Voters were asked to compare the relative risks of 30 activities and technologies. Some of the rankings were quite similar. Both experts and League members rated the danger associated with motor vehicles as high (number 1 and 2, respectively). There were relatively close agreements on other risks as well. League members rated handguns 3, smoking 4, motorcycles 5, and alcoholic beverages 6. Their expert counterparts rated these same activities and technologies as 4, 2, 6, and 3, respectively. However, there were some striking differences. Notably, League members rated nuclear power as the top risk; whereas the experts rated it number 20. League members rated police work 8 and firefighting 11, whereas the experts accorded them lower risk (17 and 18).

Similar differences between experts and the public have been found in Canada. Slovic et al. (1995) examined the risk perceptions of the Canadian toxicologists and members of the public and found that, similar to their U.S. counterparts; the

Canadian public was less sensitive to the effect of dose of exposure on level of risk and was less afraid of natural chemicals than human-made ones. Elsewhere in the world, marked differences in lay and expert risk perceptions have also been found (Purvis-Roberts et al. 2007; Siegrist and Gutscher 2006).

Overall, differences in training and experience and the public's application of outrage factors make it difficult for experts and public to comfortably interact. Many experts view the public's perceptions and behaviors as a distraction from scientific reality. Many argue that the public's lack of scientific knowledge and reliance on feelings and interpersonal information channels rather than deliberative reasoning and scientific data lead people to make irrational decisions that ultimately hurt the public (Bond 2009; Jasanoff 1993). However, the desire to dismiss public concerns is not advisable because public agreement or at least acquiescence to policy is important (Slovic 1999). Furthermore, the kind of public responses and reasons for those responses have been observed in scores of studies around the world. Public reactions appear to be psychologically driven protective mental guides that people use to weigh information and react to stressful information. Indeed, they comprise a part of risk science, but a part that is difficult for some classically trained scientist to appreciate.

16.6 Narrowing the Gap

The gap separating experts from the public cannot be fully bridged. However, it can be spanned in places to allow better public policy decisions that combine science and public input. The biggest requirement is mutual respect, and the bigger burden is on the experts.

We begin with suggestions for the public. First, the public should not rush to judgment about experts because of appearance, accent, age, gender, ethnicity/race, and affiliation. Not every expert who works for a business is out to pillage the community. Nor is every not-for-profit group willing to align their objectives with public needs. And the role of government experts to defend the community position cannot be assumed because of deep staff cuts.

Our second suggestion is that the public should ask difficult questions of experts (Fig. 16.3). Some questions should be about experts' financial support. Others should address experts' qualifications; are the people presenting the work those who did it? Questions should also address the quality of the information presented. Here are some of our favorite questions for ascertaining research quality. Is this the first study of this kind? If not, how are these results different and similar to others and why? What are the key assumptions (e.g., about exposure, diffusion of hazard)? What are the assumptions made to base the economic impacts being asserted? How exactly are the conclusions presented justified by the data gathered and the analyses? If more time and money were available, how would this study have been done? If an expert cannot answer all or most of these questions, we would be concerned. There would be less need for these kinds of questions if public representatives were included in the process of building the research, choosing the experts, and participating in all facets of the study.

Fig. 16.3 Agency for toxic substances and disease registry public meeting participant posing question (Martin 2009b)

If decision makers want the public to trust their findings, then they must take communication seriously. It should not be viewed as the dreaded part of the project and as something to be relegated to public relations persons. The public relations staff may be more eloquent, but they did not do the work, and the public becomes frustrated when they cannot get answers to questions (such as those in the prior paragraphs) because the expert is not present.

In practical terms, this means completing ten steps that elsewhere the first author has called ten commandments for communicating with the public and the media (Greenberg 2008). Experts must assume responsibility for what they want to present to the public. This means narrowing down the presentation to key points and concerns, since even an interested audience can only retain two or three key messages before they become overloaded with information. It is better that you edit your key points to the essentials, rather than letting your audience do the editing for you (Sandman 2008). It is also important to avoid presenting long lists of points, since many publics will perceive presenting too much information as an effort to obfuscate the issue (Fig. 16.4).

Do not make assumptions about what the audience knows and does not know. Almost everyone has heard about the H1N1 virus, but before you speak about how your group is trying to reduce the risk, make sure that you describe what it is, how it is different and the same as other viruses, and offer basic context. If you are talking about nuclear energy, if possible, take a few minutes to explain the difference between the different kinds of nuclear materials and types of nuclear waste.

Experts must be prepared for personal questions. The public may ask about your source of funding and allude to, if not directly challenge, that you are biased. Be prepared for these probes and then be prepared for questions about risk and your own family, such as "Would you let your family live near this nuclear power plant" or "Did you have your house tested for radon?" The first author has been asked

Fig. 16.4 Agency for toxic substances and disease registry scientist presenting to the public (Martin 2009a)

Fig. 16.5 Agency for toxic substances and disease registry public meeting participant posing question (Martin 2009c)

these questions. The audience asking these questions is trying to find the expert's risk-related bottom line. If you allow your family to live near a nuclear power plant, then you must not think it is very dangerous. If your house has been tested for radon, then you must be concerned about it (Fig. 16.5).

We strongly advise experts to use simple language and avoid jargon. Facing a potentially hostile audience, experts often resort to statistics, technical data, risk

Fig. 16.6 President Carter leaving three mile island (Still Picture Records Section, Special Media Archives Services Division 1979)

comparisons, and other jargon. This can make the public feel that they are being patronized or that the expert is out of touch with their concerns. If you need to use scientific or technical terms, explain them. Once you have chosen a term, stick with it, since alternating between equivalent terms can be confusing. You may know that influenza, flu, H1N1, and swine flu can all refer to the novel H1N1 influenza virus, but the public may think you are discussing four different illnesses. Avoid using complicated statistics when communicating with the public since they are poorly understood (Gigerenzer et al. 2007) and do little to calm. For instance, one of the examples described earlier was of a senior staff member trying to reassure the public that their risk was minimal. He did it with a statement of probability that to him conveyed negligible risk, but it frightened some members of the audience.

If experts insist on using numbers, they must explain them and place them in context. Use whole numbers instead of percentages, since these are more intuitive; for example, state that a risk will affect one in 100 people rather than 1%. Avoid risk comparisons whenever possible, as these often widen the divide between experts and the public (Fischhoff 1995). The classic example of a risk comparison gone awry is when experts compare the risk of smoking or driving with the risk of a chemical exposure. Even if the risk from the chemical exposure is orders of magnitude smaller than the risk of driving or smoking, this kind of risk comparison invariably backfires. Both smoking and driving a car are voluntary, familiar risks, while a chemical exposure is perceived as unfamiliar and involuntary. Because this comparison does not account for the outrage factors (see above), it creates more outrage and makes the public feel like their concerns are being trivialized. If you must compare risks, ensure that you are comparing similar types of risks. The safest kind of risk comparison is to compare a risk to itself, for example comparing the risk from a chemical at two points in time (e.g., this year's risk is lower than last year's risk due to safety improvements).

It is essential that the expert be aware of what s/he is saying and what the public is hearing. If the audience is not paying attention or seems confused and uncomfortable, it is likely to be the presentation, probably too much jargon and too much material that has not been connected to the issue in question. This monitoring of meetings includes correcting any mistakes you have made. Experts are not supposed to make mistakes in their presentations, but they do. Correct them, and your credibility will increase; do not, and it will decrease if the public realizes your mistake. Also, be sure to be up-front about any limitations of the work.

Another lesson learned is that experts with a great deal of public experience will ask a colleague to be present to take notes and to debrief them, so they can learn from their mistakes and improve their presentation. In order to avoid contradictions and misunderstandings, the experienced expert will determine who else is speaking and what they are speaking about.

Our last suggestion is to avoid the temptation to tell the publics what their values and ethics should be. Experts rarely directly challenge the public on ethical grounds. But by challenging inconsistencies in the public's positions, the expert is indirectly challenging their values and ethics. For example, over 90% of the public wants more solar and wind power. But many advocates of renewable energy do not want the facilities, including transmission lines, in their neighborhoods. If the public feels that their ethics and values are being challenged by outsiders, there will be confrontation with some and icy silence from others. Turning a meeting about information into a contest to see who sits on the highest moral ground is a lose–lose outcome. It will widen an expert–public gap, not narrow it.

16.7 Communications During a Crisis: A Special Case

Sandman (2006) has identified four essential risk communication tasks that differ based on the level of public outrage a hazard engenders and the danger it poses; they are precaution advocacy, outrage management, stakeholder relations, and crisis communication. Precaution advocacy takes place when the public lacks sufficient outrage about a hazard that presents a legitimate danger. This form of risk communication is often the purview of health educators, and radon is a good example of the kind of hazard these communications address. Outrage management occurs when the public experiences a lot of outrage in response to a hazard that poses little danger. The task in this case is to lessen the outrage and to make the public's risk perception more congruent with experts'. An example of a hazard that requires outrage management is public concern regarding the relationship between childhood vaccinations and the development of autism. Stakeholder relations are the ideal situation, where discourse takes place with an engaged public who are adequately concerned about a moderately dangerous hazard. The final risk communication paradigm, crisis communication, occurs when addressing an audience that is legitimately outraged about an extremely dangerous hazard. The task during crisis communication is to help an alarmed public make the best choices they can to protect themselves under nearly impossible circumstances (Reynolds and Seeger 2005). An example of a hazard requiring crisis communication would be the detonation of a dirty bomb.

Communication during a crisis brings a whole new set of challenges to the expert–public interaction, namely, heightened emotions, tight time constraints, and the possibility of catastrophic consequences if the wrong choices are made. During a crisis the public more than ever wants certainty, but this is often the time when experts are least able to provide it. The divide between the scientist's conceptualization of risk (as the effect of a hazard on a population) and the public's (the impact of the hazard on themselves and their loved ones) can serve to make their perspectives seem even more at odds. In addition to understanding the qualities of a risk that heighten public outrage, it is important to understand some other cognitive and emotional processes that affect people during a crisis as these will also affect your communications.

Earlier in this chapter, we mentioned that the public trusts interpersonal sources of health information. However, there are also institutions that are consistently identified as trustworthy as well as qualities that individuals communicating about risk possess that can make them seem more or less trustworthy.

Institutions and individuals perceived as most trustworthy to address health, safety, and environmental risks include medical personnel, university scientists, consumer advisory groups, and quality media (Frewer et al. 1996; Wray et al. 2006). Less trusted information sources include industry scientists, government (with the exception of the Centers for Disease Control and Prevention, which are consistently ranked as highly trustworthy (Wray et al. 2006), and activist groups. We encourage experts to develop partnerships with trusted institutions that share their views prior to a crisis, as it has been shown that these institutions confer their trustworthiness to those who share their position (Covello et al. 2001).

While the type of organization experts belong to affects their trustworthiness, there are interpersonal qualities that can also undermine or enhance the effectiveness of their crisis communications. The public assesses four domains when determining how trustworthy an individual expert is; they are how caring and empathetic the expert seems; the expert's dedication and commitment; competence and expertise; and finally the expert's honesty and openness (Peters et al. 1997). Caring and empathy account for 50% of the public's perception of an expert's trustworthiness as a source of risk information, while the other dimensions are weighed roughly equally. Consequently, while experts may think that the public will consider their expertise most heavily when contemplating their advice, they may instead find themselves entirely ignored if they are perceived as uncaring. It is also important to note that experts have a narrow window available to convey that they care, as the public often decides whether they are caring and empathetic within the first 30 s of an encounter (Covello 1993). As a consequence, ensure that when you speak with the public during a crisis you express caring and empathy right away. You can establish that you care simply by stating that you understand that the crisis is frightening, and by disclosing (if it is true) that you are frightened as well, before outlining your advice for addressing the environmental crisis taking place.

Dedication and commitment go hand in hand with caring, and can be conveyed by outlining how you or your organization are addressing the crisis and by following through on what you say you will do. By virtue of their expertise, most experts are perceived as knowledgeable and competent; however, as noted elsewhere, it is essential to maintain this perception by correcting any errors as soon as they are discovered. Additionally, it is vital that even when under the pressure exerted by a crisis, experts are able to state the limits of their knowledge and avoid speculation. Full and prompt disclosure of new information during a crisis (or lack there of) goes a long way towards establishing their openness and honesty. While many experts are deliberative and cautious by nature, it is vital that the public perceives that you are revealing what you know as soon as you know it. Consequently, it is better to make a tentative statement at the start of a crisis and revise it when new information is available, than it is to be perceived as being too slow to respond. While you may think that waiting until all the facts are in displays conscientiousness, the public will most often see this as a lack of openness, and this will erode your credibility.

People's emotions and thought processes are deeply affected when their safety or values are threatened. Covello et al. (2001) describe this effect in their mental noise model of cognitive processing. This model posits that when alarmed, people's ability to assimilate and act on information is severely impaired. Based on this, avoiding the use of jargon and numbers is even more imperative when communicating during a crisis, since it has also been found that mental noise reduces people's ability to process information by as much as 80% (Covello et al. 2007). However, this effect can be alleviated by the use of effective graphics and visual aids, as well as analogies and personal narratives, which can decrease mental noise by more than 50% (Hyer and Covello 2005). While some mental shortcuts used in decision making are the enemy during crisis communication (such as the availability heuristic which dictates that people view what is easily recallable as being likely (Keller et al. 2006)), others can be used to your advantage. When confronted with new information, people are best able to recall what comes first and last; therefore, put your most important talking points in these positions to help overcome mental noise (Hyer and Covello 2005).

It is also instructive to keep in mind how crises affect the weight that people give to positive and negative information. Upset and frightened people view negative information as more credible and recall it for longer than they do positive information (Slovic 1999). This negative dominance (Covello et al. 2001) makes it all the more important to maintain the public's trust by being transparent in your communications and by correcting errors as soon as possible. This principle also serves as a guideline for structuring your communications with the public during a crisis. Because positives and negatives are weighed asymmetrically, whenever you must deliver a negative message, aim to counterbalance it with a greater number of positive messages (three is a good rule of thumb) (Hyer and Covello 2005). Avoid using negative language (words such as no, not, nothing, none, etc.), focusing instead on what is being done to address the crisis (Hyer and Covello 2005).

To communicate effectively with the public, strive to understand things from their perspective whenever possible. Although you may not share the public's concerns, they reflect deeply personal values, and are consequently important considerations in the environmental health and safety arenas. Realize that the public is a legitimate partner in decision-making, not an obstacle to achieving your scientific or policy aims (Glik 2007; Jasanoff 1993; Slovic 1999). If expert–public interactions are approached with a sense of mutual respect for the differences that exist between their perspectives, this will go a long way towards fostering greater understanding between these often disparate groups.

References

Bond M (2009) Decision-making: Risk school. Nature 461:1189–1192

Bureau of Land Use Management, California (2008) Public Meeting Image BL031113OR. http://www.blm.gov. Accessed 18 April 2010

Burger J, Mayer HJ, Greenberg M, Powers C, Volz CD, Gochfeld M (2006) Conceptual site models as a tool in evaluating ecological health: The case of the department of energy's amchitka island nuclear test site. J Toxicol Environ Health: Part A 69:1217–1238

Conservation West (2009) State Budget Cuts Threaten Protections for Washington's Environment, http://www.pugetsound.org/news/news-about-people-for-puget-sound/033009cuts/. Accessed 7 August 2009

Covello VT (1993) Risk communication and occupational medicine. J Occup Med 35:18–19

Covello VT, Minamyer S, Kathy C (2007) Effective risk and crisis communication during water security emergencies-summary report of EPA sponsored message mapping workshops. EPA/600/R-07/027:US EPA

Covello VT, Peters RG, Wojtecki JG, Hyde RC (2001) Risk communication, the west nile virus epidemic: Responding to the communication challenges posed by the intentional and unintentional release of a pathogen in an urban setting. J. Urban Health: Bull. N. Y. Acad. Med. 78:382

Department of Energy (DOE) (1996) Sun Tracking Heliostats at Solar Two, Daggett, California Near Barstow. http://www.doedigitalarchive.doe.gov/ImageDetailView.cfm?ImageID=1000361&page=search&pageid=thumb. Accessed 20 May 2010

Fischhoff B (1995) Risk perception and communication unplugged: Twenty years of progress. Risk Anal 15:137

Frewer LJ, Howard C, Hedderley D, Shepherd R (1996) What determines trust in information about food-related risks? underlying psychological constructs. Risk Anal 16:473–486

Gigerenzer G, Gaissmaier W, Kurz-Milcke E, Schwartz LM, Woloshin S (2007) Helping doctors and patients make sense of health statistics. Psychol Sci Public Interest 8:53–96

Glik DC (2007) Risk communication for public health emergencies. Ann Rev Public Health 28:33–54

Greenberg M, Truelove H (2010) Right answers and right-wrong answers: Factors influencing knowledge of nuclear-related information. Socio-Economic Planning Sciences 44:130–140

Greenberg MR (2009a) How much do people who live near major nuclear facilities worry about those facilities? analysis of national and site-specific data. J Environl Plan Manag 52:19–937

Greenberg MR (2009b) Energy sources, public policy, and public preferences: Analysis of US national and site-specific data. Energy Policy 37:3242–3249

Greenberg MR (2008) Environmental Policy Analysis & Practice. Rutgers University Press, New Jersey

Greenberg MR (2009c) NIMBY, CLAMP, and the location of new nuclear-related facilities: U.S. national and 11 site-specific surveys. Risk Analysis 29:1242–1254

Greenberg MR, Schneider D (1996) Environmentally Devastated Neighborhoods: Perceptions, Policies, and Realities. Rutgers University Press, New Jersey

Hyer RN, Covello VT (2005) Effective Media Communication during Public Health Emergencies. Geneva: World Health Organization

Jasanoff S (1993) Bridging the two cultures of risk Analysis. Risk Analysis 13:123–129
Keller C, Siegrist M, Gutscher H (2006) The role of the affect and availability heuristics in risk communication. Risk Anal 26:631–639
Kraus N, Malmfors T, Slovic P (1992) Intuitive toxicology: Expert and lay judgments of chemical risks. Risk Anal 12:215–232
Lowrance WW (1976) Of Acceptable Risk: Science and the Determination of Safety. William Kaufmann Inc, California
Martin C (2009a) Public Health Image Library (PHIL) Image 11528. http://www.phil.cdc.gov/phil/details.asp. Accessed 20 May 2010
Martin C (2009b) Public Health Image Library (PHIL) Image 11602. http://www.phil.cdc.gov/phil/details.asp. Accessed 20 May 2010
Martin C (2009c) Public Health Image Library (PHIL) Image 11612. http://www.phil.cdc.gov/phil/details.asp. Accessed 20 May 2010
Nagy J (2002) State Environmental Budgets Take $200M Hit. http://www.stateline.org/live/ViewPage.action?siteNodeId=136&languageId=1&contentId=14735. Accessed 7 August 2009
NCI (2002) Making Health Communications Programs Work. U.S. Department of Health and Human Services, Washington
Peters RG, Covello VT, McCallum DB (1997) The determinants of trust and credibility in environmental risk communication: An empirical study. Risk Anal 17:43
Purvis-Roberts KL, Werner CA, Frank I (2007) Perceived risks from radiation and nuclear testing near semipalatinsk, kazakhstan: A comparison between physicians, scientists, and the public. Risk Anal 27:291–302
Reynolds B, Seeger M (2005) Crisis and emergency risk communication: An integrative approach. J. Health Commun 10:43–55
Sandman PM (1989) Hazard versus outrage in the perception of risk. In: Covello VT (ed) Effective Risk Communication: The Role and Responsibility of Government and Nongovernment Organizations. Plenum Press, New York
Sandman PM (2006) Crisis communication best practices: Some quibbles and additions. J. of Applied Commum Res 34:257–262
Sandman PM (2008) Simplification made Simple. http://www.psandman.com/col/simplify.htm. Accessed 1 Nov 2008
Siegrist M, Gutscher H (2006) Flooding risks: A comparison of lay people's perceptions and expert's assessments in switzerland. Risk Anal 26:971–979
Slovic P, Fischhoff B, Lichtenstein, S (1985) Characterizing perceived risk. In: Kates RW (ed) Perilous Progress: Managing the Hazards of Technology. Westview, Colorado
Slovic P (1999) Trust, emotion, sex, politics, and science: Surveying the risk-assessment battlefield. Risk Anal 19:689–701
Slovic P, Malmfors T, Krewski D, Mertz CK, Neil N, Bartlett S (1995) Intuitive toxicology. II. expert and lay judgments of chemical risks in Canada. Risk Anal 15:661–675
Still Picture Records Section, Special Media Archives Services Division (1979) President Jimmy Carter Leaving [Three Mile Island] for Middletown, Pennsylvania. http://www.arcweb.archives.gov/arc. Accessed 20 May 2010
Wray R, Rivers J, Whitworth A, Jupka K, Clements B (2006) Public perceptions about trust in emergency risk communication: Qualitative research findings. International Journal Mass Emerg Disasters 24:45–75

Chapter 17
Media, Local Stakeholders, and Alternatives for Nuclear Waste and Energy Facilities

Karen W. Lowrie, Amanda Kennedy, Jonathan Hubert, and Michael R. Greenberg

Contents

Abstract In the early part of the twenty-first century, it appears more likely than ever that the United States will need to consider siting additional nuclear power plants as part of its overall strategy to reduce dependence on fossil fuels. At the same time, there is a continuing need to manage legacy wastes from the nuclear weapon development era, as well as current and future high level wastes from power generation. An important determining factor in the ability to locate and build needed nuclear facilities will be the reaction of the nearby residents. As these proposals and projects are discussed in local arenas, their coverage by local media will serve to inform and possibly shape residents' views about the facts and issues that are important to consider. This chapter discusses the influence of media stories on public perceptions about hazards and risks, and then presents results of a recent content analysis of

K.W. Lowrie (✉)
Edward J. Bloustein School of Planning and Public Policy, Rutgers University,
33 Livingston Avenue, New Brunswick, NJ 08901, USA
e-mail: klowrie@rutgers.edu

J. Burger (ed.), *Stakeholders and Scientists: Achieving Implementable Solutions to Energy and Environmental Issues*, DOI 10.1007/978-1-4419-8813-3_17,

stories about proposed new or expanded projects at existing nuclear power or waste sites. Finally, we describe some implications related to media, local stakeholders, and alternatives for expanding nuclear facilities in the age of the Internet.

17.1 Introduction

Given the United States' need to reduce reliance on fossil fuels and that many existing power plants are approaching the end of their lifetimes, new facilities will need to be constructed to meet growing energy demands in the coming decades. A related issue is the need to safely manage our hazardous and radioactive wastes, a legacy left at scattered nuclear power generating sites and government-owned weapons design and construction facilities across the country (Greenberg et al. 2009). There are currently 63 nuclear power plant sites and more than a dozen large nuclear research or waste management sites in the United States. Applications for 22 new nuclear power plants to be sited at existing plant locations are under consideration by the U.S. Nuclear Regulatory Commission (2008). And a number of new large waste processing facilities are being considered or are under construction at major nuclear weapons sites owned by the U.S. Department of Energy (DOE), like the multibillion-dollar waste vitrification plant planned for the Hanford Reservation in eastern Washington state.

Surveys and observers suggest that there is increasing public support for new nuclear power plants (Venables et al. 2009; Gertner 2006). Finding locations for new nuclear power plants, nuclear waste management facilities, and research laboratories is a key part of a potential nuclear renaissance. But the majority of new applications are at sites that already have at least one site (Greenberg 2009; Venables et al. 2009). "Concentrating locations at major plants" (CLAMP) is a pragmatic policy because the utility or government already owns or controls the land; there is an existing workforce; and workers, families, and friends are likely to support facility expansion (Greenberg 2009). However, NIMBY ("not in my backyard") sentiments have dominated public opposition to siting of these large, hazardous installations (O'Hare et al. 1983; Portney 1991; Sjoberg 2004). When residents perceive that the risks they will face outweigh the benefits they will receive, NIMBY attitudes are likely to occur.

Fig. 17.1 Artist's concept of the Savannah River N River National Laboratory's New Center for Hydrogen Research (DOE Photo, 2004)

Decision-making and public policies that affect the placement, construction, and operation of new energy and waste facilities partly depends on public attitudes, which, in turn, tend to be affected by mass media coverage (McCombs and Shaw 1972; Cook et al. 1983; Kitzinger 1999; Flynn et al. 2001; Allen et al. 2000). Stories in local and regional media outlets are likely to be the primary source of information for residents (Greenberg et al. 2008). It therefore is important to ask how local nuclear projects are depicted and portrayed in media coverage. For example, a front-page newspaper story about a proposed new energy-producing plant could emphasize local jobs and income and de-emphasize negative local impacts such as traffic or environmental hazards, influencing public perceptions about the relative risks and benefits of these projects. But it could also focus on the dangers of nuclear materials.

To explore the issue of media's portrayal of new projects at existing nuclear power and waste plants, this chapter first includes a review of prior research related to three major areas: (1) public attitudes about nuclear facilities, (2) media reporting about risk and nuclear sites, and (3) influence of media on public perceptions. Then, a recent study that examines the orientation of newspaper coverage related to new projects at 11 nuclear sites in the United States regarding NIMBY or CLAMP policies is presented, followed by discussion and directions for future inquiry and research.

17.2 Background

17.2.1 Public Attitudes About Nuclear Facilities

Although logic might suggest that those who live closest to a nuclear facility should be more concerned than those who live further away (Clay and Hollister 1983), many studies have found that personal knowledge, desensitization, and employment and other local benefits may outweigh NIMBY attitudes in locations close to nuclear sites (Kivimaki and Kalimo 1993; Halpern-Felsher et al. 2001; Greenberg et al. 2007). For example, a survey of Idaho residents found that over 60% supported the Idaho National Laboratory's (INL's) nuclear research mission, which includes a plutonium project and research on nuclear reactors – over 70% had a favorable opinion of the site, and many perceived local economic benefits (Nemich 2006). Regarding nuclear power sites, the United States public appears to have become more favorably inclined to accept nuclear power than they were a decade ago. A 2007 survey by Bisconti for the Nuclear Energy Institute found that 76% of residents within ten miles of a nuclear plant said that it was acceptable to add a new nuclear reactor at the site, compared to only 22% who said it was not acceptable. Eighty-six percent had a "favorable" impression of the nearby plant, and only 11% did not (Bisconti 2007).

A great deal of literature also supports the idea that the general public typically incorporates subjective factors such as dread, unfamiliarity, and catastrophic potential into perceptions of risk (Fischhoff et al. 1981; Slovic 1987). So the public

Fig. 17.2 Workers install steel structure frame for Pit 4 Retrieval Enclosure Structure at Idaho National Laboratory (DOE Photo, 2004)

perception of a proposed project is formed from a combination of personal experience, personal attitudes and beliefs, and the social context (Sjoberg 2004). However, if the more negative aspects of a facility predominate in the public's collective mind, NIMBY complaints will result (Portney 1991; Kraft and Clary 1991). This means that perceptions about any new large project or project expansion located on a site already associated with negative factors could be subject to those same influences.

A recent survey (Greenberg 2009) focused on public preferences for locations of nuclear waste management and laboratory facilities, as well as nuclear power plants. The study found that about a third of people surveyed who live close to nuclear sites favored CLAMP policies for new nuclear plants, and just over half were in favor of CLAMP for nuclear waste management sites. Those people with more information about the site were more likely to favor CLAMP policies, a finding that has implications for information sharing and for understanding public reactions and attitudes. Yet, Rosa (2001, 2004) found that the majority of the U.S. public is still not ready to have a nuclear power plant in their jurisdiction. Rosa concludes that negative perceptions are related to concerns about reactor safety and waste disposal and low level of trust of the nuclear industry and its government regulators.

Several recent studies have tied together nuclear energy and nuclear waste management. For example, about two thirds of respondents to a recent survey said that they would support a significant expansion of nuclear power if the waste storage problem could be more effectively solved (Ansolabehere 2007). Pasqualetti (1987) noted that too much emphasis is placed on siting power plants and not enough on waste management and transportation of waste products. A recent U.K. study observed that more people were concerned about nuclear power's waste products than were concerned about nuclear power plants (Poortinga et al. 2005). In other words, it is possible that even if the U.S. public is largely convinced of the need for nuclear power plants, it might reject the waste management, transportation, and research facilities associated with nuclear power.

17.2.2 Media Reporting About Risk and Nuclear Sites

The media are important players in communication about risk, particularly regarding risks of things that are not familiar and well-known by most people in their daily lives. Greenberg et al. (2008) noted that particularly in less populated areas, a large facility will be relatively more important to news media outlets (bringing jobs and income to the region) than it might be in more urban areas. Yet journalists can be charged for exaggerating risk and distorting reality. Media have been accused of oversimplifying complex or technical issues (Breakwell 2007; Kasperson et al. 1988). For example, Boholm (2009) analyzed newspaper coverage of river valley risks in Sweden and found that articles gave simple causal explanations of risks such as water pollution, landslides, and flooding. Literature about media reporting on risk suggests a more complex picture. Kitzinger (1999) explored factors influencing media coverage of risk that included the resources of the news organization, their news-gathering routines, and the "cultural givens" about hazards in a particular area. Entman (1993) talked about a media "frame" that chooses to make some aspects of reality more salient than others to promote a certain interpretation of a problem or conflict. Some of the frames suggested by literature are economic, conflict, and human-impact frames (Neuman et al. 1992) and precaution, scientific, technocratic, and scandal frames (Vasterman et al. 2008).

The emphasis of media stories has been the subject of numerous studies. Kenix (2005) content analyzed 1,180 articles about environmental pollution and found that content was targeted to upper-socioeconomic groups. O'Donnell and Rice (2008) found that articles about environmental issues and events covered hazards with an emphasis on the categories of solutions, costs, concentration, and actual or potential mortality of nonhuman beings (i.e., animals and plants). Analysis of newspaper coverage of six major nuclear weapons research and waste sites in the United States found that environmental contamination was by far the most mentioned topic of stories (Greenberg et al. 2008).

In addition to what topics media stories emphasize, researchers have examined the scientific accuracy of media coverage and whether the media report "important" risks. (Kasperson et al. 1988; Singer and Endreny 1987, 1993; Kitzinger 1999; Vasterman et al. 2008). An abundance of research shows that risks deemed important by media do not necessarily match those identified by scientists. Risks can be either overblown or underreported. Neuman et al. (1992: 49) point out that accuracy has been evaluated differently for different kinds of media, with television more subject to charges of "sensationalism" than newspapers that are sometimes praised for in-depth news coverage. Driedger (2007) concluded that print media give better

Fig. 17.3 St. Thomas, VI, August 31, 2010 – newspaper vendor Earl Jibbs (Andrea Booher/FEMA)

coverage (analysis and process) than televised news, especially when the risk is chronic and the scale is at a local level.

It is less clear whether journalists simply avoid presenting science, or whether they tend to simplify scientific or technical issues to be understandable to the general public (Hernes 1978). But if the risks presented in media stories about nuclear sites are not based on science, it is relevant to understand the basis of risk reporting.

Lowrie et al. (2000) analyzed the content of news stories about major nuclear waste and chemical weapon facilities and concluded that discussion of risk is seldom communicated through newspaper coverage. When hazards are emphasized in a story, it can signal outrage factors in the audience, but Swain (2007) found in a study of the coverage of anthrax that explanations that put the hazard into a broader context help to mitigate negative reactions.

Regional or cultural trends or attitudes can affect how a story is portrayed. A study of newspaper coverage of the Yucca Mountain nuclear waste repository site found that the advocacy position of the journalist (suggested by their geographic region) influenced the language used to describe the site (Larsen and Brock 2005). Those advocating use of the site used words like "desolate" and "isolated," while those more critical of using the site talk about the "beauty" of the area and its proximity to Las Vegas. The Lowrie et al. study (2000) posited a "geo-cultural" explanation for the finding that newspapers in the south were less likely to talk about risk and hazard than those in the mountain's west. General moods of public optimism or pessimism were found to be reflected in media reporting of harms in a classic study by Singer and Endreny (1987), with those cycles affecting the ratio of benefits to costs associated with various harms reported by the media.

Fig. 17.4 Los Alamos, NM, May 4, 2000 – officials hold a press meeting to discuss the consequences of the devastating fire. Photo by Andrea Booher/FEMA News Photo

Wakefield and Elliott (2003) performed a content analysis of risk in regional newspaper coverage of a landfill approval process, and found that reporters rely on personal sources over written information, raising issues of the accuracy of risk communication. They also found that amount of coverage increased when a controversy arose.

17.2.3 Influence of Media on Public Perceptions

Studies of risk reporting by the media often rest on the theoretical underpinning that media depiction influences public risk perception and therefore behavior, decision-making, and policy (Breakwell 2007; Hughes et al. 2006; Kasperson et al. 1988; Kitzinger 1999; Renn 2008; Wåhlberg and Sjöberg 2000). An analysis of media coverage by Simon and Jerit (2007) affirmed the idea of "uptake" – that if certain terms are used in describing an issue, it can lead to increased or decreased support for policies. Media's impact on public perceptions can occur both because of the topics covered by media and the way that the topics are covered. In other words, the media may tell people what to think (how topics are covered and framed, positive vs. negative, risks, etc.) and may also tell people what to think *about* (what topics are covered), or the idea of agenda setting (Cohen 1973; Allen et al. 2000). Mazur (1990) goes further to say that it is not the content of coverage, but the amount of media coverage of an issue that has the stronger influence on risk perception.

Some studies have looked at whether consumers of media have individual frames that affect how they understand information (Johnson-Cartee 2005; Scheufele 1999). Most local residents around nuclear sites report feeling well informed about the site, according to the Bisconti research cited above (2007). But how is that information being presented and interpreted? As an example of individual frames, Williams et al. (1999) found that personal probability of economic or job loss can create heightened focus related to new facilities or removal or facilities from regions. Individual frames could also take the form of socioeconomic status. Media dependence theory, for example, tells us that the people most dependent on mass media for information (lower socioeconomics, older) are also most affected by its content (Ball-Rokeach and Defleur 1976), particularly if they have little or no personal experience with the topic.

An important factor affecting the extent to which media coverage impacts individual perceptions is the trustworthiness of the source. Studies have shown, for instance, that even if people go to newspaper stories for in-depth coverage, many still do not consider them trustworthy and suspect bias in reporting (Wakefield and Elliott 2003).

17.3 Newspaper Content Analysis Study: Coverage of Nuclear Sites

The authors conducted a content analysis study, building on a recent risk perception and preference survey (Greenberg 2009) of residents living near nuclear waste management and laboratory facilities and nuclear power plants. In the study, we collected local newspaper stories about the same nuclear facilities as in the Greenberg study and focused on analyzing stories about new projects or project expansions at those sites. The study was conducted under the theory that framing in media can have an agenda-setting function on the media audience (Entman 2007). Announcements and plans for major new facilities or expansions of existing facilities are likely to be judged as newsworthy by local media outlets, where these facilities can bring both new jobs and income (positives), and new hazards and risks (hazardous wastes, pollution, etc.)

Our intent was to examine the extent to which local newspapers print articles about the local nuclear facilities, and for those articles about proposed or actual new projects or expansions at the facilities, how the stories portrayed the projects and whether they were presented with a clear emphasis toward either NIMBY or CLAMP policy. That is, do these articles tend to present information that would influence readers to either want the new project placed somewhere else or to be persuaded that concentrating new projects at existing locations is preferred? We present the results of the study here.

17.3.1 Data and Methods

17.3.1.1 Article Selection and Coding

We selected two daily newspapers within each of the 11 site regions (described below). We performed Internet searches of the newspaper archives for articles published between July 2006 and June of 2008, searching for the name of the nuclear site or facility in the title or content of the articles, or the name of the site plus the word "nuclear," if the name of the site is not unique to the site. Articles were then coded for their main subject or emphasis: what is the article about? If the online data indicated it, we also coded the article for its length in words and its location within the newspaper (section and page number).

The focus of the study was to examine articles about new or expanded projects at the existing nuclear locations. So from the subset of articles that were about new or expanded projects, we then selected articles that were either located on the front page of a section or were judged to be major articles by either their length or the presence of a picture or graphic. For this subset of more prominent articles about new or expanded projects, we read the articles and then coded them on two dimensions:

1. Overall tone of article with regard to new or expanded project:
   - Positive
   - Negative
   - Neutral
2. Mention/emphasis of NIMBY or CLAMP?
   - NIMBY: Yes/No
   - CLAMP: Yes/No

The tone coding was based on the use of strongly positive or negative words or imagery in the article. The NIMBY/CLAMP coding decision was based on whether the article mentioned, implied, or provided an argument that the facility or project is not a good idea for the region (NIMBY), or that the facility or project is well suited to the region because of the presence of the existing facility (CLAMP). A single coder ensured consistency of judgment.

17.3.1.2 Site Selection

The 11 nuclear sites match those in the Greenberg study. There are six major U.S. DOE facilities: Hanford [WA], Idaho National Laboratory [ID], Los Alamos [NM], Oak Ridge [TN], Savannah River [SC], and the Waste Isolation Pilot Plant (WIPP) [NM]. Each of these has major nuclear waste management facilities. Their aggregate budget for waste management averages approximately $4.5 billion per year.[8, 50] Los Alamos, Idaho, and Oak Ridge are also major DOE research facilities. Five of

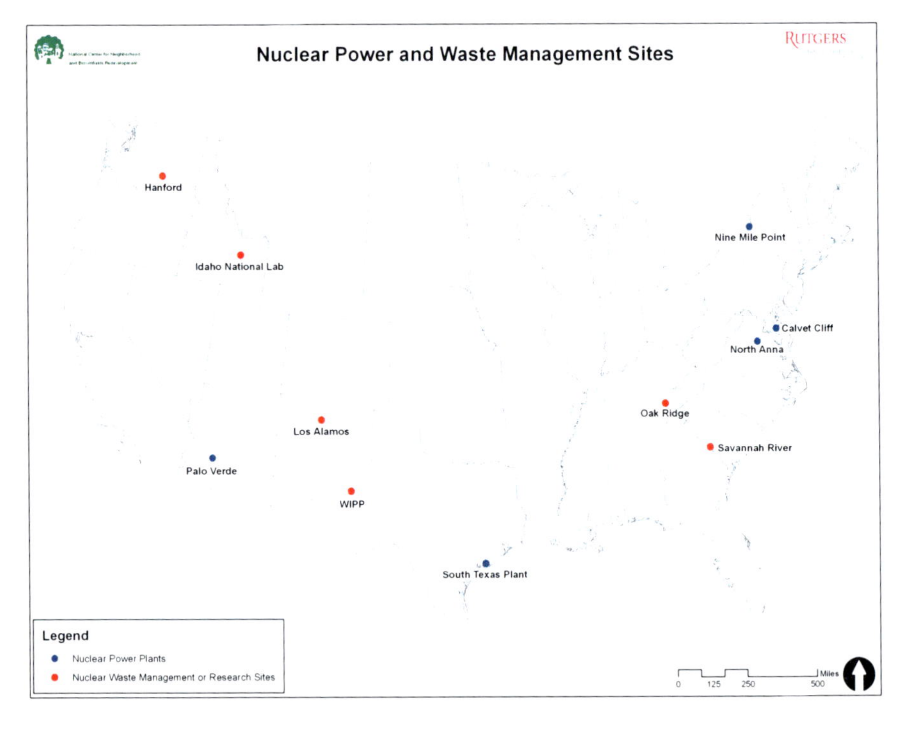

Fig. 17.5 Nuclear sites in study

the 11 sites are existing nuclear power plants that were chosen as locations for new nuclear power plants: Calvert Cliffs [MD], Nine Mile Point [NY], North Anna [VA], Palo Verde [AZ], and South Texas [TX].

We chose the most prominent regional daily newspaper that would be the most likely to be read by residents within a 50-mile radius of the facility, particularly regarding news about the facilities. Many of the areas within close proximity to the sites have had a strong economic relationship with the sites, while populations farther away are likely to be less concerned but still have a familiarity-based relationship with sites (see Table 17.1).

17.3.2 Results and Discussion

The article search yielded a total of 845 articles about the 11 sites over the course of 2 years. Of those, we looked at whether the article was primarily about a new proposed or planned project or facility, and found that for about a third of the articles about power plants, the article concerned a new proposal. For waste/weapons sites, the percentage was less, at around over one-fifth of articles (21%) (see Table 17.2).

We then looked at the 204 articles about new projects, and selected 66 that were more prominent articles, including some from each site, anywhere from 3 (Palo Verde)

Table 17.1 List of sites and newspapers

Nuclear sites	Newspaper	Daily circulation[a]
Nuclear power sites		
Palo Verde	Arizona Republic	434,000
North Anna	Richmond Times-Dispatch	125,000
South Texas Plant	Houston Chronicle	494,000
Calvert Cliffs	Washington Post	578,000
Nine Mile Point	Post Standard	90,000
Nuclear waste/research sites		
Hanford	Tri-City Herald	42,000
Los Alamos	Albuquerque Journal	95,000
Savannah River Site	Augusta Chronicle	61,000
Oak Ridge	Knoxville Sentinel	100,000
Idaho Falls	Idaho Falls Post Register	55,000
WIPP	Albuquerque Journal	95,000

[a]Rounded to nearest thousand. Based on most recent obtainable figures from newspaper Web sites

Table 17.2 Articles about new projects

Type of site	Articles about the site	Articles about new projects	%
Power	225	75	33
Waste/weapons	620	129	21
Total	825	204	25

to 14 articles (Idaho). By prominent we mean that they appeared on the front page of a section or were relatively long articles. Because these articles were longer or more visible than other articles, they presumably would be seen by more people. The tone of these articles can serve to influence public opinion and perceptions. We found that more of the articles were positive (44%) in tone than negative (32%), and that this difference was more pronounced for the set of articles about nuclear power plants, with twice as many positive articles as negative articles. The ratio was close to even for the waste management sites.

We had expected that power sites would generally be portrayed more favorably because of increased support for nuclear power as a more "clean" technology to meet our growing energy demands, despite the hazard presented by storage of spent fuel at reactor sites. Media stories may emphasize the contribution of the sites to the region's economy and the relatively good track record of the businesses running the plants in terms of safety and security. The massive DOE weapons sites, however, are a double-edged sword for the communities around them. While they bring jobs, they are also well-known for the nuclear wastes stored on site and for their role in the development of nuclear weapons, creating a stigma. Some of the sites in the study (Hanford, Savannah River, Oak Ridge) have contaminated soils and water supplies, or have had accidents that pose health and environmental dangers to surrounding populations (Pavey 2010; U.S. EPA 2009; Washington State Dept. of Ecology 2008; U.S. DOE 1996). So it is not surprising that media stories may tend to be somewhat more negative in tone when describing new initiatives at these sites (see Table 17.3).

Table 17.3 Characteristics of prominent articles about new projects

Type of site	Number of articles	Positive	Negative	Neutral	NIMBY	CLAMP	Both
Power	25	12 (48%)	6 (24%)	7(28%)	5(20%)	5 (20%)	2 (8%)
Waste/weapons	41	17 (42%)	15 (37%)	9 (21%)	4 (10%)	11 (27%)	1 (2%)

Regarding whether the articles tended to support policies of NIMBY or CLAMP, we found that in the sample of 66 more prominent articles, just over 40% of the articles were judged to convey a preference for either one or both of the policies. Others did not include any arguments or implied leaning toward either policy. Overall, there were more CLAMP-favoring articles (16) than NIMBY (9), but in articles about the weapons sites, the ration of CLAMP to NIMBY articles was almost 3 to 1, while the ratio was even for power facilities. Most of the articles about new projects for the power sites were about possible new reactors being built, while those for the weapons site ranged from new waste processing facilities or industrial plants to new reactors. The weapons/waste sites are much larger geographically, so that major structures are far from public visibility or spread out over vast areas. It makes sense that a CLAMP policy could be supported here for practical reasons like the abundance of space, while power plants are usually more openly visible and on smaller sites, so placing new facilities there could raise more public and practical issues.

Greenberg's (2009) study found that people who had mixed reactions about CLAMP policy were most likely to know little about the sites or not trust the site managers. We use caution in drawing strong conclusions from the content analysis because we used judgment in selecting the set of most prominent articles, and there was inherent judgment in coding for the tone and NIMBY/CLAMP orientation, since neither is a straightforward quantitative count or tally. However, our findings could be extrapolated to provide some evidence that local media outlets do not stress CLAMP in their coverage of nuclear power plants because they know less about them than about the nuclear weapons/waste/research sites in their regions, presumably because the private companies managing these sites may not communicate as frequently with the press. Greenberg concluded that based on current public opinion, we can no longer assume NIMBY reactions to new proposals at nuclear sites, but can expect a more tentative reaction that is willing to consider new nuclear-related projects.

17.4 Stakeholders and Media: Future Directions

It is clear that behemoth nuclear facilities that are either clearly visible or occupy a great deal of space have undeniable real and perceived impacts on surrounding communities. Because the risks posed by site operations and the potential hazards associated with nuclear power, materials, and waste management can instill fear and dread, it is important for the local public to obtain clear and accurate information about facilities that addresses their concerns and answers their questions. The local media are likely to be an important part of the process of informing the public about what is going on at these sites, particularly about how to evaluate proposals for site additions or expansions. The goal for all parties involved should be for the public to have the most accurate and unbiased information about proposals that affect their regions so that informed debate can occur, outrage can be minimized, and trust can be instilled.

What is presented by media is only part of a more complex picture, as citizens need to process information from many sources. So a larger question for both the managers and policy makers at the sites and for those concerned about citizen participation is ...what should be presented from the perspective of plant management and what have the public and journalists said that they want to know? How much data about these subjects is available and can be made public? Is the information balanced? Is it biased toward one or more policy options or deliberately positive or negative in tone? We finish with a discussion of future implications for effective risk communication and a look forward to the significance of media in the Internet age of instant communication and information overload.

17.4.1 Implications for Risk Communication

Journalists have a responsibility to cover technical and complex issues in ways that are understandable to their audience. Reporters who gather information from a variety of sources, such as key informants from both proponent and opponent groups, industry and government representatives, as well as attend public meetings and use written sources, are likely to present a well-balanced set of information. For the private companies that manage nuclear sites and for the U.S. DOE, the responsibility is to communicate with the press openly and often, and importantly, to appear before the public directly to communicate messages about risk and address controversy.

It's critical for the press and for the risk communicators and public relations personnel at the sites to consider the culture of the surrounding community; their literacy about complex economic, environmental, and health issues; and their main concerns. If messages are not presented at the correct level of detail, are not well-understood, or are perceived as one-sided, amplified levels of risk perception can occur (Feldman and Hanahan 1996). Normally, media information about new projects at nuclear sites should be provided not in a crisis context, but within the context of normal stakeholder relations. That is, when no immediate dangers or public outrage is present, it is a good opportunity to present sound balanced information. Presentation of fair and balanced information should in turn help to increase effective citizen participation in decision making and maintain trust.

17.4.2 Media Influence in the Internet Age

Recent shifts in communication technology have served to compress the time and space of information flow. It is now routine for members of the general public to see

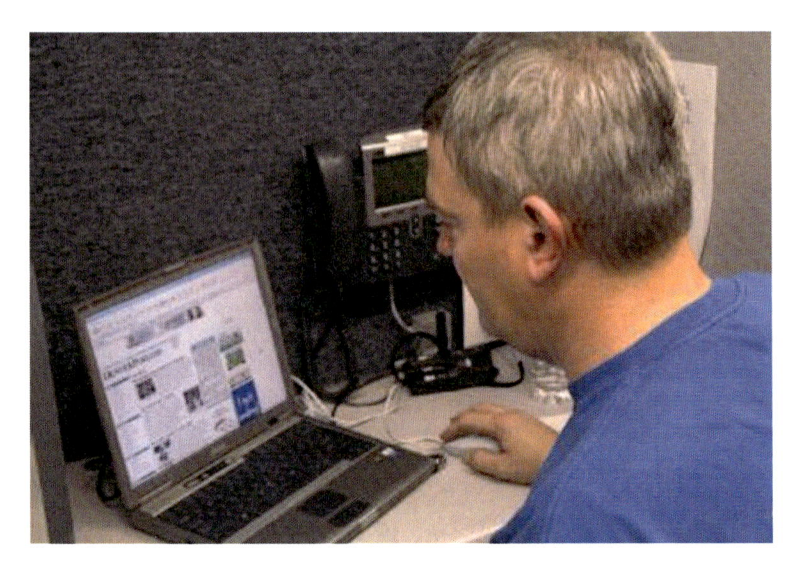

Fig. 17.6 Cameron, LA, 11-10-05 – user of Internet connections provided by FEMA-sponsored MCI phone/communication trailer (MARVIN NAUMAN/FEMA photo)

instantaneous video feeds from news events on their home computer screens or remote hand-held remote devices. Many people get most of their information from Internet-based sources such as blogs and news sites like Yahoo, etc. Further, the rise of social networking sites means that the exchange of information occurs at such a rapid pace that fact checking may take a backseat.

Conventional print media is constrained and requires greater physical and mental effort to use as information source. But because most major national and regional newspapers post their news stories on their websites, the influence of the newspapers is not completely lost to the Internet. Stories are still getting out there. However, stories designed for web pages can often be cut in length and may be tailored to be more sensationalized and colorful to maintain reader interest. Further research is needed to determine which of these formats – conventional print media, electronic versions of newspapers, blogs, other news services, social networking sites – has the strongest influence on public perceptions as we enter the second decade of the third millennium. Managers of nuclear sites, and all kind of facilities, will need to adjust public messages to reach target audiences in the right format with the right messages, whether it is Twitter, Facebook, or other popular electronic formats. It will be even more important to analyze media messages to understand the focus and slants of various news delivery methods, how they are used, and what influence they have on public opinion.

The widespread use of Internet sites, blogs, and forums as avenues by which people obtain their news has created a two-way relationship between mass media and public opinion. It is now very easy for individuals to comment on stories, post their own interpretations of stories, and share them with people around the globe at the push of a button (Wong and Fryxell 2004). This involvement allows stakeholders greater opportunity to shape news, or to at least force responses from news media outlets that are challenged in a public forum. In a sense, it could also reduce the relative importance of official news media stories in shaping or influencing public views or agendas, as many different stakeholders have the opportunity to publish messages (Ryan 2003; Leading Futurist: Online Influence is Transforming Society 2009; Opinion: Internet conversations can have valuable print influence (Viewpoint essay) 2007; van der Merwe et al. 2005). In the age of the Internet, stakeholders will have more and more strength, visibility, and ability to call out news media sources for ignoring important issues or portraying only a partial picture. Media companies will need to recognize the "word of mouth" power of blogs and social media. Further analysis is needed to examine to what extent the presence and circulation of an issue through social media, forums, and blogs finds its way back to the media, who respond in turn with stories that address those issues. Our analysis focused on a one-way process, but democratization of media influence is a phenomenon that should result in stories that seek to inform the public about risks and impacts that concern them the most.

References

Allen S, Adam B, Carter C (eds) (2000) Environmental risks and the media. Routledge, London & New York

Ansolabehere S (2007) Public attitudes toward America's energy options: Insights for nuclear energy MIT-NES-TR-08

Ball-Rokeach S, Defleur ML (1976) A dependency model for mass-media effects. Comm Res 3:3–21

Bisconti Research, Inc. (2007) National survey of nuclear power plant communities. For nuclear energy institute. Available at: http: www.nei.org/newsandevents/newsreleases/nuclearpower-plantneighborsaccept.html

Boholm M (2009) Risk and causality in newspaper reporting. Risk Anal 29:1566–1577

Breakwell GM (2007) The psychology of risk. Cambridge University Press, Cambridge

Clay P, Hollister R (eds) (1983) Neighborhood policy and planning. Lexington Books, Lexington

Cohen BC (1973) The press, the public and foreign policy. Princeton University Press, Princeton

Cook FL, Tyler TR, Goetz EG, Gordon MT, Protess D, Leff DR, Molotch,HL (1983) Media and agenda-setting: Effects on the public, interest group leaders, policy makers and policy. Pub Op Quart 47:16–35

Driedger SM (2007) Risk and the media: A comparison of print and televised news stories of a Canadian drinking water risk event. Risk Anal 27:775–786

Entman RF (1993) Framing: Toward clarification of a fractured paradigm. Journal of Communication 43:51–58

Entman R (2007) Framing bias: Media in the distribution of power. J Comm 57:163–173

Feldman DL, Hanahan RA (1996). Public perceptions of a radioactively contaminated site: Concerns, remediation preferences and desired involvement. Environl Health Persp 104:1344–1352

Fischhoff B, Lichtenstein S, Slovic P, Derby SL, Keeney RL (1981) Acceptable risk. New York: Cambridge University Press

Flynn J, Slovic P, Kunreuther H (eds) (2001) Risk, media and stigma: Understanding public challenges to modern science and technology. Earthscan, London and Sterling

Gertner J (2006) Atomic balm? New York Times July 16:36–47

Greenberg MR (2009) NIMBY, CLAMP, and the location of new nuclear-related facilities: US national and 11 site-specific surveys. Risk Anal 29:1242–1254

Greenberg M, Lowrie K, Hollander J, Burger J, Powers C, Gochfeld M (2008) Citizen board issues and local newspaper coverage of risk remediation, and environmental management: six U.S. nuclear weapons facilities. Remediation Journal 18:72–90
Greenberg M, Lowrie K, Burger J, Powers C, Gochfeld M, Mayer H (2007) The ultimate LULU? Public reaction new nuclear activities at major weapons sites. J Am Plan Ass 173:346–351
Greenberg M, West B, Lowrie K, Mayer H (2009) The Reporters Handbook on Nuclear Materials, Energy, and Waste Management. Nashville: Vanderbilt University Press
Halpern-Felsher B, Millstein S, Ellen J, Adler N, Tschann J, Biehl M (2001) Role of behavioural experience in judging risks. Health Psychol 20:120–126
Hernes G (1978) Det mediavridde samfunn. In: Hernes G (ed) Forhandlingsøkonomi og blandingsadministrasjon, Universitetsforlaget, Bergen
Hughes E, Kitzinger J, Murdock G (2006) The media and risk. In: Taylor-Gooby P, Zinn J (eds) Risk in social science. Oxford University Press, Oxford
Johnson-Cartee KS (2005) News narratives and news framing: Constructing political reality. Rowman & Littlefield Publishers, Lanham
Kasperson RE, Renn O, Slovic P, Brown HS, Emel J, Goble R, Kaperson JX, Ratick SJ (1988) The social amplification of risk: A conceptual framework. Risk Anal 8:178–187
Kenix LJ (2005) A comparison of environmental pollution coverage in the mainstream, african american, and other alternative press. Howard J Comm 16:49–70
Kitzinger J (1999) Researching risk and the media. Health, Risks, Soc 1:55–69
Kivimaki M, Kalimo R (1993) Risk perception among nuclear power plant personnel: a survey. Risk Anal 13:421–425
Kraft ME, Clary BB (1991) Citizen participation and the NIMBY syndrome: Public response to radioactive waste disposal. West Polit Quart 44:299–328
Larsen SC, Brock TJ (2005) Great Basin imagery in newspaper coverage of Yucca Mountain. Geo Rev 95:517–536
Leading Futurist: Online Influence is Transforming Society (2009). PR Newswire. PR Newswire Association LLC. http://www.highbeam.com/doc/1G1-206783545.html. Accessed September 24, 2010
Lowrie K, Greenberg M, Waishwell L (2000) Hazards, risk and the press: Newspaper coverage of US nuclear and chemical weapons sites. Risk: Health, Safety, Environ 11:49–67
Mazur A (1990) Nuclear power, chemical hazards, and the quantity of reporting. Minerva 28:294–323
McCombs ME, Shaw DL (1972) The agenda-setting function of mass media. Pub Op Quart 36:176–187
Nemich C (2006) Boise State survey shows statewide support for INL. http://news.boisestate.edu/newsrelease/032006/0302INLrelease.html. Accessed 26 August 2006
Neuman WR, Just MR, Cringler AN (1992) Common knowledge: News and the construction of political meaning. The University of Chicago Press, Chicago and London
O'Donnell C, Rice RE (2008) Coverage of environmental events in US and UK newspapers: frequency, hazard, specificity, and placement. Int J Environ Studies 65:637–654
O'Hare M, Bacow L, Sanderson D (1983) Facility siting and public opposition. Van Nostrand and Reinhold, New York
Opinion: Internet conversations can have valuable print influence (Viewpoint essay) (2007) PR Week (US). Haymarket Media, Inc. http://www.highbeam.com/doc/1G1-167247367.html. Accessed 24 Sept 2010
Pasqualetti MJ (1987) Decommissioning as a neglected element of nuclear power plant siting policy in the US and UK. In Blowers A, Pepper D (eds) Nuclear power in crisis: Politics and planning for the nuclear state. Nichols Publishing Company, New York
Pavey R (2010) SRS contamination may lead to new procedures. Augusta Chronicle, July 29, 2010, http://chronicle.augusta.com/latest-news/2010-07-29/srs-contamination-may-lead-new-procedures. Accessed 24 Sept 2010
Poortinga W, Pidgeon N, Lorenzoni I, et al. Public perceptions of nuclear power, climate change and energy options in Britain; summary findings of survey conducted during October and November 2005. Understanding risk working paper 06–02

Portney K (1991) Siting hazardous waste treatment facilities: the NIMBY Syndrome. Auburn House, New York
Renn O (2008) Risk governance: coping with uncertainty in a complex world. Earthscan, London and Sterling
Rosa E (2001) Public acceptance of nuclear power: Déjà vu all over again? Physics Soc 30:1–5
Rosa E (2004) The future acceptability of nuclear power in the United States. Institute Francias des Relations Internationales, Paris
Ryan M (2003) Public relations and the web: organizational problems, gender and institutional type. Pub Relat Rev 29:335–349
Scheufele DA (1999) Framing as a theory of media effects. J Comm 49:103–122
Simon AF, Jerit J (2007) Toward a theory relating political discourse, media, and public opinion. J Comm 57:254–271
Singer E, Endreny P (1987) Reporting hazards: their benefits and costs. J Comm 37:10–26
Singer E, Endreny P (1993) Reporting on risk. Russell Sage Foundation, New York
Sjoberg L (2004) Local acceptance of a high-level nuclear waste repository. Risk Anal 24:737–749
Slovic P (1987) Perception of risk. Science 236:280–285
Swain KA (2007) Outrage factors and explanations in outrage factors and explanations in news coverage of the anthrax attacks. J&MC Quarterly 84:335–352
U.S. Department of Energy, Environmental Management (1996) Baseline environmental management report. Available at: http://www.em.doe.gov/bemr/pages/bemr96.aspsx
U.S. Environmental Protection Agency, Region 4, Superfund (2009) U.S. DOE Oak Ridge Reservation, Site summary profile. http://www.epa.gov/region4/waste/npl/npltn/oakridtn.htm. Accessed 24 Sept 2010
U.S. Nuclear Regulatory Commission (2008) Expected new nuclear power plant applications, updated March 19, 2008. http://www.nrc.gov/reactors/new-licensing/new-licensing-files/expected-new-rx-applications.pdf. Accessed 27 March 2008
van der Merwe R, Pitt LF, Abratt R (2005) Stakeholder Strength: PR Survival Strategies in the Internet Age. Public Relations Quarterly. http://www.highbeam.com/doc/1P3-853788301.html. Accessed Sept 24 2010
Vasterman P, Scholten O, Ruigrok N (2008) A model for evaluating risk reporting: the case of UMTS and fine particles. Europ J Comm 23:319–341
Venables D, Pidgeon N, Simmons P, Henwood K, Parkhill K (2009) Living with nuclear power: A Q-method study of local communities' perceptions. Risk Anal 29:1089–1104
Wåhlberg A, Sjöberg L (2000) Risk perception and the media. J Risk Res 3:31–50
Wakefield S, Elliott S (2003) Constructing the news: the role of local newspapers in environmental risk communication. Profess Geo 55:216–226
Washington State Deparatment of Ecology (2008) Cleaning Hanford's groundwater. Ecol Pub 08-05-001
Williams B, Brown S, Greenberg M, Kahn M (1999) Risk perception in context: the Savannah River site stakeholder study. Risk Anal 19:1019–1035
Wong LT, Fryxell GE (2004) Stakeholder influences on environmental management practices: A study of fleet operations in Hong Kong (SAR), China. Transportation Journal, American Society of Transportation and Logistics, Inc. from HighBeam Research: http://www.highbeam.com/doc/1G1-125229257.html. Accessed 24 Sept 2010

Chapter 18
Science and Stakeholders: A Synthesis

Joanna Burger

Contents

Abstract Stakeholders are all the interested and affected parties, and include (but are not limited to) Tribal nations, U.S. governmental agencies (federal, state, local), nongovernmental groups (conservation groups, recreational groups, hunting and fishing groups, citizens' groups), industry and their representative organizations, the media and information organizations, and the public. Governmental agencies include not only regulators, but human and ecological health groups. There are several levels of stakeholder involvement, including informational, acquisitional, dialogue, intragovernmental, stakeholder involvement stakeholder-driven, and stakeholder collaborative. In all cases, however, a range of stakeholders is involved in different phases of decision making. I suggest combining stakeholder models of involvement and collaboration during all phases from

J. Burger (✉)
Division of Life Sciences, Environmental and Occupational Health Sciences Institute (EOHSI), Consortium for Risk Evaluation with Stakeholder Participation (CRESP), and Rutgers University, 604 Allison Road, Piscataway, NJ 08854, USA
e-mail: burger@biology.rutgers.edu

J. Burger (ed.), *Stakeholders and Scientists: Achieving Implementable Solutions to Energy and Environmental Issues*, DOI 10.1007/978-1-4419-8813-3_18,

problem formulation to solutions and decision making, with an adaptive management approach. This would involve an adaptive management approach of a structured, iterative process of optimal decision making, with stakeholder involvement at all phases.

18.1 Introduction

As the Nation and World move forward in the twenty-first century, we must find solutions to complex environmental and energy-related problems. Many of the problems are those that remain from human population increases, concentration of populations in cities and along coastal zones, intense industrialization, and the legacy from the Cold War, and chemical and toxic waste facilities. These issues and problems are complex, are ongoing, and require consultation and collaboration among many different agencies, organizations, and individuals, as well as Tribal Nations. The complexities will require solutions that may change over time as conditions change, as the nature of the problem changes, and as societal and cultural needs and requirements change. Iteration and modifications to established methods and practices will be needed to move forward.

Solutions to environmental and energy problems will only come with interactions, collaborations, and decision making that includes governments, Tribal nations, federal, state and local governments, regulators, environmental and human health agencies, conservation and service organizations, relevant commissions or committees, public policy makers, and the general public. In current language, all of these agencies, nations, and people are considered stakeholders (Table 18.1). In short, "stakeholder" indudes all the agencies, Tribes, or other people who are interested and affected by the decisions and environmental conditions that are created by those decisions (Boiko et al. 1996). People with interests can include local people, but also those who live far away, but are interested in the region or problem, or who merely appreciate knowing that the environment exists. Such existence values should not be discounted, although they are often difficult to quantify or evaluate (Diamond and Hausman 1994; Costanza et al. 1997; Chambers and Whitehead 2003; Efroymson et al. 2008).

18.2 Approaches and Solutions Described in the Book

The chapters in this book were aimed at understanding ways of finding solutions or making decisions about environmental and energy-related problems, with stakeholder participation. They were selected to provide a range of problems (both environmental and energy-related), stakeholders, solutions, and completed projects, as well as ongoing processes. Some were successful in that solutions were reached that were accepted by all interested and affected parties, others were less successful, and still others are ongoing and involve continued information gathering and research aimed at filling knowledge gaps. Stakeholder involvement is both necessary and essential to solving complicated environmental and energy-related problems (DOE 1994, 1997; NRC 2008).

Table 18.1 Types of stakeholders to include for environmental and energy-related decision making

1. Tribal Nations – all federally and state-recognized Tribal nations with interests and/or treaty rights
2. Tribal Nation Members – all individuals belonging to Tribal nations with legal status or treaty rights, as well as those unrecognized by the Bureau of Indian Affairs
3. Federal Agencies – all federal agencies with environmental, human or ecological health, or regulatory authority
4. State Agencies – all state agencies with environmental, human or ecological health, or regulatory authority
5. Federal or State Commissions or Committees – all such commissions or committees with environmental, human or ecological health, or regulatory or legal authority
6. Conservation, Human Health, Toxic groups, or other Organizations – all such organizations aimed at protecting the environment, ecological or human health, specific ecological receptors or groups of receptors, watersheds, and other organizations. Also to include special interest groups for specific environmental problems
7. Economic Interests and Organizations – all organizations involved with industry and economic interests, of both individuals and companies
8. Interested and Affected Parties – all individuals (or organizations) that are interested in the problem or issue, and all those potentially affected, even if they have expressed no interest
9. Neighbors, Regional Interests, and Those Farther Away – different levels of geographical and economic interest, based on distance from the site or the problem
10. Existence Values – the values placed by individuals who are not affected, live far away, and may never visit the place, but nonetheless have an interest in the existence of the place (and its associated condition)

While this list is by no means complete, it gives a picture of the levels of stakeholders to consider. Within each category, there may be several individuals or groups to consider

All of the case studies have several things in common: (1) a commitment to stakeholder involvement, (2) a commitment to meaningful stakeholder involvement where a range of stakeholders directly affect the process from problem formulation to workable solutions, and (3) a realization that stakeholder participation may require more time and money, but that the final solutions or process developed is better and more likely to be accepted by the interested and affected parties. Participation must be real, and not just a means to placate an angry public. Such participation will lead to making better, more cost-effective management decisions (Burger et al. 2008, 2009; Brody 2009).

Several different types of problems and processes have been presented, and they represent a beginning typology for stakeholder participation. All of the cores presented in this book, however, involve multiple stakeholders.

Stakeholders and Energy

General description of stakeholders (Burger et al. 2011a)
Energy diversification (Gochfeld 2011)

Single Environmental or Energy-related Problems:

Dealing with residual waste from mill tailings (Waugh et al. 2011)
Reducing bird mortality from wind power (Bartlett 2011)
Reducing human risk from mercury in fish (Burger 2011b)
Reducing the risk from decommissioning (Clarke et al. 2011)

Complex and Multiple-issue Problems

Consensus building in a uranium plant (Morgan 2011)
Evaluating risk to wildlife from different energy sources (Zillioux et al. 2011)
Melding the needs of salmon, watershed protection, and hydropower (Opperman et al. 2011)
Understanding the risk from radionuclides to ecosystems, native peoples, and commercial fisheries (Burger et al. 2011)
Estuary enhancement as a solution to an industry's problem (Baletto and Teal 2011)
Evaluating wind energy in Vermont (English 2011)

Environmental Equity Approaches and Legal Mandates

Native American viewpoints and Treaty Rights (Bohnee et al. 2011)
Inclusion of minorities in decisions (Johnson 2011)

Institutional/Communication Approaches

Institutional and stakeholder approaches (Fremeth and Marcus 2011)
Communication (Greenberg et al. 2011)
Media and Stakeholders (Lowrie et al. 2011)

It is clear that the complexity of environmental problems relates to the complexity of the component parts, which include the problem, the stakeholders, the processes, and the solutions. Special attention needs to be directed to groups that may not be included (minorities, Native Americans, low-income groups), and to institutional and communication/media approaches. Native Americans have different rights affirmed in treaties with the U.S. government, which recognizes their Sovereign Nation Status and the right to participate in decisions (Nez Perce 2003; Federal Register 2008).

Two types of processes bear special mention: ongoing issues, and research-based solutions, and the relationship among them. Ongoing issues are those that may have no permanent or static solution. That is, the problem may continue to exist, and what changes is the practice or methods of dealing with the issue. Three examples from this book deal with bird mortality at wind facilities, Native American rights in environmental decision making, and inclusion of minorities in these decisions. All three will continue as important national issues, although for different reasons. As we move forward with development of wind power, dealing with the threat to birds and bats that use the same air space will be a continuing problem, and research aimed at reducing this risk will change and shift with new techniques and technologies. Protecting the rights of Native Americans, minorities, and low-income Americans will continue to be important, legally mandated, and an environmental justice issue that is key to solving complex environmental problems.

Other issues, as well as those described above, have a research-based (or data gap) requirement to move forward. Such issues will require multidimensional dialogues (and collaboration) among different stakeholders (Tribal, U.S. governmental, nongovernmental, scientists, citizens) to move forward. In this regard, science is moving toward increased collaboration among scientists, among disciplines, and among scientists and other stakeholders. Several of the chapters in this book describe interaction, where scientists and other stakeholders design and implement research projects designed to fill data gaps.

For all problems, reality and perceptions play a key role, and information gaps (or research needs) are critical aspects of moving forward. It is essential to understand not only the science base, but the perception of that science. For the purposes of solving environmental problems, perceptions are reality for the people involved. The science base must include traditional science (as practiced by university and governmental scientists), and native and traditional science (as practiced by native Tribes and subsistence peoples; Burger et al. 2008). In a stakeholder-driven process, all points of view have validity for the purposes of discussion.

18.3 Iteration and Reconsiderations

The solving of environmental and energy-related problems requires an approach that involves iteration and reconsideration at each and every step in the process (Fig. 18.1). That is, it is generally an illusion that an environmental problem can be addressed, and then is completed or "solved." There will be continued interest in most environmental problems, and both the problems and the solutions will change with time. This is particularly true with energy issues as we develop new technologies and approaches to energy production, transportation, and use.

And at every step, all interested and affected parties must be considered, and all interested and affected parties must be involved. Being considered and being involved are not the same thing. I alone could consider the needs, wishes, or viewpoints of several different groups or stakeholders (i.e., ecosystems, hunters or fishers, industry, regulators), but I alone cannot represent these interests. They themselves must do this, and to represent themselves, they must be identified.

18.4 Stakeholders: Including the Broadest Range of Tribes, Agencies, and Groups

Determining who the appropriate stakeholders are is the most critical step in the process, and should be in itself an iterative and ongoing process. New stakeholders may wish to participate at any point in the process. New stakeholders may include young people who grow into an interest, others who were not previously interested or affected, and still others whose interests were not previously identified or understood. For example, for many years the major stakeholders involved in addressing mortality at wind facilities were the companies, landowners, ornithologists, and conservation organizations. Then recently, bat mortality was recognized as an important negative effect of wind turbines (Durr and Bach 2004), and thus mammalogists and bat conservation groups became important stakeholders.

Stakeholders are all the interested and affected parties, and include (but are not limited to) Tribal nations, U.S. governmental agencies (federal, state, local), non-governmental groups (conservation groups, recreational groups, hunting and fishing groups, citizens' groups), industry and their representative organizations, the media and information organizations, and the public. Governmental agencies include not only regulators, but human and ecological health groups.

The process of stakeholder identification should include (1) listing all possible interested and affected parties, (2) using the initial list to solicit other interested or affected parties, (3) identifying spokespersons from each group, (4) continuing to solicit or identify new stakeholders as the process unfolds, and (5) assuring that all stakeholders remain committed or seeking new representatives. In essence, the process is one of continued commitment to including all interested stakeholders, at all stages, early and often.

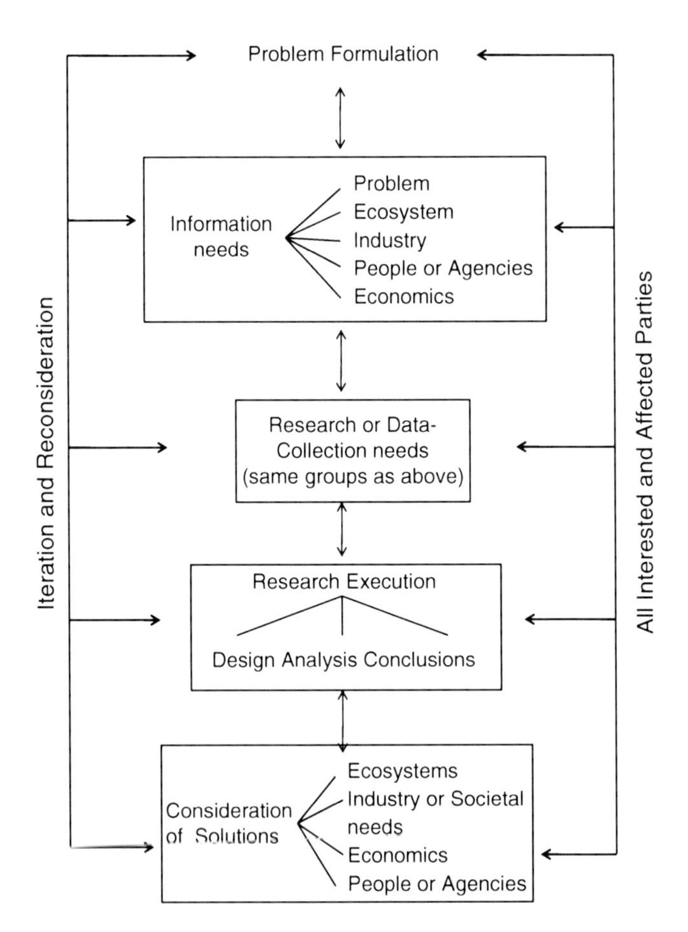

Fig. 18.1 Schematic of stakeholder involvement in environmental decision making from problem formulation to consideration of solutions. The model involves both iteration and reconsideration of options and the involvement of all interested and affected parties (the full range of stakeholders)

18.5 Stakeholder Inclusion

There are many models of approaching environmental problems, but two seem particularly critical to solving environmental and energy-related problems: (1) adaptive management and (2) stakeholder models, which include community-based participatory research. In traditional ecological or wildlife management, agencies consider the data and options and make a decision. This decision is then implemented and followed until there is another problem with the resource or ecosystem, and then another solution is devised. Wildlife management has moved forward with adaptive management to recognize that problems, conditions, options, and solutions change with time, and with management.

18.5.1 Adaptive Management

Recently, adaptive management has emerged as a viable approach to wildlife and ecosystem management (Lee 1999). Adaptive Management is a structured, iterative process of optimal decision making. It involves continually improving management decisions and practices by learning from the outcomes of observations, experiments, and management measures.

In this model, relevant environmental agencies and other scientists formulate the problem, consider different solutions, select a possible solution to implement, implement it, and implement it in such a way that a hypothesis is tested. When the results of the "management experiment" are available, the group reconsiders this and other possible solutions to implement in the future. In other cases, very specific experiments are designed to provide data that will address data gaps, and these experiments then inform future management decisions.

The adaptive management process is never truly completed since all solutions are examined for efficacy, usefulness, and success on a continual basis. It allows managers to move forward with imperfect information, select a solution, and move forward with the assumption that adjustments will be made in the future when new information becomes available. It is generally a process that involves agency personnel and nongovernmental scientists, and not the general public. To some extent adaptive management recognizes that decisions need to be made without perfect information, and that the decisions should not be written in stone, but can be modified when new data become available.

18.5.2 Stakeholder Models

Interestingly, stakeholder models were first developed in business, where management began to consider the interests of a wide range of stakeholders, as opposed to the usual emphasis on shareholders (Jansson 2005). The business stakeholder model started to

Table 18.2 Types of public meetings with stakeholder participation

Type	Objective	Direction of information flow	Outcome
Informational	Inform the public	One-way; leader provides information	Public obtains information and data from agency, responsible party, or other leader
Acquisitional	Solicit information from public	Two-way; but leader mainly solicits information or data	Agency, responsible party, or other leader listens to the concerns and perceptions of stakeholders
Two-way dialogue	Exchange of information, data, and concerns	Two-way, but leader and stakeholders exchange information, data, and concerns	Agency, responsible party, or other leader exchanges information with a range of stakeholders in an open forum
Multidimensional dialogue	Exchange of data, information, concerns, and concepts	From and to all participants. Multidirectional communication	All parties are considered stakeholders. The views, concerns, data, and science of all participants are considered, and all work toward improving the process or practice leading to solutions

consider the views and interests of consumers, workers, suppliers, advertisers, transport personnel, and the communities in which they operated. This changed the way companies viewed their role in the business world and within their communities.

Stakeholder models for ecosystem management usually assume that stakeholders are involved in the process of solving environmental problems, but the degree of involvement varies. The first step in any stakeholder process is having public meetings, but the nature of those meetings can vary, from purely informational to a multidimensional dialogue where all present are considered stakeholders, have valid viewpoints and data to offer, and are involved in meaningful dialogue that can lead to better decisions (Table 18.2). It is essential at the outset to make sure that all participants understand the objectives, potential outcomes, and nature of the public meeting being held. Ground rules need to be clear to everyone; within reason everyone should be able to contribute viewpoints and data; and everyone should expect clear notes following the meeting.

Burger (2009, 2011c) suggested that there are different levels of stakeholder involvement, including informational, acquisitional (information gathering), dialogue, intragovernmental, stakeholder involvement, stakeholder-driven, and stakeholder collaborative. In all cases, however, a range of stakeholders are involved in different phases of decision making. Whereas the Presidential/Congress Commission on Risk Assessment and Management (PCCRAM 1997, see Chap. 1) suggested that stakeholders should be engaged in considering the risks and options that lead to evaluations, actions, and decisions, they did not suggest that stakeholders should be involved in all phases, including information gathering and data collection (either in design or implementation), and I do so here. Stakeholders should be involved in the science itself, not just as directors in problem formulation or solution consideration (Burger in press).

Stakeholder conceptual models or analyses usually attempt to include stakeholders in as many phases of the decision-making process as possible, from problem definition or formulation to the final decision. While for a given environmental problem or issue, it may not be possible to involve the full range of stakeholders in all phases, it is desirable to involve them in as many as possible, in as many aspects as possible (Fig. 18.2). Further, the types of stakeholders involved vary, depending upon their legal role. For example, Tribes are mandated by treaty rights to be part of environmental decisions, U.S. governmental agencies are mandated by laws and regulations to initiate or take part in some of the processes, and governmental regulators have legal responsibility for other aspects of environmental problems.

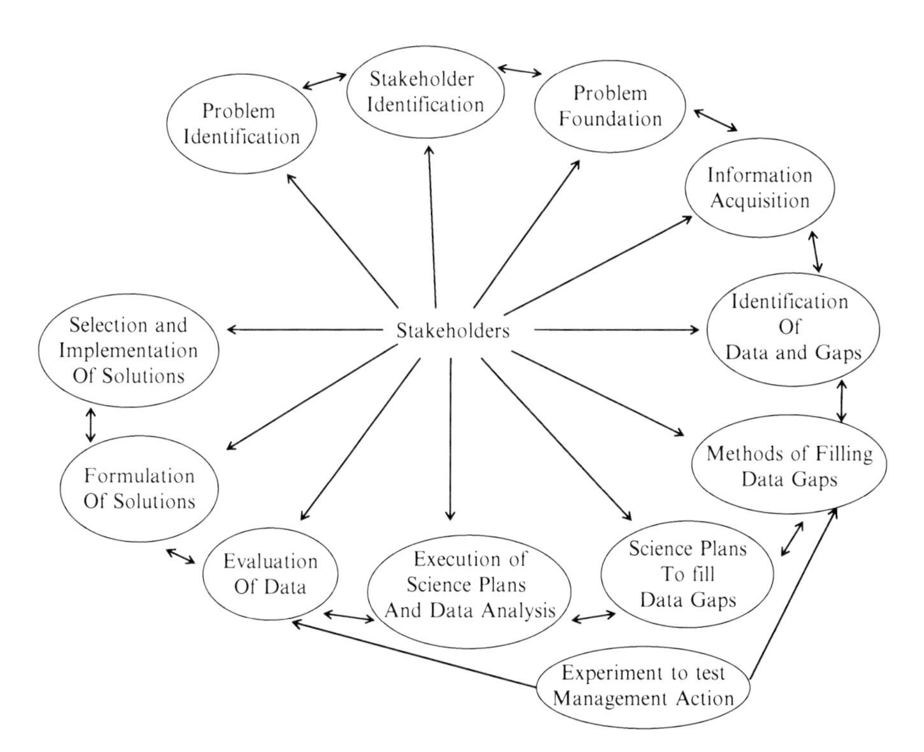

Fig. 18.2 Schematic of the phases of environmental decision making from problem identification to selection and implementation of solutions. Note that stakeholders should be involved in as many of the phases as possible, although the particular stakeholders involved will vary

18.5.3 Community-based Participatory Research

As is clear from the above descriptions of "stakeholder-involvement" and "adaptive management," may different fields have independently come to the realization that involving a wide range of people and interests in research is critical to finding solutions to difficult, contentious, and continually-changing environmental problems. Social scientists are using the term "community-based participatory research" to describe the slightly different process of including the public or community in research (O'Fallon and Dearry 2002; Wallerstein and Duran 2006; Munoz and Fox 2011, NIESH 2011). Community participatory research is of growing importance in defining, studying, and resolving complex exposure and risk issues, and this many of the case studies described in this book are at the intersection of traditional stakeholder approaches and community based participatory research. And to some extent, both use an adaptive management approach in that the problem definition, methods, and solutions are continually being evaluated and changed as necessary.

18.5.4 Combining Adaptive Management with Stakeholder Models

Adaptive management models are structured and iterative, and lead to decision making that takes advantage of changing levels of information and data, allowing for the improvement of policies and practices of management. It is, however, usually an agency approach that may involve several agencies, as well as outside scientists. It is not usually a wide-ranging stakeholder approach. I suggest that combining an adaptive management model with a stakeholder approach will lead to better, more cost-effective decisions regarding environmental and energy-related problems (Fig. 18.3). This approach combines an iterative, continually evolving management strategy with the widest inclusion of stakeholders at as many levels as possible. In this manner, a wide range of stakeholders can be involved at different phases, and the decisions are modifiable, depending on the outcome of experiments, and additional data and analyses.

Each of the phases in Fig. 18.2 would therefore be open to continued modification as the results of the other phases became known and examined in light of current understanding of the problem (e.g. Fig. 18.3). The problem may be reformulated, new data gaps identified, and new research initiated. These in turn would lead to new solutions that reflect new understandings of the issue or problem. This approach has the advantage of incorporating new approaches, practices, and knowledge to reach consensual solutions that take into account the views of a wide range of environmental and human needs.

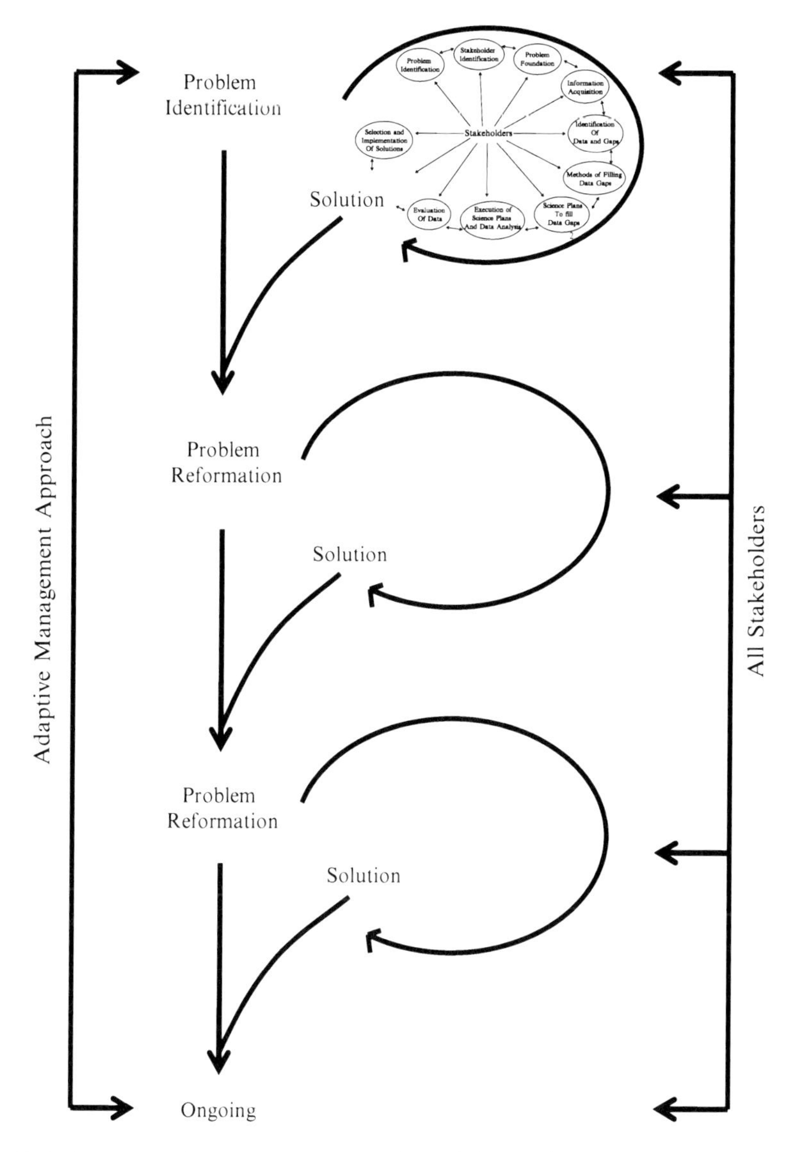

Fig. 18.3 Schematic of an evolving strategy for solving environmental and energy-related problems by combining an adaptive management strategy with a full range of stakeholders

18.6 Conclusions and Recommendations

Including the full range of stakeholders in the decision-making process involved in environmental and energy-related problems is neither simple nor fast. It takes time to identify the stakeholders, contact them and solicit their involvement, consider the phases for involvement, consider the places where different stakeholders can influence both the process and the decisions, and conduct the process in such a way to maximize stakeholder involvement. As Brody (2009) and the cases presented in this book suggest, however, inclusion of stakeholders leads to better and more cost-effective solutions. The cost of the stakeholder process may be relatively large, but often it pales in light of the ongoing cost of not solving the problem for many years (see Burger et al. 2011).

Some general suggestions about stakeholder involvement and behavior emerge from the case studies in this book, from the literature, and from my work (Table 18.3). Whether the stakeholder process is aimed at deciding where to cite a wind facility or a nuclear power plant, deciding how to move forward with chemical or radiological cleanup, or developing plans to manage or recover an endangered species, the tenets are the same. Mainly these include (1) treating all views of all stakeholders with respect, including all interested and affected parties and agencies as stakeholders, (2) including stakeholders early and often, (3) including a range of stakeholders in as many phases leading to decision making as possible, (4) making clear which aspects of the problem or decisions stakeholders can influence, and (5) defining clear timelines for actions and decisions. In some cases, one or more stakeholder groups will be more involved in the entire process leading to decision making than others. In all cases, however, it must be clear what decisions are required, and how different stakeholders can contribute and collaborate.

Table 18.3 General suggestions for stakeholder involvement

Stakeholder identification

1. Include all interested and affected parties
2. Use a broad approach to identify stakeholders
3. Identify the primary stakeholders early, and use this network to identify other stakeholders
4. Identify key Tribal, church, or other community leaders who can help identify individual or group stakeholders
5. Make it known to potential stakeholders that true collaboration, rather than information provision, is the primary goal (for scientists as well as all others)

Stakeholder Participation

1. Treat every point of view as valuable and worth considering (and listening to)
2. Recognize the importance of Tribal science as well as the knowledge base of all indigenous and subsistence people
3. Involve stakeholders early, often, and wherever possible
4. Consider information and data flow to be multidimensional and flowing in all directions
5. Consider the time and money constraints of stakeholders to participate (particularly Tribal and general public members), and plan accordingly
6. Consider monetary compensation for prolonged stakeholder involvement, much as scientists, governmental personnel, or others are compensated for their expertise and time
7. Be appreciative of stakeholder involvement, making data and written material available in sufficient time for distribution and review, and setting meeting times to meet the needs of the primary stakeholders (and others, where possible)
8. Develop interactive web-based dialogue tools that are available to a broad range of people (a web page is only the first step)

9. Where necessary, responsible parties (governmental agencies, tribes, companies, others) should identify a key person to answer questions and seek information for stakeholders, This person should be readily reached via the phone and electronic means, and should respond rapidly to such inquiries
10. Include a range of stakeholders as authors of reports or scientific papers, where it is warranted

Stakeholder involvement in decision making

1. Define the decisions to be made as early as possible
2. Make the time line for decision-making process clear
3. Make it clear that stakeholders will be involved in some of the decisions involved in the process (see Fig. 18.2 for the phases)
4. Define which decisions stakeholders can be involved in
5. Determine at the start which phases require or would benefit the most from stakeholder involvement, and which stakeholders should be approached
6. Make it clear which phases stakeholders can influence, and which are clearly mandated by treaties, laws, or regulations to involve particular agencies (i.e., when are stakeholder views or data merely advisory)
7. Make it clear which phases or decisions stakeholders cannot make, especially when a commission, agency, or other entity is mandated to make it
8. Solicit stakeholders' input regarding how they can take part in the process

Fig. 18.4 View of the Brookfield Power Plant in Cohoes (near Albany, New York) – although this plant was established on the Mohawk River long before site neighbors or a full range of stakeholders were involved, it now solicits such participation when making new decisions

In many cases, environmental or energy-related decisions were made decades ago, before the full environmental implications of such actions were known, or a range of stakeholders participated in these decisions – for example with hydroelectric facilities (Fig. 18.4). Even so, with continued operations of hydroelectric plants, the permitting of new nuclear plants, and the establishment of wind power facilities, a fuller range of stakeholders (including Tribes, site neighbors, and the general public) are being included. A more open, transparent, public debate about environmental issues is now possible, and this leads to better, more cost-effective solutions that protect humans and the environment.

Finally, true collaboration among stakeholders involves participation in research and data gathering, participation in report writing and scientific papers (as full authors), and participation in decision making. To some extent it requires that we all, regardless of our position (in agencies, in Tribes, in universities, in organizations, as the public), recognizes that we are all stakeholders, with a knowledge and a view to contribute, and that collaboration will lead to better decisions and to an adaptive management strategy that will allow us to solve today's environmental problems and be able to face those of tomorrow.

Acknowledgments I particularly thank Michael Gochfeld, Charles W. Powers, Caron Chess, Michael Greenberg, David S. Kosson, James Clarke, Lisa Bliss, Larry Niles, and Mandy Dey for valuable discussions about science, stakeholders, and environmental health problems. I also thank Chris Jeitner and Taryn Pittfield for technical support. This research was funded mainly by the Consortium for Stakeholder Participation (CRESP) through a grant from the Department of Energy (DE-FC01-06EW07053) to Vanderbilt University and Rutgers University, as well as the Nuclear Regulatory Commission (NRC 38-07-502M02), NIEHS (P30ES005022), and Rutgers University. The conclusions and interpretations reported herein are the sole responsibility of the author, and should not in any way be interpreted as representing the views of the funding agencies.

References

Baletto JH, Teal JM (2011) PSEG's Estuary Enhancement Program: An Innovative Solution to an Industry Problem. In: J Burger (ed) Science and stakeholders: Achieving implementable solutions to energy and environmental issues. Springer: New York

Bartlett G (2011) Joint Fact-finding and Stakeholder Consensus Building at the Altamont Wind Resource Area in California. In: J Burger (ed) Science and stakeholders: Achieving implementable solutions to energy and environmental issues. Springer: New York

Bohnee G, Matthews J, Pinkham J et al (2011) Nez Perce Involvement with Solving Environmental Problems: History, Perspectives, Treaty Rights and Obligations. In: J Burger (ed) Science and stakeholders: Achieving implementable solutions to energy and environmental issues. Springer: New York

Boiko PE, Morrill RL, Flynn J et al (1996) Who holds the stakes? A case study of stakeholder identification at two nuclear weapons sites. Risk Anal 16:237–249

Brody SD (2009) Measuring the effects of stakeholder participation on the quality of local plans based on principles of collaborative ecosystem management. J Plan Ed Res 22:407–419

Burger J (2009). Stakeholder involvement in indicator selection: case studies and levels of participation. Environ Bioindicat 4:170–190

Burger J (2011a) Introduction: Stakeholders and Sciences. In: J Burger (ed) Science and stakeholders: Achieving implementable solutions to energy and environmental issues. Springer: New York

Burger J (2011b) Stakeholders, Risk from Mercury, and the Savannah River Site: Iterative and Inclusive Solutions to Deal with Risk from Fish Consumption. In: J Burger (ed) Science and stakeholders: Achieving implementable solutions to energy and environmental issues. Springer: New York

Burger J (2011c) Science and Stakeholders: A Synthesis. In: J Burger (ed) Science and stakeholders: Achieving implementable solutions to energy and environmental issues. Springer: New York

Burger J (in press) Protecting human health and the environment around nuclear facilities: Native Americans, stakeholders, and environmental justice. NovaScience Press, New York

Burger J, Gochfeld M, Pletnikoff K (2009) Collaboration versus communication: the epartment of Energy's Amchitka Island and the Aleut community. Environ Res 109:503–510

Burger J, Gochfeld M, Pletnikoff K et al (2008) Ecocultural attributes: evaluating ecological degradation: ecological goods and services vs subsistence and Tribal values. Risk Anal 28:1261–1271

Burger J, Gochfeld M, Powers CW et al (2011) Amchitka Island: Melding Science and Stakeholders to Achieve Solutions at a Former Department of Energy Nuclear Site. In: J Burger (ed) Science and stakeholders: Achieving implementable solutions to energy and environmental issues. Springer: New York

Chambers C, Whitehead J (2003) A contingent valuation estimate of the value of wolves in Minnesota. Environ Res Econ 9:225–238

Clarke J, Burger J, Powers CW et al (2011) Decommissioning of Nuclear Facilities and Stakeholder Concerns In: J Burger (ed) Science and Stakeholders: Achieving implementable solutions to energy and environmental issues. Springer: New York

Costanza R, d'Arge R, deGroot RS et al (1997) The value of the world's ecosystem services and natural capital. Nature 387:253–260

Department of Energy (DOE) (1994) How to design a public participation program. Battelle Pacific Northwest Labs (James L. Creighton for EM-22)

Department of Energy (DOE) (1997) Linking legacies: Connecting the Cold War Nuclear Weapons Production Processes To Their Environmental Consequences. Washington DC: Office of Environmental Management, Department of Energy http://www.em.doe.gov/Publications/linklegacy.aspx. Accessed 3 Aug 2010

Diamond P, Hausman J (1994) Contingent valuation: is some number better than no number? J Econom Perspect 8:45–64

Durr T, Bach L (2004) Bat deaths and wind turbines – a review of current knowledge, and the information available in the database for Germany. Brem Beirage Naturkunde Naturschutz 7:743–744

Efroymson RA, Peterson MJ, Welsh CJ et al (2008) Investigating habitat value to inform contaminant remediation options: approach. J Environ Manage 88:1436–1451

English MR (2011) Wind Energy in Vermont: The Benefits and Limitations of Stakeholder Involvement. In: J Burger (ed) Science and stakeholders: Achieving implementable solutions to energy and environmental issues. Springer: New York

Federal Register (2008) Indian entities recognized and eligible to receive services from the Unites States Bureau of Indian Affairs. Federal Register 73:18553–18557

Fremeth AR, Marcus AA (2011) Institutional Void and Stakeholder Leadership: Implementing Renewable Energy Standards in Minnesota. In: J Burger (ed) Science and stakeholders: Achieving implementable solutions to energy and environmental issues. Springer: New York

Gochfeld M (2011) Energy Diversity: Options and Stakeholders. In: J Burger (ed) Science and stakeholders: Achieving implementable solutions to energy and environmental issues. Springer: New York

Greenberg M, Babcock-Dunning L (2011) Communication between the Public and Experts: Predictable Differences and Opportunities to Narrow Them. In: J Burger (ed) Science and stakeholders: Achieving implementable solutions to energy and environmental issues. Springer: New York

Jansson E (2005) The stakeholder model: the influence of ownership and governance structures. J Business Ethics 56:1–13

Johnson J (2011) Minority Participants in Environmental and Energy Decision Making Process. In: J Burger (ed) Science and stakeholders: Achieving implementable solutions to energy and environmental issues. Springer: New York

Lee KN (1999) Appraising adaptive management. Conserv Ecol 3(2):3

Lowrie K, Greenberg M, Kennedy A et al (2011) Media, Local Stakeholders, and Alternatives for Nuclear Waste and Energy Facilities. In: J Burger (ed) Science and stakeholders: Achieving implementable solutions to energy and environmental issues. Springer: New York

Morgan K (2011) How Clean is Clean? Stakeholders and Consensus-Building the Fernald Uranium Plant. In: J Burger (ed) Science and stakeholders: Achieving implementable solutions to energy and environmental issues. Springer: New York

Munoz R, Fox MD (2011) Research impacting social contexts: the moral import of community-based participatory research. American Bioethics 11:37–38

National Research Council (NRC) (2008) Public participation in environmental assessment and decision making. Nat Acad Press, Washington

National Institute of Environmental Health Research (NIEHS) (2011) Environmental justice and community-based research. http://www.niehs.nih.gov/research/supported/programs/justice/. Accessed 5 May 2011

Nez Perce Tribe (2003) Treaties: Nez Perce perspectives. US DOE and Confluence Press

O'Fallon LR, Dearry A (2002) Community-based participatory research as a tool to advance environmental health science. Envir Health Perspect 110:155–159

Opperman JJ, Apse C, Ayer F et al (2011) Hydropower, Salmon, and the Penobscot River (Maine, USA): Pursuing Improved Environmental and Energy Outcomes through Participatory Decision-making and Basinscale Decision Context In: J Burger (ed) Science and stakeholders: Achieving implementable solutions to energy and environmental issues. Springer: New York

President's Commission (PCCRAM) (1997) Presidential/Congressional Commission on risk assessment and management. U.S. Government Printing Office, Washington DC

Wallerstein NB, Duran B (2006) Using community-based participatory research to address disparities. Health Promot Pract 7:312–323

Waugh J, Glenn EP, Carroll MK et al (2011). Helping Mother Earth Heal: Dine' College Collaboration on Enhanced Attenuation Pilot Studies at U.S. Department of Energy Uranium Processing Sites on Navajo Land. In: J Burger (ed) Science and stakeholders: Achieving implementable solutions to energy and environmental issues. Springer: New York

Zillioux Z, Newman JR, Lampman GG et al (2011) Using Stakeholder Input to Develop a Comparative Risk Assessment for Wildlife from the Life Cycles of Six Electrical General Fuels. In: J Burger (ed) Science and stakeholders: Achieving implementable solutions to energy and environmental issues. Springer: New York

Index

J. Burger (ed.), *Stakeholders and Scientists: Achieving Implementable Solutions to Energy and Environmental Issues*, DOI 10.1007/978-1-4419-8813-3,

Printed by Publishers' Graphics LLC USA
MO20120424-022
2012